Biochemistry and Cell Biology of *Artemia*

Editors

Thomas H. MacRae, Ph.D.
Associate Professor
Department of Biology
Dalhousie University
Halifax, Nova Scotia
Canada

Joseph C. Bagshaw, Ph.D.
Professor and Head
Department of Biology and Biotechnology
Worcester Polytechnic Institute
Worcester, Massachusetts

Alden H. Warner, Ph.D.
Professor
Department of Biological Sciences
University of Windsor
Windsor, Ontario
Canada

CRC Press
Taylor & Francis Group
Boca Raton London New York

CRC Press is an imprint of the
Taylor & Francis Group, an **informa** business

First published 1989 by CRC Press
Taylor & Francis Group
6000 Broken Sound Parkway NW, Suite 300
Boca Raton, FL 33487-2742

Reissued 2018 by CRC Press

Library of Congress Cataloging-in-Publication Data

Biochemistry and cell biology of Artemia/editors, Thomas H. MacRae,
Joseph C. Bagshaw, Alden H. Warner.
p. cm
Bibliography: p.
Includes index.
ISBN 0-8493-4897-8
1. Artemia — Physiology. 2. Artemia — Cytology. 3. Crustacea—Physiology. 4. Crustacea — Cytology. I. MacRae, Thomas H.
II. Bagshaw, Joseph C., 1943- III. Warner, Alden H. (Alden Howard), 1937-
QL444.B815B56 1989
595.3'843 — dc19 88-14562

A Library of Congress record exists under LC control number: 88014562

Publisher's Note
The publisher has gone to great lengths to ensure the quality of this reprint but points out that some imperfections in the original copies may be apparent.

Disclaimer
The publisher has made every effort to trace copyright holders and welcomes correspondence from those they have been unable to contact.

ISBN 13: 978-1-315-89105-7 (hbk)
ISBN 13: 978-1-351-07015-7 (ebk)

Visit the Taylor & Francis Web site at http://www.taylorandfrancis.com and the
CRC Press Web site at http://www.crcpress.com

INTRODUCTION

The unusual life history of the brine shrimp, *Artemia*, and the relative ease with which it can be experimentally manipulated have long made this crustacean a favorite system for biological studies. Over the years, descriptive morphological work has given way to a rigorous analysis of biochemical and cellular aspects of the organism. The underlying theme of the work has often been developmental in nature.

We have attempted, with this book, to bring together a rather wide spectrum of topics currently under study in the shrimp. Analyses of gene structure and protein synthesis are combined with descriptions of protein interactions characteristic of functional cells. Although we have not restricted the contents of the book to developmentally related processes, several chapters are more or less concerned with this aspect of the brine shrimp. It will become apparent to the reader that significant contributions to our appreciation of eukaryotic cell function are being made by the study of *Artemia*, but the potential of the system is far from being fully exploited. We, the editors, will thus consider the book a success if our efforts not only provide researchers working with *Artemia* a useful framework into which to integrate their findings, but also prompt others to consider the study of *Artemia* as a way to answer the questions associated with their specific interests.

The editors gratefully acknowledge the efforts of all individuals who have so willingly contributed their time and expertise to this book.

T. H. MacRae
J. C. Bagshaw
A. H. Warner

THE EDITORS

Thomas H. MacRae, Ph.D. is an Associate Professor in the Department of Biology at Dalhousie University in Halifax, Nova Scotia.

Dr. MacRae obtained his B.Sc. in Biology from Mount Allison University in New Brunswick. He received an M.Sc. in 1973 and a Ph.D. in 1976 from the University of Windsor, Ontario. His graduate work, in microbiology and electron microscopy, was supported by several awards including a National Research Council of Canada Postgraduate Scholarship. Three years of postdoctoral research in biochemistry at the University of Sherbrooke and University of Mississippi Medical Centers, where Dr. MacRae was holder of a Medical Research Council of Canada Postdoctoral Research Fellowship, were followed by a year at the University of Ottawa in the Biology Department. Dr. MacRae has been Assistant Professor (1980—1987) and Associate Professor (1987—present) at Dalhousie University and was a Visiting Research Fellow at the University of Kent, England in 1987.

Dr. MacRae is currently a member of the American Society for Biochemistry and Molecular Biology, Biochemical Society (U.K.), American Society for Cell Biology, Canadian Society for Cell Biology, and the Canadian Biochemical Society. His major research interests are tubulin gene organization and expression in *Artemia,* tubulin assembly and function, and the effects of metals on brine shrimp development. In addition to his research activities, Dr. MacRae is a very active teacher of cell/molecular biology at Dalhousie University.

Joseph C. Bagshaw, Ph.D., is Professor and Head of the Department of Biology and Biotechnology at Worcester Polytechnic Institute, Worcester, Massachusetts. He earned his B.A. degree from the Johns Hopkins University in 1965, and his Ph.D. degree from the University of Tennessee, Oak Ridge Graduate School of Biomedical Sciences in 1969. Dr. Bagshaw was the first graduate from this innovative program at the Oak Ridge National Laboratory. Following post-doctoral research at Massachusetts General Hospital, he was appointed Assistant Professor of Biochemistry and then Associate Professor of Biochemistry at Wayne State University School of Medicine, Detroit, Michigan, before joining the faculty of Worcester Polytechnic Institute in 1984. Dr. Bagshaw is also Director of the Worcester Consortium Ph.D. Program in Biomedical Science.

Dr. Bagshaw is an author of over 60 research papers and reports and seven book chapters, and has presented his research at many national and international scientific meetings. He has been a co-organizer of two International Symposia on *Artemia,* and co-editor of the resulting proceedings. He is a member of the American Society for Biochemistry and Molecular Biology, the American Society for Cell Biology, the American Association for the Advancement of Science, and the New England Biotechnology Association. His research interest is in molecular aspects of development in *Artemia,* focusing on gene structure, organization, and expression.

Alden H. Warner, Ph.D., is Professor of Biology at the University of Windsor, Windsor, Ontario, Canada. He received his Bachelor of Arts degree in Zoology from the University of Maine in 1959, his Master of Science degree in Physiology at Southern Illinois University in 1961, and the Ph.D. degree in Physiology/Biochemistry from Southern Illinois University in 1964 under the joint sponsorship of a U.S.P.H. Pre-doctoral Fellowship and the Biology Division of the Oak Ridge National Laboratory in Tennessee. After the Ph.D. he was appointed as Research Associate (P.D.F.) in the Developmental Biology Group at Oak Ridge. Since 1965 he has held successive appointments as Assistant Professor of Biology, Associate Professor of Biology and, currently, Professor of Biology in the Department of Biological Sciences at the University of Windsor.

Dr. Warner is currently President of the Canadian Society for Cell Biology and a member of the Board of Directors of the Canadian Federation of Biological Societies. He has also served as Associate Editor of the *Canadian Journal of Biochemistry and Cell Biology* since 1986. Between 1979 and 1985 he was Head of the Department of Biology at the University of Windsor. He is a member of the American Society for Biochemistry and Molecular Biology, Society for Developmental Biology, and the Canadian Society for Cell Biology. He is an author of over 50 articles dealing with the biochemistry of *Artemia* development and several others dealing with biochemical regulation in dystrophic muscle. In addition to his research interest, he is an active teacher, community volunteer, and sports enthusiast.

CONTRIBUTORS

Myriam Aerden
Scientific Collaborator
Department of Biochemistry
University of Antwerp
Wilrijk, Belgium

Matthew T. Andrews, Ph.D.
Assistant Professor
Department of Genetics
North Carolina State University
Raleigh, North Carolina

Hubert Backhovens
Industrial Engineer
Department of Biochemistry
University of Antwerp
Wilrijk, Belgium

Joseph C. Bagshaw, Ph.D.
Professor and Head
Department of Biology and Biotechnology
Worcester Polytechnic Institute
Worcester, Massachusetts

Julius Clauwaert, Dr. in Sciences
Professor
Department of Biochemistry
Biophysics Research Group
University of Antwerp
Antwerp, Belgium

E. De Herdt, Ph.D.
Scientific Collaborator
Department of Biochemistry
University of Antwerp
Wilrijk Belgium

Andre P. De Leenheer, Ph.D.
Professor
Department of Medical Biochemistry
State University of Gent
Faculty of Pharmaceutical Sciences
Ghent, Belgium

K. De Smet
Scientific Collaborator
Department of Biochemistry
University of Antwerp
Antwerp, Belgium

John A. Freeman, Ph.D.
Associate Professor
Department of Biology
University of South Alabama
Mobile, Alabama

D. Geelen
Scientific Collaborator
Department of Biochemistry
University of Antwerp
Antwerp, Belgium

Patrick Lavens, Licentiate Biology
Researcher
Artemia Reference Center
State University
Ghent, Belgium

Thomas H. MacRae, Ph.D.
Associate Professor
Department of Biology
Dalhousie University
Halifax, Nova Scotia, Canada

Luc Moens, Dr. Sc.
Professor
Department of Biochemistry
University of Antwerp
Wilrijk, Belgium

Hans J. Nelis, Ph.D.
Research Associate
Department of Medical Biochemistry
State University of Gent
Faculty of Pharmaceutical Sciences
Ghent, Belgium

Paul Nieuwenhuysen, Dr. Sc.
Library
Free University of Brussels
Brussels, Belgium

Brian A. Perry
Department of Zoology
Arizona State University
Tempe, Arizona

E. Piot, Ph.D.
Department of Biochemistry
Janssen Pharmaceuticals
Beerse, Belgium

Erwin Roggen, Ph.D.
Scientific Collaborator
Department of Biochemistry
University of Antwerp
Wilrijk, Belgium

Antonio Sillero, Ph.D.
Professor and Chairman
Department of Biochemistry
Faculty of Medicine
University of Extremadura
Badajoz, Spain

Maria A. Günther Sillero, Ph.D.
Investigator
Instituto de Investigaciones Biomédicas (C.S.I.C.)
Faculty of Medicine
University of Extremadura
Badajoz, Spain

Herman Slegers, Ph.D.
Associate Professor
Department of Biochemistry
University of Antwerp
Wilrijk Belgium

Patrick Sorgeloos, Ph.D.
Research Coordinator
Artemia Reference Center
State University of Gent
Ghent, Belgium

Chris Thoen, Ph.D.
Scientific Collaborator
Department of Protein Engineering
Plant Genetic Systems
Ghent, Belgium

J. Van Beeumen, Ph.D.
Associate Professor
Department of Biochemistry
University of Antwerp
Wilrijk, Belgium

M.-L. Van Hauwaert
Technician
Department of Biochemistry
University of Antwerp
Wilrijk, Belgium

Lauris Van Hove, Ph.D.
Scientific Collaborator
Department of Biochemistry
University of Antwerp
Wilrijk, Belgium

G. Verpooten
Scientific Collaborator
Department of Biochemistry
University of Antwerp
Wilrijk, Belgium

Albert Wahba, Ph.D.
Professor and Chairman
Department of Biochemistry
University of Mississippi Medical Cent
Jackson, Mississippi

Alden H. Warner, Ph.D.
Professor
Department of Biological Sciences
University of Windsor
Windsor, Ontario, Canada

Charles L. Woodley, Ph.D.
Associate Professor
Department of Biochemistry
University of Mississippi Medical Center
Jackson, Mississippi

TABLE OF CONTENTS

Chapter 1

STRUCTURE AND EXPRESSION OF HISTONE GENES IN *ARTEMIA*

Joseph C. Bagshaw, Matthew T. Andrews, and Brian A. Perry

TABLE OF CONTENTS

I. INTRODUCTION

Of all the genetic information encoded in DNA, only a fraction is expressed in a given cell type. Development is characterized by a succession of precisely programmed changes in selective gene expression, and developing organisms and differentiating tissues have provided many useful models for studies of eukaryotic gene expression. Technological advances have now made possible studies of regulation of specific genes, including single-copy genes. Recombinant DNA technology has provided means to identify and amplify any gene for which an appropriate probe is available. It is now possible to put an entire genome into a "library" of cloned sequences, and then retrieve the genes of interest at any time. These and other advances have brought us to the point where virtually any eukaryotic gene can be cloned and isolated, and its structure and function can be studied in detail at the molecular level. Despite all this, the molecular biologist must still work within the limitations imposed by the biological system. Sometimes the investigator can manipulate the biological system; more often, the biological system manipulates the investigator. Obviously if one wants to study regulation of a specific gene it is advantageous to have a source of material that is expressing that gene, and another source that is not. Ideally, expression should be synchronized in the population of cells or organisms, and it should be possible to obtain these cells in large quantity (tens or hundreds of grams), to isolate and identify the molecules involved in regulation. Postgastrula development of *Artemia* provides an advantageous biological system, and the regulated expression of histone genes provides an appropriate model.

A. Histone Genes and Messenger RNAs

This subject has been extensively reviewed by Kedes;[1,2] only the most pertinent aspects and recent developments will be discussed in this section. Histones have been identified as the principal or sole translation products of 7-10S polysomal RNA from sea urchins[3-5] and HeLa cells[6] and Kedes[1] has suggested in vitro translation as one criterion for identifying histone mRNAs. Histone mRNAs generally lack 3′-polyadenylate sequences[4,7,8] although a partial exception occurs in *Xenopus* oocytes.[9] Since most nonhistone messengers are polyadenylated, RNA preparations can be enriched in histone mRNAs by selecting poly(A)$^-$ molecules. In *Artemia*, the presence of histone messengers has been demonstrated by translation of poly(A)-RNA[10] as well as total polysomal RNA,[11] provided the RNA was prepared from hatched larvae; RNA from dormant or developing encysted embryos does not contain translatable histone messengers.

Histone genes are reiterated and clustered within the genome. Reiteration frequencies range from 20 to 40 copies per haploid genome in mammals[12,13] to several hundred in sea urchins.[14] Early evidence that histone genes are clustered[14-16] has been confirmed by mapping histone genes with restriction endonucleases. In most of the cases, the repeat unit within the cluster contains one gene for each of the five major types of histone. Nucleotide sequences of most of the histone gene region of two species of sea urchin (*P. sammechinus milliaris* and *Strongylocentrotus purparatis*) and of *Drosophila* have been determined. In sea urchins, the genes for histones H1, H2A, H3, H2B, and H4 are arranged in that order on the chromosome.[12-20] All five genes are coded by the same DNA strand, i.e., have the same transcriptional polarity. It is hypothetically possbile that all five genes could be transcribed in a single polycistonic messenger, but no such molecule has ever been found. The histone genes of *Drosophila* share some characteristics with those of urchins, but they differ in certain important respects. *Drosophila* histone genes are reiterated only about 100 times, and clustered in the genome, but the order of the histone genes differs from the order in sea urchins, and the transcriptional polarity is opposite for adjacent genes.[21-23]

Not all copies of the histone genes are identical, and there is good evidence that different sets of genes are expressed at different stages of development in *Strongylocentrotus*[24-27] and in *Lytechinus*.[28] Sea urchin histone genes of the "early" (preblastula) type are more abundant than the "late" type, as indicated by the hybridization data of Kunkel and Weinberg.[25] Appearance of different histone mRNA subtypes does not always coincide with synthesis of the corresponding histone subtypes.[26,27] Late mRNAs are present early in development, and early mRNAs are transcribed even during gastrulation. Lifton et al.[21] identified two types of *Drosophila* histone gene sets which differ in that one type contains an extra 220 base pairs, including a site for the restriction enzyme EcoRI. These two types of repeat units are interspersed nonrandomly in the genome, and there are about 75 copies of the longer type and 25 of the shorter. Whether these units are differentially expressed during development is not known.

Organization of histone genes in blocks of tandemly repeated units is not a universal characteristic of eukaryotes, especially higher vertebrates. In *Xenopus*, some of the histone genes are arranged in tandem repeats while others are scattered through the genome in a pattern that varies from species to species and from individual to individual.[29,30] Histone genes in the newt *Notophthalmus* are clustered in a repeat unit containing all five genes, but repeat units are separated by variable lengths of satellite DNA.[31] In rainbow trout there appears to be a conserved block of 5 histone genes with approximately 145 such blocks scattered throughout the genome.[32] Histone genes are clustered in varying groups of two to five genes but not tandemly linked in chicken,[33,34] mouse,[35] and man.[36] Similarly, the late histone genes of the sea urchin *Lytechinus pictus* are clustered but not tandemly repeated.[37]

B. Histone Synthesis and DNA Replication

In most cases, histones are synthesized only during the S phase (DNA replicative period) of the cell cycle, and histone mRNAs appear on polysomes only during periods of DNA replication. Borun et al.[38] found no translatable histone mRNAs in the cytoplasm of HeLa cells after inhibition of DNA replication or during the G1 phase of the cell cycle. Stein et al.,[39] using hybridization of RNA to DNA complementary to histone mRNAs (cDNA), confirmed the absence of histone mRNA sequences in the G1 phase of HeLa cells. Similarly, Parker and Fitschen[40] found histone mRNA sequences only during S phase in synchronized 3T6 cells. Stein et al.,[41] presented evidence that the histone genes of HeLa cells are available for transcription only during S phase, and Detke et al.,[42] showed that nuclei from S phase HeLa cells, partially synchronized by double thymidine block, contain and synthesize histone mRNA sequences. In nuclei from G1 cells highly synchronized by mitotic detachment, Detke et al.[42] found no histone mRNA sequences and no synthesis thereof, as judged by the hybridization assay with histone cDNA. The best evidence indicates that the primary mode of histone gene regulation in cultured cells is transcriptional. Whether this is true in other types of tissue remains unknown, due to the lack of suitable experimental systems. *Artemia* provides an opportunity to answer this question in a developing eukaryote.

Coincidence of histone synthesis and DNA replication does not always occur during the earliest stages of development. This has been most clearly shown in sea urchin species and in the amphibian *Xenopus laevis*. Arceci and Gross[43] examined the timing of both histone and DNA synthesis during the first two (synchronous) cleavage cycles in fertilized sea urchin eggs. They found that histone synthesis was not restricted to the period of DNA replication, and continued at an accelerating rate during the first G2 phase. Adamson and Woodland[44,45] found that histones were synthesized during *Xenopus* oogenesis and in enucleated oocytes; in both cases, no DNA synthesis occurs. During early cleavage in *Xenopus* embryos, the timing of histone synthesis was not coordinated with DNA synthesis. This noncoincidence of histone and DNA synthesis in early embryos is not difficult to rationalize, but it illustrates a tactical problem. Sea urchin and amphibian eggs can be obtained in reasonable quantity, and the first few DNA replicative periods are naturally synchronized, but histone gene expression is not coupled to DNA replication during early cleavage. At later stages (e.g., blastula), when DNA and histone synthesis are probably coupled in the individual cell, DNA synthesis is no longer synchronous in all cells. A population of embryos may be synchronous with respect to developmental stage, but histone gene expression has become randomized, considering all cells in the population. Sea urchin or amphibian embryos at this stage are analogous to randomly dividing cultures of HeLa cells, and their utility for studies of histone gene regulation are somewhat limited. It would be a great advantage to be able to obtain reasonable quantities of developing organisms in which the histone genes are expressed only during certain predictable periods. This is clearly not possible with sea urchins and amphibia, and available evidence indicates it is not possible with *Drosophila*. It is not only possible but very easy with *Artemia*.

C. *Artemia* as a Biological System for Studies of Histone Genes

Artemia offers unique opportunities for studies of gene expression in a developing eukaryote. Fertilized eggs develop within the ovisac to a late blastula or early gastrula stage, then development is arrested and the now dormant embryos are encysted in a chitinous shell and expelled from the ovisac. These so-called "winter eggs" are dehydrated osmotically after being washed ashore by wind and sunlight. The encysted embryos are completely metabolically dormant, but will resume development immediately upon rehydration at a suitable temperature. The rehydrated embryos undergo a period of synchronous development within their shells, and about 16 hr after immersion of embryos, swimming nauplius larvae hatch out. The dormant cysts, commonly but erroneously called brine shrimp "eggs", are collected commercially and sold as fish food. Thus, they are available at any time of year in essentially unlimited quantity. They can be stored indefinitely in the freezer and hatched synchronously at the investigator's convenience. The larvae can be kept up to 3 days in sterile artificial sea water without feeding, and all the larvae remain within a few hours of one another in development. Thus, relatively large quantities of synchronously developing nauplius larvae can easily be obtained.

A number of interesting biochemical events can be studied during embryonic and early larval development of *Artemia*.[11] The most relevant, for the purposes of this chapter, are DNA and histone synthesis. During the period of pre-emergence development, no cell division occurs, and the number of nuclei per embryo and the mass of DNA per embryo remain constant.[46,47] Shortly after hatching, DNA synthesis resumes, continues for about 24 to 36 hr, then stops.[47] We have independently confirmed these results, using the San Francisco strain of *Artemia*; McClean and Warner[47] used the Salt Lake strain. In both strains, DNA synthesis in early larvae is confined to a single period and results in a doubling of the mass of DNA per larva. Because the larvae are far beyond the early cleavage stage of development, we reasoned that histone synthesis should be coordinated with DNA replication. This prediction was verified by our studies of in vivo histone synthesis, which showed that histones are synthesized in newly hatched larvae but not in developing encysted embryos. Thus, we can easily obtain tens of grams of synchronously developing organisms that are or are not (our choice) expressing their histone genes. This led us to clone and isolate the *Artemia* histone genes in order to study their structure, organization, and expression.

II. COORDINATE SYNTHESIS OF DNA AND HISTONES IN *ARTEMIA* LARVAE

For reasons outlined above, it seemed probable that histone synthesis would be temporally coordinated with the wave of DNA synthesis in newly hatched *Artemia* larvae. However, it was necessary to prove this point experimentally by in vivo radiolabeling. The impermeability of the cyst shell to phosphate precluded in vivo labeling of DNA with ^{32}P, and these studies were thus confined to the postemergence stages of development. Fortunately for our work, the cyst shell is permeable to bicarbonate ions, and more than half of the ^{14}C taken up as $[H^{14}CO_3]^-$ is rapidly incorporated into amino acids.[48] Thus, we were able to examine directly the synthesis of specific proteins, including histones, even during the pre-emergence stage.

A. DNA and Protein Radiolabeling In Vivo

Dormant encysted embryos of the San Francisco (Metaframe brand) variety were obtained from a local supplier, surface sterilized with antiformin, and cultured in sterile artificial sea water at 30°C. For radiolabeling of DNA, hatched nauplius larvae in l-g samples were collected by filtration and quickly tranferred to 10 mℓ of sterile sea water containing 0.1 mCi of carrier-free ^{32}P-orthophosphate and incubated a further 4 hr. For labeling of proteins the $NaHCO_3$ in the sea water was replaced by NaH $^{14}CO_3$ at 40 to 60 mCi/mol and 25μCi/mℓ. Larvae were harvested after a further 8 hr of incubation. Incorporation of ^{32}P into DNA was measured by liquid scintillation counting after hot acid hydrolysis. Incorporation of ^{14}C into protein was measured by scintillation counting after acid extraction of nuclear proteins. A comparison of incorporation of ^{14}C into acid-soluble nuclear protein and of ^{32}P into DNA at various times during embryonic and larval development is shown in Figure 1. As previously described by McClean and Warner,[47] radiolabeling of DNA increased rapidly after hatching, reaching a maximum about 12 hr after hatching, then declined to a level of less than 10% of the maximum. Synthesis of acid-soluble nuclear proteins also increased immediately after hatching, then declined to about 10% of the maximal level. There was also a small but measurable amount of ^{14}C incorporation into acid-soluble nuclear proteins during pre-emergence development (0 to 8 hr). Similar but quantitatively less dramatic changes in radiolabeling of total protein were seen. Specific radioactivities of total protein in nauplius larvae were in the range of 7 to 10 × 10^4 cpm/mg, and total protein labeled in encysted embryos (0 to 8 hr) had a specific radioactivity of 3.3 × 10^4 cpm/mg. The very low level of incorporation of ^{14}C into acid-soluble nuclear protein in pre-emergence embryos was clearly not due to a general absence of, or failure to detect, protein synthesis.

B. Specific Labeling of Histones

In order to verify the apparent coordination of DNA and histone synthesis, it was essential to identify histones unequivocally among the acid-soluble nuclear proteins, and to examine incorporation of ^{14}C into histones at different stages of development. Acid-soluble proteins extracted from *Artemia* nuclei were fractionated by the method of Johns[49] and analyzed by polyacrylamide gel electrophoresis in the presence of sodium dodecyl sulfate (SDS-PAGE). Fractionation on acid-urea gels showed the expected microheterogeneity, and each of the five major histones could be readily identified. More importantly, all of the histones could be distinguished among total acid-soluble nuclear proteins fractionated by SDS-PAGE.[50] In order to examine in vivo radiolabeling of histones, developing *Artemia* embryos or larvae were incubated for 8-hr periods in the presence of $NaH^{14}CO_3$. Total acid-soluble nuclear proteins were fractionated by SDS-PAGE and analyzed for incorporated ^{14}C by fluorography, as shown in Figure 2. It is clear that protein radiolabeling during the 0 to 8-hr period (developing encysted embryos) was limited to nonhistone proteins, and no labeling of histones during this period was detected. Neither unlabeled nor labeled histones could be detected in the cytoplasm or in any other sub-cellular fraction except nuclei.[60] Subsequent experiments (not shown) demonstrated that no labeling of histones occurred even after incubation of embryos with ^{14}C from 0 to 12 hr of development.

The data presented in Figures 1 and 2 indicate that histone synthesis is, as expected, coordinated with DNA synthesis in early larvae. A point that should be considered is whether we could have detected histone synthesis if it had occurred in developing encysted embryos. Although radiolabeling of acid-soluble nuclear proteins was extremely low at this stage, radiolabeling of total cell protein was about one third of the highest level seen in larvae (3.3

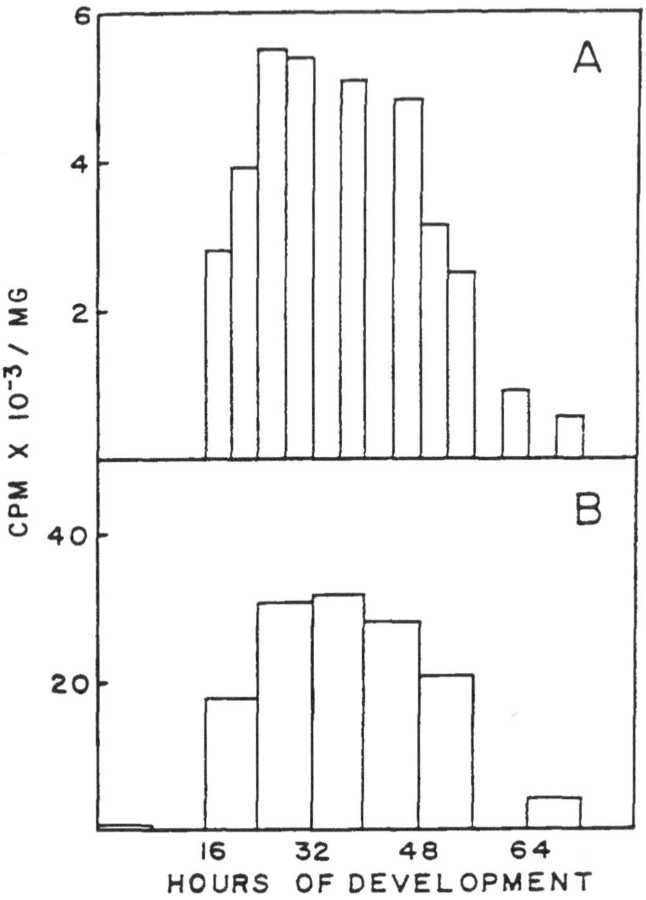

FIGURE 1. Radiolabeling in vivo of DNA and acid-soluble nuclear proteins. Development of *Artemia* embryos was initiated by adding dormant encysted embryos to artificial sea water at zero time. Free-swimming nauplius larvae were harvested at 16 hr of development and further incubated in sea water. Panel A: nauplius larvae were incubated with ^{32}P-orthophosphate for periods of 4 hr indicated by the bars. At the end of each labeling period, incorporation of ^{32}P into DNA was measured. Panel B: developing embryos or larvae were incubated with ^{14}C-bicarbonate for periods of 8 hr indicated by the bars. At the end of each labeling period, incorporation of ^{14}C into acid-soluble nuclear proteins was measured. (From Bagshaw, J. C., *Dev. Genet.*, 3, 41, 1982. With permission.)

$\times$ 10^4 vs 10 $\times$ 10^4 cpm/mg). Synthesis of some of the nonhistone acid-soluble nuclear proteins in developing embryos was clearly seen as bands in the original fluorograph. Moreover, labeling of abundant soluble proteins in the same embryos was readily detected by the same analytical techniques.[11] Given that the mass of histone protein in each gel lane shown in Figure 2 was approximately the same, synthesis of histones should have been detected if it occurred at any time during the labeling period. It is apparant that histones are not synthesized in developing Artemia embryos for at least the first 12 hr after dormant gastrulae resume development.

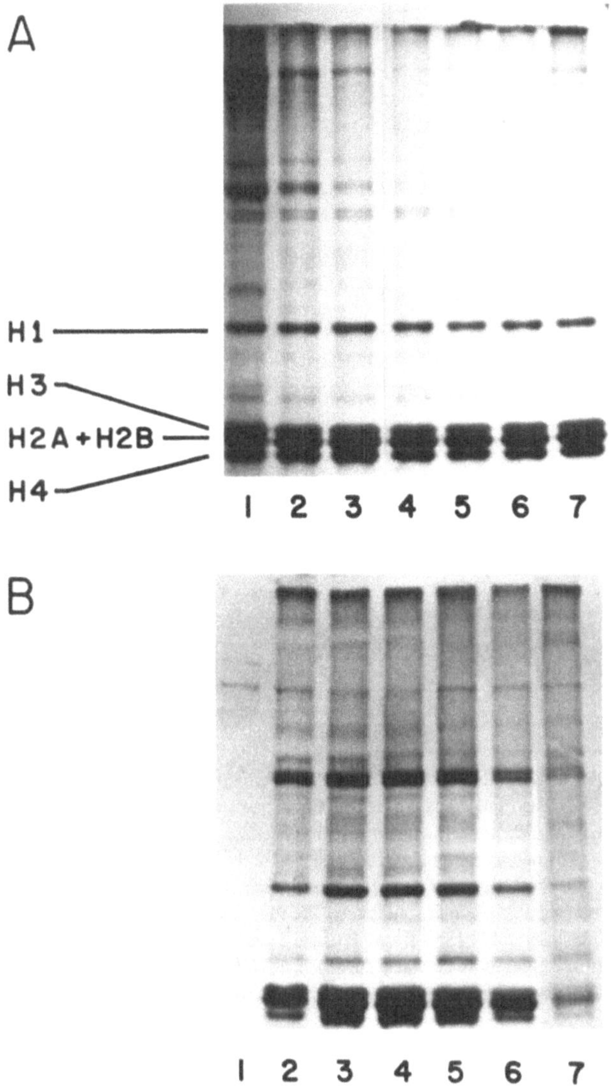

FIGURE 2. Analysis of radiolabeled histones. Developing embryos or larvae were incubated with $NaH^{14}CO_3$ for periods of 8 hr at 0 to 8 hr (lane 1), 16 to 24 hr (lane 2), 24 to 32 hr (lane 3), 32 to 40 hr (lane 4), 40 to 48 hr (lane 5), 48 to 56 hr (lane 6), or 64 to 72 hr (lane 7) of development. Acid-soluble nuclear proteins were fractionated by SDS-PAGE (A). The same gel was then subjected to fluorography (B). (From Bagshaw, J. C., *Dev. Genet.*, 3, 41, 1982. With permission.)

III. GENOMIC ORGANIZATION OF HISTONE GENES IN *ARTEMIA*

A. Cloning and Isolation of *Artemia* Histone Genes

The first step in the isolation of the *Artemia* histone genes was construction of a "library" of *Artemia* genomic DNA sequences. The library was constructed by first isolating *Artemia* nuclear DNA, subjecting it to partial digestion with the restriction enzyme EcoRI, and selecting fragments of 15 to 25 kilobase (kb) pairs by sucrose gradient centrifugation. These fragments were then cloned in the bacteriophage vector lambda Charon 4A. To screen the library for histone genes, we used as a hybridization probe the cloned *Drosophila* genes from the plasmid cDm500,[21] a gift of Dr. David Hogness. A DNA fragment containing one

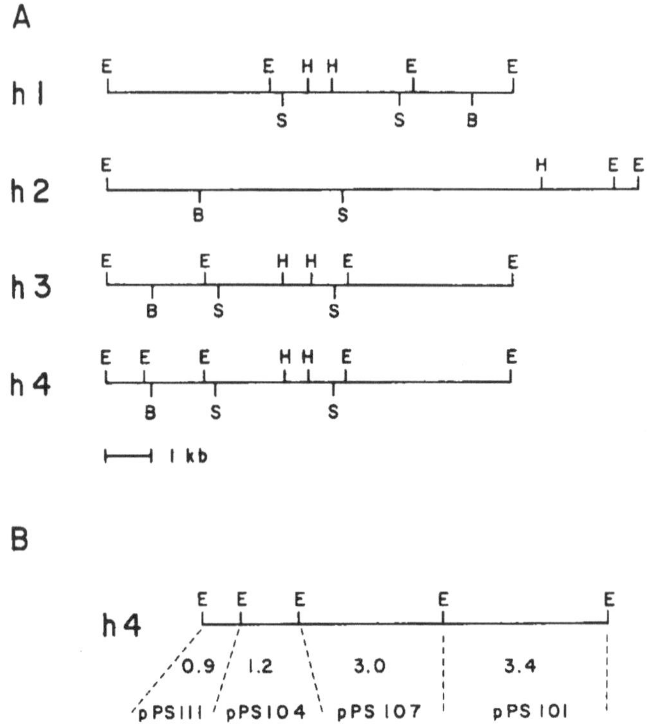

FIGURE 3. Structure of bacteriophage recombinants and plasmid subclones. (A) Restriction maps of recombinant bacteriophage λBSh1, λBSh2, λBSh3, and λBSh4. Only insert regions (h1, h2, h3, h4, respectively) are shown, oriented as they exist in the phage genomes. Sites for the following restriction enzymes are shown: EcoRI (E), BamHI (B), HindIII (H), SmaI (S). (B) Orgin of plasmid subclones. The four EcoRI fragments that comprise the insert of λBSh4 were separately subcloned into the EcoRI site of pBR322. Numbers between dashed lines are sizes (in kb) of the fragments. (From Andrews, M. T. et al., *Gene*, 51, 61, 1987. With permission.)

copy of each of the five histone genes was excised from cDm500 by cleavage with Hind III, isolated by agarose gel electrophoresis, and radiolabeled by nick translation. Approximately 10,000 phage from the *Artemia* gene library were plated, and the resulting plaques, transferred to nitrocellulose, were hybridized with the histone gene probe. Four probe-positive recombinants were found, each of which proved to be authentic after rescreening at lower phage densities. The four isolates were designated λBShl, λBSh2, λBSh3 and λBSh4. To facilitate further structural analysis, each of the four EcoRI fragments from λBSh4 was subcloned into the EcoRI site of pBR322, creating plasmids named pPS101, pPS104, pPS107, and pPS111.

Restriction maps of the inserts of these four isolates, oriented as they exist in the recombinant phage genomes, are shown in Figure 3. It is clear that these four phage isolates arose from four independent cloning events. Recombinants λBSh1 and λBSh3 each have the same inserted EcoRI fragments, with lengths of approximately 3.4, 3.0, and 2.1 kb, but the fragments are arranged in opposite orientations in these two isolates. In λBSh4, the 2.1-kb fragment is divided by an additional EcoRI site into fragments of 1.2 and 0.9 kb. The EcoRI fragments of λBSh2, which measure 10.9 and 0.6 kb, are very different from the other three, but hybridization studies have shown that the inserted DNA sequences in λBSh2 are homologous to specific regions of the other isolates. The restriction maps of the inserted *Artemia* DNA in λBSh1, λBSh3, and λBSh4 are identical except for the additional EcoRI

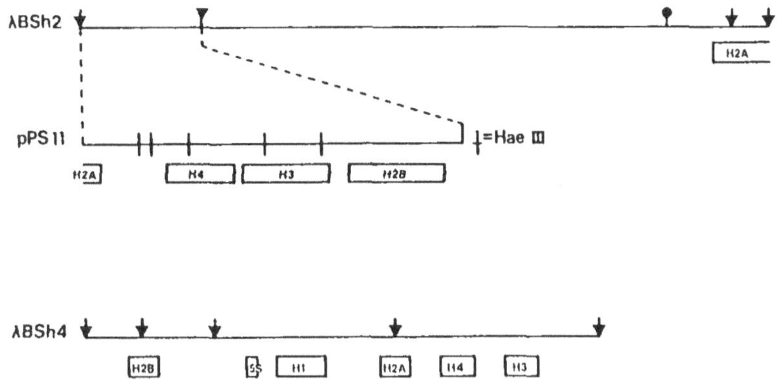

FIGURE 4. Histone gene order in bacteriophage recombinants λBSh2 and λBSh4. Plasmid pPS11 contains a fragment subcloned from λBSh2 as shown. The following restriction sites are indicated: EcoRI, downward arrows; BamHI, triangle; HindIII, circle; HaeIII, vertical lines.

site in λBSh4. The restriction maps suggest that the structures of λBSh1, 3, and 4 represent the most common form of histone gene repeat unit in *Artemia*, and this has been confirmed by hybridization of histone gene subclones to restriction fragments of *Artemia* genomic DNA (see below). The insert in λBSh2 appears to represent a rare variant of histone gene structure.

B. Histone Gene Order in *Artemia*

The clustering of histone genes in different species occurs in a wide variety of patterns of gene order, spacer distances, and transcriptional polarities. In higher organisms containing a relatively low repetition of genes, different histone gene orders within the same species are common, often varying greatly from clone to clone. The invertebrates, with their high copy number, tend to have more conserved histone gene organization with all five genes clustered in a tandemly repeated unit. It was therefore reasonable to suppose that *Artemia* has a relatively constant histone gene order containing all five histone genes, but experimental verification of this hypothesis was necessary.

Restriction fragments derived from λBSh2 and λBSh4 DNA were hybridized with a probe containing one complete set of *Drosophila* histone genes to determine the location of histone genes in recombinants. We also obtained from Dr. Carl Parker a set of recombinant plasmids, each containing portions of one or two *Drosophila* histone genes, from which we isolated probes specific for individual histone genes. We then hybridized these with Hae III fragments of the insert region isolated from plasmid pPS11, subcloned from λBSh2, and with appropriate restriction fragments of λBSh4. The results of these experiments are illustrated in Figure 4. Hybridization of probes for histone genes H2B, H3, and H4 to specific Hae III fragments derived from pPS11 indicated that these three genes are arranged in the order shown. Placement of the H2A gene is based on the observation that a probe for this gene hybridizes weakly with the left end of the pPS11 insert and strongly with the small EcoRI fragment at the right end of the λBSh2 insert. We have not detected any cross hybridization between a cloned *Drosophila* HI gene and any *Artemia* DNA sequence, probably owing to greater sequence divergence in this gene than in other histone genes. The location of the HI gene was based on hybridization of subclone pPS107 from λBSh4 to *Artemia* RNA (see below).

C. Tandem Repetition of *Artemia* Histone Genes

The similarity in structure among λBSh1, 3, and 4 suggested that these three recombinants contained examples of the most common histone gene repeat unit in the *Artemia* genome.

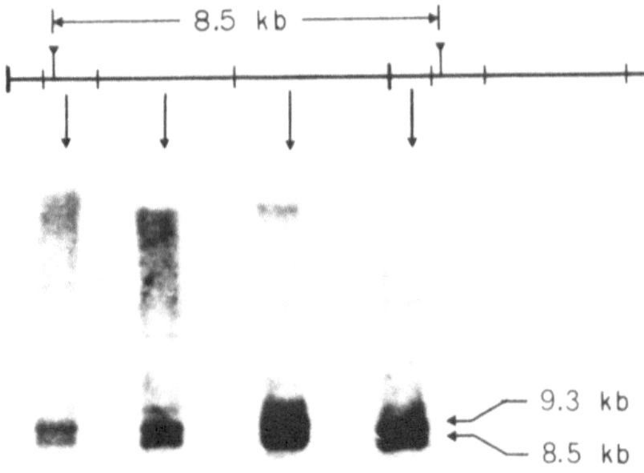

FIGURE 5. Southern blot analysis of *Artemia* genomic DNA. *Artemia* DNA was digested with BamHI, fractionated on a 0.8% agarose gel, and blotted onto nitrocellulose. Identical lanes were separately hybridized with each of the plasmid subclones identified in Figure 3B. At the top is a hypothetical restriction map comprising two tandem repeats of the insert found in λBSh4 (Figure 3A), showing locations of EcoRI sites (vertical lines) and BamHI sites (stalked inverted triangles). Below each EcoRI fragment is shown (see downward arrows) the result of hybridizing the corresponding plasmid to the BamHI-digested *Artemia* DNA. Sizes (in kb) of the two major probe-positive bands are indicated. The probe-positive band at 8.5 kb is the BamHI fragment predicted by the restriction map at the top. (From Andrews, M. T. et al., *Gene*, 51, 61, 1987. With permission.)

The presence of a single Bam HI site in each isolate made it possible to test the hypothesis that *Artemia* histone genes are arranged in a tandem repetition of this structure. DNA isolated from *Artemia* nuclei was digested to completion with Bam HI and replicate samples were fractionated electrophoretically and transferred by blotting to nitrocellulose. Individual lanes of the blot were probed with each of the four subclones from λBSh4. The Bam HI digest (Figure 5) shows two prominent bands of approximately 9.3 and 8.5 kb with each of the four probes. These results verify that the *Artemia* genome contains a tandem repetition of the structure represented by the 8.5-kb inserts of λBSh1, λBSh3, and λBSh4. The probe-positive 9.3-kb band is probably due to an insertion of about 800 bp into the 8.5-kb repeat unit, which resembles the case for the tandem repeats in *Drosophila* where two classes of units differ by a 200-bp insertion in the spacer between H1 and H3.[21] The lack of any other probe-positive bands suggests that the single Bam HI site is highly conserved, probably due to its location in a histone coding sequence.

IV. HISTONE GENE EXPRESSION IN EARLY LARVAE

As noted before, DNA replication and histone synthesis are coordinated in newly hatched nauplius larvae of *Artemia*, but neither occurs during pre-emergence development of encysted embryos.[46,47,50] Amaldi et al.[10] detected histone mRNAs by translation of poly(A)$^-$ RNAs from hatched nauplius larvae, but found no histone mRNAs in dormant embryos by the same test. The apparent absence of functional histone messengers, as judged by translation in vitro, cannot rule out the possible presence of inactive stored forms or "masked" histone mRNAs in dormant or developing embryos. The possible presence of such mRNAs can be tested by employing molecular hybridization with a suitable probe, which will react with the putative mRNAs regardless of their functional status.

Having cloned and isolated the *Artemia* histone genes, we used molecular hybridization to determine whether the dormant grastrulae contain cryptic or "masked" histone mRNA.

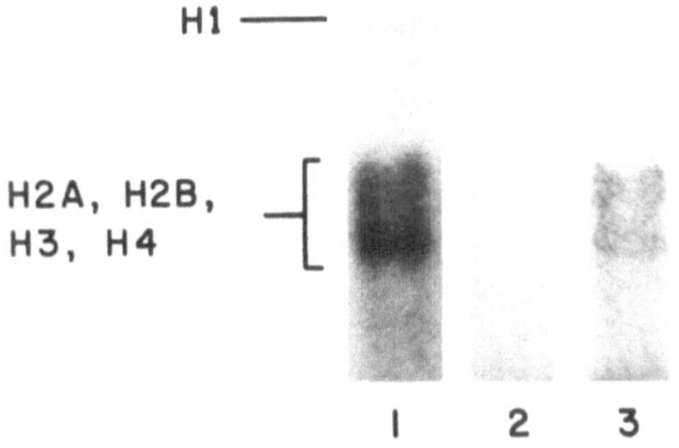

FIGURE 6. Hybridization analysis of histone mRNA sequences. Total RNA from cleavage stage *Drosophila* embryos (lane 1), dormant *Artemia* embryos (lane 2), and *Artemia* larvae at 48 hr of development (lane 3) was fractionated by agarose gel electrophoresis and transferred to nitrocellulose filter material. Filter strips were hybridized with DNA containing *Drosophila* (lane 1) or *Artemia* (lanes 2 and 3) histone genes. Washed filters were then subjected to autoradiography. (From Bagshaw, J. C., *Dev. Genet.*, 3, 41, 1982. With permission.)

The probe used was a recombinant bacteriophage DNA containing *Artemia* histone genes. Whole cell RNAs from dormant *Artemia* gastrulae and from nauplius larvae were fractionated by agarose gel electrophoresis, blotted onto nitrocellulose, and hybridized with the recombinant DNA probe. As a control, *Drosophila* RNA was fractionated on the same gel, blotted, and hybridized with a probe containing *Drosophila* histone genes.[21] As shown in Figure 6, the core histone mRNAs were readily detected but not fully resolved by this procedure. Histone H1 mRNA was well separated from the core histone messengers, but the latter formed only two poorly resolved bands. Nevertheless, the results showed that the *Artemia* histone gene probe hybridized to larval RNAs having the electrophoretic mobility of the *Drosophila* core histone mRNAs (Figure 6, lane 3). Neither *Artemia* nor *Drosophila* histone genes hybridized detectably to RNAs from dormant encysted embryos (Figure 6, lane 2). In a separate experiment not shown here, total RNA from developing (8-hr) embryos also gave no hybridization signal with either histone gene probe. The relatively low abundance of histone mRNAs in *Artemia* larval RNA compared with *Drosophila* embryo RNA probably reflects the relatively leisurely pace of histone synthesis in *Artemia* larvae compared with *Drosophila* cleavage stage embryos. From these results we conclude that the histone mRNAs present in newly hatched *Artemia* larvae arise by transcription of the histone genes, rather than recruitment from a pool of pre-existing "masked" mRNAs.

By the criteria of translation in vitro and molecular hybridization with a cloned *Artemia* histone gene probe, dormant or developing encysted embryos do not contain histone mRNAs. Functional histone mRNAs can be recovered from hatched larvae, and can be detected among total cell RNAs by molecular hybridization. The simplest interpretation of these results is that histone genes are transcriptionally inactive during the pre-emergence period, and become active at or just prior to the period of DNA replication.

For reasons discussed above, the observation that histone and histone mRNA synthesis is coordinated with DNA replication in *Artemia* is not surprising. The significance of this observation lies in the identification of a specific set of genes, the expression of which is regulated during the readily available stages of *Artemia* development. At present this phenomenon should be viewed as an example of developmental regulation, rather than cell cycle regulation. If the cells of dormant *Artemia* embryos are arrested in G1, then developmental stage-specific and cell cycle stage-specific regulation become synonymous in this case. The appearance of histone mRNAs in early larvae will provide a useful model system for studies of transcriptional regulation in *Artemia*. The synchrony of development in the laboratory makes it possible to obtain tens of grams of embryos or larvae in which the histone genes are either transcriptionally active or inactive. This is not possible with the more familiar sea urchins or amphibia, in which histone gene expression is not coordinated with DNA replication during the synchronous early cleavage cycles. Moreover, *Artemia* embryos undergo extensive differentiation and morphogenesis in the absence of cell division (and histone synthesis). If histone gene transcription does not begin until about the time of DNA replication in newly hatched larvae, it will be possible to compare the structure and function of these quiescent genes with other genes actively transcribed in the same developing embryos.

V. INTERSPERSION OF HISTONE AND 5S RNA GENES IN *ARTEMIA*

A. Discovery of the 5S Gene

Early blotting experiments produced the startling result that each of our recombinant phage clones, in addition to hybridizing to histone mRNAs, also hybridized to a highly abundant RNA too small to be any histone mRNA. This was more clearly demonstrated, and the site of the hybridizing DNA sequence was approximately located, by using each of the four plasmid sub-clones depicted in Figure 3B to probe identical samples of *Artemia* polysomal RNA on "Northern" blots. The results are shown in Figure 7. Each plasmid depicted in Figure 3B hybridized to one or more discrete species of RNA. Plasmid pPS107, which contains the 3.0 EcoRI fragment from λBSh4, hybridized to the small RNA and to another RNA of about 850 bases. Based on its size in comparison with histone mRNAs of other invertebrates we assume that this 850-base RNA is the *Artemia* histone Hl mRNA. Plasmids pPS101, pPS104, and pPS111 collectively hybridize to additional RNAs which are presumably the mRNAs for *Artemia* histones H2A, H2B, H3, and H4. The small, highly abundant RNA was subsequently identified as 5S ribosomal RNA on the basis of its size and location in the large (60S) ribosomal subunit.

The presence of a 5S-related sequence in plasmid pPS107 was further confirmed and its location within this subclone determined more precisely by hybridization and restriction mapping experiments. Identification of the small RNA species as 5S RNA was further confirmed by the observation that plasmid 12D1, which contains *Drosophila* 5S RNA genes,[51] hybridizes to the same RNA.[52] The sequence of *Artemia* 5S RNA, and thus the restriction map of a 5S RNA gene, is known.[53] This information enabled us to pinpoint the location of the 5S gene in pPS107 and led to further detailed restriction mapping around this site. As shown in Figure 8, sites for BstE II, Apa I, Hae III, Hpa II, and Sau 3AI were found in pPS107 in the positions predicted by the 5S RNA sequence. Fourteen other restriction enzymes that have no recognition site in the 5S sequence also fail to cut pPS107 within this region. Thus, the restriction map for 20 different enzymes matches exactly that predicted from the sequence of *Artemia* 5S RNA.

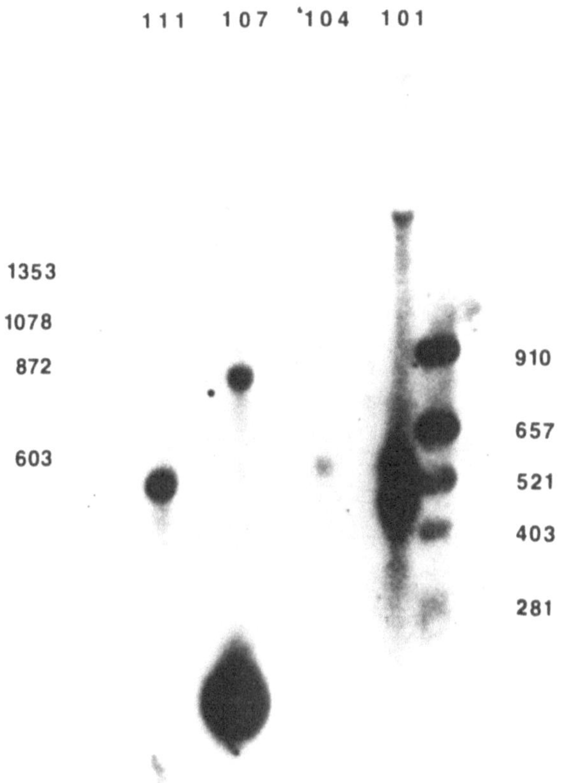

FIGURE 7. Hybridization of λBSh4 subclones to polysomal RNA. Polysomal RNA from nauplius larvae was denatured with glyoxal and dimethylsulfoxide, fractionated on a 1.7% agarose gel, and blotted. Separate lanes were hybridized with pPS101, pPS104, pPS107, and pPS111.ΦX174 RF DNA digested with Hae III (unlabeled, left lane) and pBR322 digested with AluI (end-labeled, right lane) were used as markers. Base lengths of marker fragments are shown.

The insert from plasmid pPS107 was recloned into bacteriophage M13 and the DNA sequence in the region of the 5S gene was determined by the dideoxy method. The result is shown in Figure 9. This region contains a 120-base sequence synonymous with the known sequence of *Artemia* 5S RNA. Within this sequence are several pairs of G residues spaced four or five residues apart, a sequence element identified by Rhodes and Klug[54] as essential for transcription of 5S genes and probably other genes transcribed by RNA polymerase III. We conclude that this is an authentic and functional 5S RNA gene, as opposed to a ''pseudogene''. The gene is flanked by numerous 5 to 7 base direct and inverted repeats, suggesting that the 5S gene may have been transposed into the histone gene cluster some time early in the evolution of *Artemia*.

B. Copy Number of the Histone and 5S Genes

In light of the juxtaposition of histone and 5S genes in our isolated recombinants, it was of special interest to determine the copy number for histone and 5S gene sequences in *Artemia* genomic DNA. Reiteration frequencies of both genes typically number in the hundreds for a majority of invertebrate species. In general, there are many times more copies of 5S

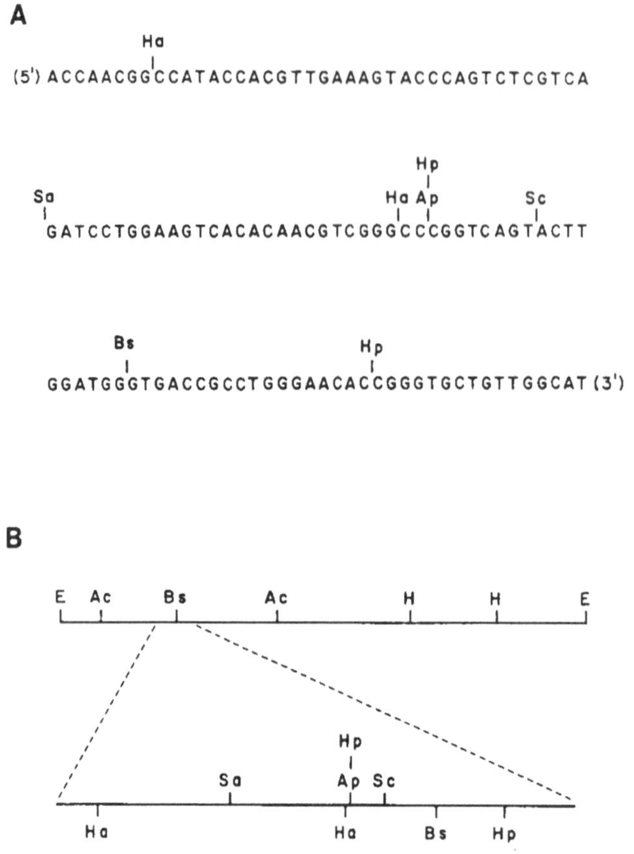

FIGURE 8. Sequence of 5S DNA and structure of plasmid pPS107. (A) Sequence of DNA synonymous with the sequence of *Artemia* 5S RNA.[53] The following predicted restriction sites are indicated: ApaI (Ap), BstEII (Bs), HaeIII (Ha), HpaII (Hp), ScaI (Sc). (B) Restriction map of plasmid pPS107 with an expanded view of one region. pPS107 was digested with EcoRI and ^{32}P-labeled by filling in the ends with the Klenow fragment of DNA polymerase I. The end-labeled fragments were digested with HindIII and a fragment of approximately 2 kb extending from the EcoRI border of the insert to the first HindIII site of the insert was isolated by electrophoresis on a 0.8% agarose gel. This fragment was further digested with additional restriction enzymes and fractionated by electrophoresis on 8% polyacrylamide gels to determine the restriction sites shown here. Restriction sites in addition to those listed in A above are AccI (Ac), EcoRI (E), HindIII (H). (From Andrews, M. T. et al., *Gene*, 51, 61, 1987. With permission.)

RNA genes than of histone genes. Thus, it appeared possible that the *Artemia* genome might contain one or more additional sets of 5S genes. To answer this question we examined the kinetics of hybridization between *Artemia* genomic DNA (driver) and two different tracer DNAs, one containing only histone genes and the other only the 5S gene.

Tracer DNA fragments were excised from plasmids, purified by agarose gel electrophoresis, and ^{32}P-labeled by nick translation. *Artemia* genomic driver DNA was sheared by sonication to an average size of 400 bp. Tracer and driver DNAs were mixed, denatured by boiling in sealed glass capillaries in 0.12 *M* PB (sodium phosphate buffer, pH 6.8) and incubated at 60°C to appropriate values of C_0t. DNA samples were diluted to 0.03 *M* PB, brought to 60°C, and loaded onto hydroxylapatite columns at 60°C. Single-stranded and

571 base pair sequence of ARTSSREG

```
         10          20          30          40          50          60
TCTAG AGGGA GAAGG TGAGG AGGTG GGTAC TTTAA AATAC CTTCC CGGGC ATACT TTAGC
         70          80          90         100         110         120
CTGTA GACCC ATCCC TGAAA GTTTC ATTTT TCTAA CCTAA ACCCT TTCTG AGATA TCAAG
        130         140         150         160         170         180
AAGTC AATTC ACTAG AATTT TACCA AAACA ATTAA TAAAC ATATT TTGAT TCTTT TTAAA
        190         200         210         220         230         240
TATTT GTCTA CTAAA AACCT TGGTT CACGA CTCTA AAACT AGGCA TTTTA TTTCT TTTCA

        250         260         270         280         290         300
GTTTG TGGAC CAACA ACTAT TTTGT TACAA AATTT CTCAG ACCAA CGGCC ATACC ACGTT
        310         320         330         340         350         360
GAAAG TACCC AGTCT CGTCA GATCC TGGAA GTCAC ACAAC GTCGG GCCCG GTCAG TACTT
        370         380         390         400         410         420
GGATG GGTGA CCGCC TGGGA ACACC GGGTG CTGTT GGCAT TTTTT TTGTT TTTGT TTTTT
        430         440         450         460         470         480
TTTCT TATAT ATATA TTTAT ATATT ATAAT GAATT ATTTT CAAGT AAAAT AATAT TCTTT

        490         500         510         520         530         540
AACAA ACTGC TTTCC TTATT CTAAT AGCGC CAAGC TGATT ACATC TTTTA TAAGA AAAAA
        550         560         570         580         590         600
GAAAA ACAGG AAGAG 'ACTAA GTTAC ATCTC C
```

FIGURE 9. DNA sequence in the vicinity of the *Artemia* 5S gene. A 120-base sequence synonymous with *Artemia* 5S RNA[53] is underlined.

double-stranded DNAs were separated on hydroxylapatite columns at 60°C and analyzed by liquid scintillation counting and absorbance at 260 nm. The histone gene probe was the 3.4-kb insert of pPS101 which contains histone genes, H2A, H3, and H4[52] and the 5S probe was the 960-bp Acc I fragment from pPS107 which contains no histone gene sequences. Kinetic values for the *Artemia* genome determined in this experiment were plotted in order to construct the genomic renaturation curve shown in Figure 10. The single-copy $C_0t_{1/2}$ of 1500 was used for the determination of the histone gene and 5S gene reiteration frequencies. Tracer DNA C_0t values and their corresponding percent reassociations are shown in Figure 10, fitted to computer-generated second reassociation curves. The nearly identical profiles for both curves result in nearly identical $C_0t_{1/2}$ values: 15.5 for the Acc I fragment containing the 5S gene and 16.0 for the histone genes. These values correspond to 97 copies of the Acc I fragment and 94 copies of histone genes per haploid genome, indicating that genes for histones 5S RNA gene have the same reiteration frequency in *Artemia*. It is especially significant that none of the tracer containing the 5S RNA gene was driven into hybrids by any other component of genomic DNA. On the basis of kinetic analysis, we find no evidence for any 5S RNA genes in *Artemia* other than the ones interspersed with histone genes.

C. Functional and Evolutionary Significance

The presence of multigene families organized in tandem arrays of DNA sequence is a common feature of eukaryotic genomes. The repeat unit may contain a single functional gene, as in the case of both nucleolar and extranucleolar ribosomal RNA genes, or it may include a number of related genes as in the histone gene repeats of *Drosophila* and sea urchins.[55,56] Tandemly repeated genes are separated by "spacer" DNA of unknown function. Although repeat units may contain pseudogenes clearly related to the functional genes, as in the 5S RNA genes of *Xenopus*, regular interspersion of genes of two apparently unrelated repeated families has not previously been reported. We have shown that the genes for all five histone proteins in *Artemia* are clustered and flank a single gene for 5S RNA and that this novel structure is repeated in tandem with reiteration frequency of approximately 95 to 100 copies per haploid genome. It has been proposed that the generation and maintenance of tandemly reiterated multigene families is due to strong selection for expression of their gene products.[57] The rapid cell division of early embryogenesis requires large amounts of histone proteins to package and condense the newly synthesized DNA, and the accompanying

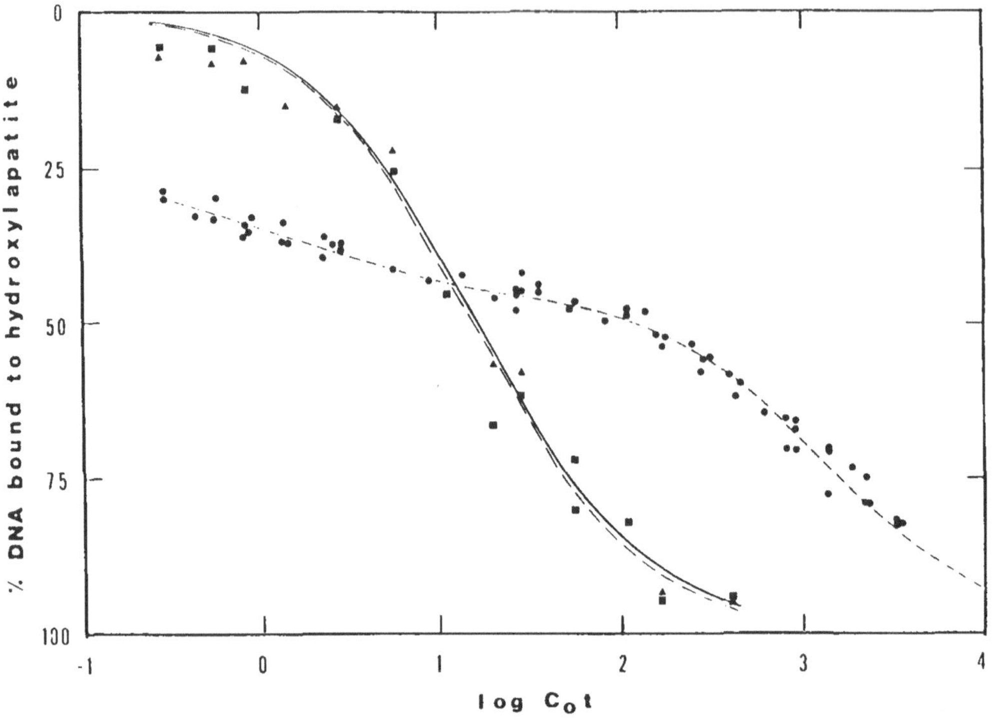

FIGURE 10. Hybridization kinetic analysis of *Artemia* DNA. Histone gene tracer was the complete insert from plasmid pPS101 (see Figure 3) excised by digestion with EcoRI. 5S gene tracer DNA was the 960-bp AccI fragment from the insert of plasmid pPS107 (See Figure 8). These fragments were purified by electrophoresis on 1.0% agarose gels and ^{32}P-labeled by nick translation. Driver DNA was *Artemia* genomic DNA sheared to an average fragment size of 400 bp. Circles, driver DNA; squares, histone gene tracer; triangles, 5S gene tracer. (From Andrews, M. T. et al., *Gene*, 51, 61, 1987. With permission.)

amplification of protein synthesis requires increased production of 5S RNA for ribosome formation. The occurrence of histone and 5S RNA genes clustered together in the same repeat unit could allow simultaneous transcriptional control of both genes at a common locus. An alteration in chromatin structure proximal to the repeat unit, such as the derepression of *Xenopus* oocyte 5S RNA genes by removal of histone H1,[58] could place both genes in a conformation accessible to gene-specific transcription factors and RNA polymerases. Our results also raise some interesting questions regarding the evolution of histone and 5S genes in crustacea. Repeated gene families are believed to have arisen by tandem duplication of an existing structure, sometimes followed by dispersal throughout the genome (e.g., mammalian histone genes). In *Drosophila* and in sea urchins, histone genes and 5S RNA genes are organized in tandemly repeated but separated blocks in the genome. In *Artemia* these genes are interspersed. This suggests that the histone and 5S RNA genes were already adjacent in the ancestor of *Artemia* prior to duplication of the ancestral gene cluster. A further implication is that the 5S gene was transposed to the vicinity of the histone genes prior to the tandem duplication of all these genes as a unit.

REFERENCES

1. **Kedes, L. H.**, Histone messengers and histone genes, *Cell*, 8, 321, 1976.
2. **Kedes, L. H.**, Histone genes and histone messengers, *Annu. Rev. Biochem.*, 48, 837, 1979.
3. **Gross, K. W., Jacobs-Lorena, M., Baglioni, C., and Gross, P. R.**, Cell-free translation of maternal messenger RNA from sea urchin eggs, *Proc. Natl. Acad. Sci. U.S.A.*, 70, 2614, 1973.
4. **Grunstein, M., Levy, S., Schedl, P., and Kedes, L.H.**, Messengers RNAs for individual histone proteins: fingerprint analysis and *in vitro* translation, *Cold Spring Harbor Symp. Quant. Biol.*, 38, 717, 1973.
5. **Levy, S., Wood, P., Grunstein, M., and Kedes, L. H.**, Individual histone messenger RNAs: identification by template activity, *Cell*, 4, 239, 1975.
6. **Borun, T. W., Gabrielli, F., Ajiro, K., Zweidler, A., and Baglioni, C.**, Further evidence of transcriptional and translation control of histone messenger RNA during the Hela S3 cell cycle, *Cell*, 4, 59, 1975.
7. **Adesnik, M. and Darnell, J. E.**, Biogenesis and characterization of histone messenger RNA in HeLa cells, *J. Mol. Biol.*, 67, 397, 1972.
8. **Skoultchi, A. and Gross, P. R.**, Maternal histone messenger RNA: detection by molecular hybridization, *Proc. Natl. Acad. Sci. U.S.A.*, 70, 2840, 1973.
9. **Levenson, R. G. and Marcu, K. B.**, On the existence of polyadenylated histone mRNA in *Xenopus laevis* oocytes, *Cell*, 9, 311, 1976.
10. **Amaldi, P., Felicetti, L., and Campioni, N.**, Template activity for histones of a poly(A)-minus RNA fraction from different developmental stages of *Artemia salina* embryos, *Biochim. Biophys. Acta*, 518, 518, 1978.
11. **Bagshaw, J. C. and Warner, A. H., Eds.**, Biochemistry of *Artemia*, University Microfilms International, Ann Arbor, Mich., 1979.
12. **Wilson, M. C., Melli, M., and Birnsteil, M. L.**, Reiteration frequency of histone coding sequences in man, *Biochem. Biophy. Res. Commun.*, 61, 354, 1974.
13. **Wilson, M. C. and Melli, M.**, Determination of the number of histone genes in human DNA, *J. Mol. Biol.*, 110, 511, 1977.
14. **Kedes, L. H. and Birnstiel, M.**, Repetition and clustering of DNA sequences complimentary to histone messenger RNA, *Nature (London) New Biol.*, 230, 165, 1971.
15. **Weinberg, E. S., Birnstiel, M. L., Purdom, I. F., and Williamson, R.**, Genes coding for polysomal 9S RNA of sea urchins: conservation and divergence, *Nature (London)*, 240, 225, 1972.
16. **Birnstiel, M. L., Telford, , J., Weinberg, E. S., and Stafford, D.**, Isolation and some properties of the genes coding for histone proteins, *Proc. Natl. Acad. Sci. U.S.A.*, 71, 2900, 1974.
17. **Schaffner, W., Gross, K., Telford, J., and Birnstiel, M.**, Molecular analysis of the histone gene cluster of *Psammechinus miliaris*. II. The arrangement of the five histone-coding and spacer sequences, *Cell*, 8, 471, 1976.
18. **Cohn, R., Lowry, J. C., and Kedes, L. H.**, Histone genes of the sea urchin *(S. purpuratus)* cloned in *E. coli:* order, polarity and strandedness of the five histone-coding and spacer regions, *Cell*, 9, 147, 1976.
19. **Gross, K., Schaffner, W., Telford, J., and Birnstiel, M.**, Molecular analysis of the histone gene cluster of *Psammechinus miliaris*. III. Polarity and asymmetry of the histone-coding sequences, *Cell*, 8, 479, 1976.
20. **Sures, I., Lowry, J., and Kedes, L. H.**, The DNA sequence of sea urchin *(S. purpuratus)* H2A, H2B and H3 histone coding and spacer regions, *Cell*, 15, 1033, 1978.
21. **Lifton, R. P., Goldberg, M. L., Karp, R. W., and Hogness, D. S.**, The organization of the histone genes in *Drosophila melanogaster:* functional and evolutionary implications, *Cold Spring Harbor Symp. Quant. Biol.*, 42, 1047, 1978.
22. **Golberg, M. L.**, Sequence Analysis of *Drosophila* Histone Genes, Ph.D. thesis, Stanford University, Stanford, Calif., 1979.
23. **Karp, R.**, The Organization of the Histone Genes in *Drosophila melanogaster*, Ph.D. thesis, Stanford University, Stanford, Calif., 1980.
24. **Newrock, K. M., Cohen, L. H., Hendricks, M. B.,, Donnelly, R. J., and Weinberg, E. S.**, Stage-specific mRNAs coding for subtypes of H2A and H2B histones in the sea urchin embryo, *Cell*, 14, 327, 1978.
25. **Kunkel, N. S. and Weinberg, E. S.**, Histone gene transcripts in the cleavage and mesenchyme blastula embryo of the sea urchin, *S. purpuratus*, *Cell*, 14, 313, 1978.
26. **Childs, G., Levy, S.,, and Kedes, L. H.**, Rapid purification of biologically active individual histone mRNAs by hybridization to cloned DNA linked to cellulose, *Biochemistry*, 18, 208, 1979.
27. **Childs, G., Maxson, R., and Kedes, L. H.**, Histone gene expression during sea urchin embryogenesis: isolation and characterization of early and late messenger RNAs of *Strongylocentrotus purpuratus* by gene specific hybridization and template activity, *Dev. Biol.*, 73, 153, 1979.
28. **Grunstein, M.**, Hatching in the sea urchin *Lytechinus pictus* is accompanied by a shift in histone H4 gene activity, *Proc. Natl. Acad. Sci. U.S.A.*, 75, 4135, 1978.

29. **Old, R. W., Woodland, H. R., Ballantine, J. E. M., Aldridge, T. C., Newton, C. A., Bains, W. A., and Turner, P. C.**, Organization and expression of cloned histone gene clusters from *Xenopus laevis* and *X. borealis*, *Nucleic Acids Res.*, 10, 23, 1982.
30. **Van Dongen, W., de Laaf, L., Zall, R., Moorman, A., and Destree, O.**, The organization of the histone genes in the genome of *Xenopus laevis*, *Nucleic Acids Res.*, 9, 10, 1981.
31. **Stephenson, E. C., Erba, H. P., and Gall, J. G.**, Histone gene clusters of the newt *Notophthalmus* are separated by long tracts of satellite DNA, *Cell*, 24,, 639, 1981.
32. **Connor, W., Mezquita, J., Winkfein, R. J., States, J. C., and Dixon, G. H.**, Organization of the histone genes in the rainbow trout *(Salmo gairdnerii)*, *J. Mol. Evol.*, 20, 227, 1984.
33. **D'Andrea, R. J., Coles, L. S., Lesnikowski, C., Tabe, L., and Wells, J. R. E.**, Chromosomal organization of chicken histone genes: preferred associations and inverted duplications, *Mol. Cell. Biol.*, 5, 3108, 1985.
34. **Engel, J. D. and Dodgson, J. B.**, Histone genes are clustered but not tandemly repeated in the chicken genome, *Proc. Natl. Acad. Sci. U.S.A.*, 78, 2859, 1981.
35. **Sittman, D. B., Chiu, I., Pan, C., Cohn, R. H., Kedes, L. H., and Marzluff, W. F.**, Isolation of two clusters of mouse histone genes, *Proc. Natl. Acad. Sci. U.S.A.*, 78, 4078, 1981.
36. **Sierra, F., Lichtler, A., Marashi, F., Rickles, R., Van Dyke, T., Clark, S., Wells, J., Stein, G., and Stein, J.**, Organization of human histone genes, *Proc. Natl. Acad. Sci. U.S.A.*, 79, 1795, 1982.
37. **Childs, G., Nocente-McGrath, C., Lieber, T., Holt, C., and Knowles, J. A.**, Sea urchin *(Lytechinus pictus)* late-stage histone H3 and H4 genes: characterization and mapping of a clustered but nontandemly linked multigene family, *Cell*, 31, 383, 1982.
38. **Borun, T. W., Gabrielli, F., Ajiro, K., Zweidler, A., and Baglioni, C.**, Further evidence of transcriptional and translational control of histone messenger RNA during the HeLa cell cycle, *Cell*, 4, 59, 1975.
39. **Stein, G., Stein, J., Shaphard, E., Park, W., and Phillips, I.**, Evidence that the coupling of histone gene expression and DNA synthesis in HeLa S_3 cells is not mediated at the transcriptional level, *Biochem. Biophys. Res. Commun.*, 77, 245, 1977.
40. **Parker, I. and Fitschen, W.**, Histone mRNA metabolism during the mouse fibroblast cell cycle, *Cell Differentiation*, 9, 23, 1980.
41. **Stein, J. L., Thrall, C. L., Park, W. D., Mans, R. J., and Stein, G. S.**, Hybridization analysis of histone messenger RNA: association with polyribosomes during the cell cycle, *Science*, 189, 557, 1975.
42. **Detke, S., Stein, J. L., and Stein, G. S.**, Synthesis of histone mRNAs by RNA polymerase II in nuclei from S-phase HeLa S3 cells, *Nucleic Acids Res.*, 5, 1515, 1978.
43. **Arceci, R. J. and Gross, P. R.**, Noncoincidence of histone and DNA synthesis in cleavage cycles of early development, *Proc. Natl. Acad. Sci. U.S.A.*, 74, 5016, 1977.
44. **Adamson, E. D. and Woodland, H. R.**, Histone synthesis in early amphibian development: histone and DNA syntheses are not co-ordinated, *J. Mol. Biol.*, 88, 263, 1974.
45. **Adamson, E. D. and Woodland, H. R.**, Changes in the rate of histone synthesis during oocyte maturation and very early development of *Xenopus laevis*, *Dev. Biol.*, 57, 136, 1977.
46. **Nakanishi, Y. H., Okigaki, T., Kato, H., and Iwasaki, T.**, Cytological studies of Artemia salina II. Deoxyribonucleic acid (DNA) content and chromosomes in encysted dry eggs and nauplii, *Annot. Zool. Jpn.*, 35, 223, 1962.
47. **McClean, D. K. and Warner, A. H.**, Aspects of nucleic acid metabolism during development of the brine shrimp Artemia salina, *Dev. Biol.*, 24, 88, 1971.
48. **Clegg, J. S.**, Interrelationships between water and cellular metabolism in *Artemia* cysts, *J. Cell Physiol.*, 89, 369, 1976.
49. **Johns, E. W.**, Studies on histones. VII. Preparative methods for histone fractions from calf thymus, *Biochem. J.*, 92, 55, 1964.
50. **Bagshaw, J. C.**, Developmentally regulated gene expression in *Artemia:* histone gene expression in newly hatched larvae, *Dev. Genet.*, 3, 41, 1982.
51. **Artavanis-Tsakonas, S., Schedl, P., Tschudi, C., Pirrotta, V., Steward, R., and Gehring, W.**, The 5S genes of *Drosophila melanogaster*, *Cell*, 12, 1057, 1077.
52. **Andrews, M. T.**, The Organization of Histone Genes in *Artemia*, Ph.D. thesis, Wayne State University, Detroit, Mich., 1983.
53. **Diels, L., DeBaere, R., Vandenberghe, A., and DeWatcher, R.**, The sequence of the 5S ribosomal RNA of the crustacean *Artemia salina*, *Nucleic Acids Res.*, 9, 5141, 1981.
54. **Rhodes, D. and Klug, A.**, An underlying repeat in some transcriptional control sequences corresponding to half a double helical turn of DNA, *Cell*, 46, 123, 1986.
55. **Long, E. O. and Dawid, I. B.**, Repeated genes in eukaryotes, *Annu. Rev. Biochem.*, 49, 727, 1980.
56. **Maxson, R., Cohn, R., Kedes, L., and Mohun, T.**, Expression and organization of histone genes, *Annu. Rev. Genet.*, 17, 239, 1983.
57. **Fedoroff, N. V.**, On spacers, *Cell*, 16, 697, 1979.

58. **Schlissel, M. S. and Brown, D. D.,** The transcriptional regulation of *Xenopus* 5S RNA genes in chromatin: the roles of active stable transcription complexes and histone H1, *Cell*, 37, 903, 1984.
59. **Andrews, M. T., Vaughn, J. C., Perry, B. A., and Bagshaw, J. C.,** Interspersion of histone and 5S RNA genes in *Artemia, Gene*, 51, 61, 1987.
60. **Bagshaw, J. C.,** unpublished data.

Chapter 2

ACTIVATION OF STORED MESSENGER RIBONUCLEOPROTEINS: IDENTIFICATION AND FUNCTION OF THE PROTEINS ASSOCIATED WITH NONPOLYSOMAL POLY(A)-CONTAINING MESSENGER RIBONUCLEOPROTEINS OF CRYPTOBIOTIC GASTRULAE OF *ARTEMIA* SP.

H. Slegers, E. De Herdt, E. Piot, H. Backhovens, C. Thoen, L. Van Hove, E. Roggen, and M. Aerden

TABLE OF CONTENTS

I. INTRODUCTION

In the cytoplasm of eukaryotic cells messenger ribonucleoproteins (mRNP) exist as free mRNP, mRNP associated with ribosomal subunits into preinitiation complexes, and as polysomal mRNP. Free mRNP was detected in the cytoplasm of fish extracts in 1964.[1] Since then, mRNP has been characterized in many eukaryotic cells and in different cell compartments and it has become generally accepted that mRNAs and their precursors are associated with proteins.[2,3] One of the main protein binding sites is the poly(A)-sequence located at the 3′- end of the majority of mRNA. Apparently the latter segment is not needed for translation but is almost universal in eukaryotes and enables purification of mRNA and mRNP by affinity chromatography on oligo(dT)- or poly(U)-matrices.

The free mRNP forms a heterogeneous population of translationally active and repressed mRNP. The repressed mRNP may be stored mRNA to be translated at a later stage in development, DNA transcripts in excess of what can be translated, incompletely processed mRNP, or "run off" mRNP released from polysomes between two translation cycles. The repression of mRNP is mainly due to the reversible association of translation inhibitors such as proteins or small ribonucleoproteins.[3] mRNP may also become untranslatable by irreversible modifications including degradation by cellular ribonucleases.

Although a discrete number of proteins is associated with mRNA, the composition varies with its location in the cell and an exchange of proteins has been observed after its transport from the nucleus to the cytoplasm and after entry into polysomes.[2-7] Furthermore, the protein composition of mRNP is dependent on the cellular origin and no uniform and universal composition has been demonstrated. The published data are difficult to interpret due to the various techniques used in the isolation of mRNP, to tissue specificity, to the existence of different functional forms of mRNP, and to nonspecific binding of proteins to mRNA. As such, the study of the protein composition of mRNP is even more complicated than the study of the protein composition of ribosomes some 20 years ago. The lack of suitable standard conditions to purify mRNP and the existence of different functional classes of mRNP certainly delayed further studies on the functional significance of mRNP and mRNP proteins. Despite these difficulties it is generally assumed that mRNP proteins have a function in mRNA-related phenomena. Several functions have been postulated in the past and some evidence has been presented which demonstrates the functional significance of mRNA-associated proteins in the mechanism of mRNA biogenesis,[8,9] stability of mRNA,[8,10] transport from nucleus to cytoplasm,[11-14] and regulation of translation.[8,15-19] Up to now a small number of enzymatic activities, such as initiation and elongation factors of protein synthesis, protein kinases, and alkaline phosphatases[20-25] have been demonstrated to be associated with mRNA. It is clear that all these activities are not associated with mRNP at the same time but become associated according to a dynamic sequence. The latter sequence and the proteins involved in it have to be identified in order to understand the pathway from precursor heterogeneous nuclear (hn)RNP to polysomal mRNP active in protein synthesis. This fundamental problem may only be solved in systems which contain a few mRNPs in abundance.

Artemia sp. is one of these systems. Under certain conditions the embryos of the brine shrimp enter a cryptobiotic phase.[26,27] In this phase no polysomes are observed but their rapid formation after reinitiation of development is due to the presence of preinitiation complexes and stored mRNP.[28-30] The stored mRNP is present in abundance and its protein composition is very discrete. The latter property makes *Artemia* sp. cryptobiotic embryos a valuable system for the study of the function of mRNP proteins and for the study of the pathway from free repressed mRNP to polysomal mRNP active in protein synthesis.

II. POLY(A)-CONTAINING mRNP OF CRYPTOBIOTIC GASTRULAE

A. Distribution of Poly(A)-Containing mRNP

The cellular distribution of poly(A)-containing mRNP was measured by hybridization with (^{3}H)-poly(U).[30] Sixty-two percent of the poly(A)-sequences was localized in the postmitochondrial supernatant. Of the remaining mRNP 19% was associated with the endoplasmatic reticulum and was released with 1% Triton X-100 and 19% was associated with the cytoskeleton. From the latter network only 8% of the mRNP was released with 2% sodium deoxycholate. Eleven percent of the poly(A)-sequences was not released even after repeated washes with detergent.[31] The poly(A)-containing mRNP of the cytoplasm and membrane fraction had a different sedimentation pattern.[30-31] The cytoplasmic mRNP revealed a heterogeneous distribution with at least six discrete classes sedimenting at approximately 15S, 40S, 60S, 70S, 85S, and 110S. Less than 12% of the poly(A)-containing mRNP sedimented faster than the 80S ribosome. The sedimentation pattern was unaltered when the postmitochondrial supernatant was prepared in the presence of Triton X-100 or the RNAse inhibitor bentonite. No extensive aggregation was observed as estimated from the amount of poly(A)-sequences sedimenting to the bottom of the gradient. The mRNPs sedimenting slower than 45S constituted approximately 50% of the poly(A)-containing mRNP and was separated by a prolonged centrifugation.[32] Forty one percent of this fraction sedimented at 17S, 28% between 20 to 28S, and 31% at 34S, respectively. The mRNP sedimenting faster than 45S were characterized as mRNP-ribosome preinitiation complexes by EDTA treatment of the postmitochondrial supernatant.[30] The latter treatment converted mRNP with sedimentation coefficients above 45S to particles sedimenting between 10 to 30S, concomitant with a dissociation of ribosomes into slow sedimenting subunits of 40S (large subunit) and 23S (small subunit).[33] The mRNA-40S ribosome preinitiation complexes sediment at 48S and the higher sedimenting particles arise from multiple attachment of small ribosomal subunits to mRNP.[34]

Only two mRNP classes sedimenting at 36S and 17S were associated with the membrane fraction and mRNP-ribosome preinitiation complexes were absent.[31] The lack of these complexes suggests developmental differences in the translation of membrane-associated and free mRNP from cryptobiotic gastrulae. As observed in other systems, our results support the existence of a ribosome-independent functional mRNP class directly associated with the endoplasmic reticulum and the cytoskeleton.[35] Although membrane association by proteins located near the 3′-end of mRNA is often proposed, the actual binding is speculative.

B. Purification and Protein Composition

The main step in the purification of mRNP is affinity chromatography on oligo(dT)-cellulose. High salt concentrations (500 m*M*) are normally used in the binding of poly(A)-containing mRNP on oligo(dT)-cellulose but this may result in the removal of functional

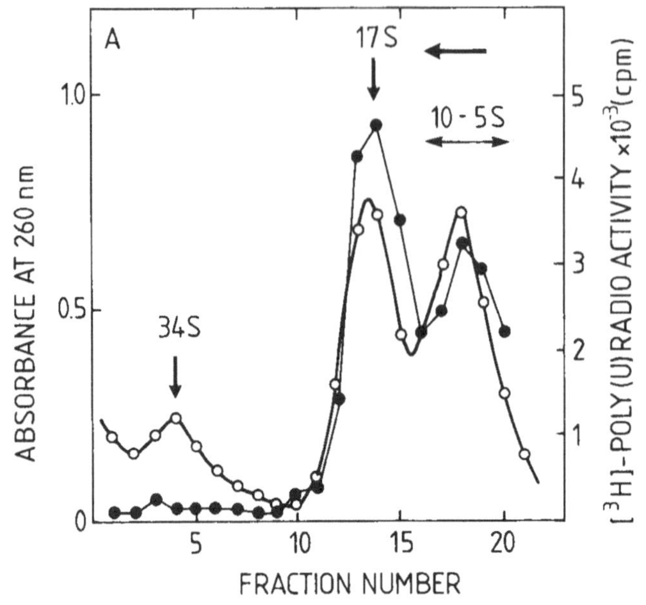

FIGURE 1A. Size distribution of free poly(A)-containing mRNP. Poly(A)-containing mRNP was purified by affinity chromatography on oligo(dT)-cellulose in 10 m*M* Hepes pH 7.2, 250 m*M* KCl, and 0.1 m*M* DTT. mRNP was eluted with 10 m*M* Hepes pH 7.2, 0.1 m*M* DTT, and at 36°C, concentrated by ultrafiltration and analyzed by 10 to 30% (w/v) sucrose density gradient centrifugation in 10 m*M* Hepes pH 7.2, 100 m*M* KCl, 0.1 m*M* DTT, and 0.1 mg/mℓ heparin. Centrifugation was in a Beckman SW27 rotor at 82,500 *g* for 15 hr and at 4°C. Poly(A)-sequences were located by hybridization with (^{3}H)-poly(U).[30] The direction of centrifugation is indicated. Radioactivity (●——●); absorbance at 260 nm (○——○).

proteins from mRNP. Therefore, a moderate salt concentration (100 to 250 m*M*) was used to bind mRNP. The latter was eluted in the absence of salt and at 36°C. Different buffer and salt conditions were used for the preparation of mRNP without striking differences in protein composition.[32]

The purified cytoplasmic mRNP is composed of a discrete number of proteins with M_r of 115000, 87,000, 76,000, 65,000, 50,000, 45,000, 38,000, and 23,500. Only in the presence of $MgCl_2$ an additional protein with an M_r of 21,000 was observed.[32] The main mRNP protein has a molecular weight of 38,000 and represents 60 to 70% of the total amount of protein. The association of mRNP to oligo(dT)-cellulose is mediated by the poly(A)-sequence as shown by rechromatography of mRNP on oligo(dT)-cellulose after hybridization with poly(U). In the latter conditions less than 5% of the mRNP was rebound to oligo(dT)-cellulose. As the M_r 38,000 protein is the poly(A)-binding protein (Section IV. A) it is surprising that the presence of 60 to 70% of the total amount of mRNP protein on the poly(A)-sequence still allows the poly(A)-tail to interact with oligo(dT).

The size distribution of the mRNP purified on oligo(dT)-cellulose was determined on a 10 to 30% (w/v) linear sucrose gradient (Figure 1A). The absorbance at 260 nm and the hybridization with (^{3}H)-poly(U) was coincident and revealed the presence of three discrete mRNP classes sedimenting at 34S, 17S, and 5 to 10S. More than 60% of the poly(A)-sequences were associated with the 17S mRNP. Compared to the distribution of mRNP before oligo(dT)-cellulose, the amount of the 34S mRNP was reduced to 3%. Thirty-seven percent of the poly(A)-sequences sedimenting at the top of the gradient was degraded mRNP.[32] It is likely that the decrease in the 34S mRNP was due to inefficient binding on

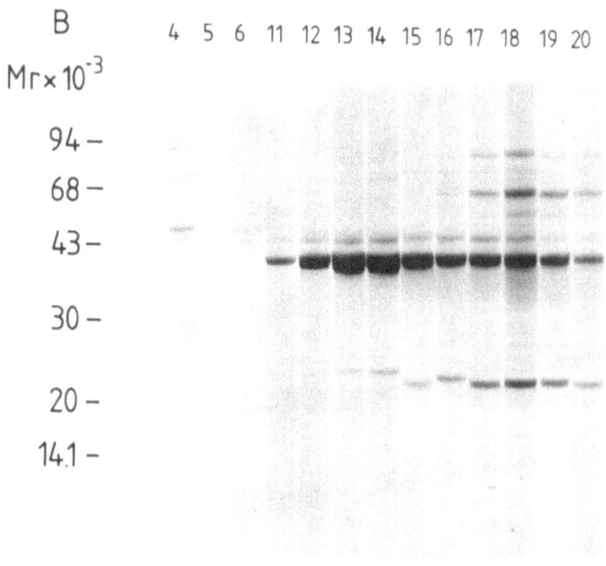

FIGURE 1B. Protein composition of poly(A)-containing mRNP. Gradient fractions were analyzed by sodium dodecylsulfate polyacrylamide gel electrophoresis. Electrophoresis was as described.[32] Standards used for molecular weight estimation were phosphorylase b (94,000), serum albumin (68,000), ovalbumin (43,000), carbonic anhydrase (30,000), trypsin inhibitor (20,000) and α-lactalbumin (14,000).

oligo(dT)-cellulose.[37] The protein composition of the discrete sedimenting particles was determined by sodium dodecylsulfate polyacrylamide gel electrophoresis (Figure 1B). The 34S mRNP is composed of proteins with M_r of 50,000, 45,000, 42,000, and 38,000. The M_r 38,000 protein is the poly(A)-binding protein and is only a minor component of the latter mRNP (Section IV.A.1.) The 17S mRNP is composed of proteins with M_r of 65,000, 45,000, and 38,000 and proteins with M_r of 76,000, 57,500, 50,000, 45,000, 38,000 and 22,000 were observed at the top of the gradient. The integrity of mRNP was further demonstrated by isopycnic centrifugation in CsCl after fixation of the particles with formaldehyde.[32] The buoyant density of mRNP is 1.4 g/cm^3 and is a characteristic property of stored mRNP.[2] From the density a protein content of 78% was measured, compared to 71% calculated from RNA and protein concentration measurements. The high protein content is in agreement with the results of others.[16,39-41]

The protein composition of cytoplasmic mRNP was compared with those of mRNP purified from preinitiation complexes and from the membrane fraction by affinity chromatography on oligo(dT)-cellulose (Figure 2). mRNP from preinitiation complexes is composed of the same proteins as free mRNP although some differences in the relative amounts of the proteins were observed. With the exception of two proteins with a molecular weight of 60,000 the same proteins are present in mRNP associated with the membrane fraction. The proteins with M_r of 60,000 may be involved in the binding of mRNP to membranes. By chromatography on poly(A)-Sepharose 4B the latter proteins were identified as membrane proteins with affinity for RNA.[38] In amphibian oocytes a 10-nm intermediate filament-forming protein of similar molecular weight has been isolated from hnRNP[42] and from stored mRNP.[43]

III. THE REPRESSION OF POLY(A)-CONTAINING mRNP

A. Translationally Active and Repressed mRNP

Although the concept of "masked" mRNA has been generally accepted, controversial

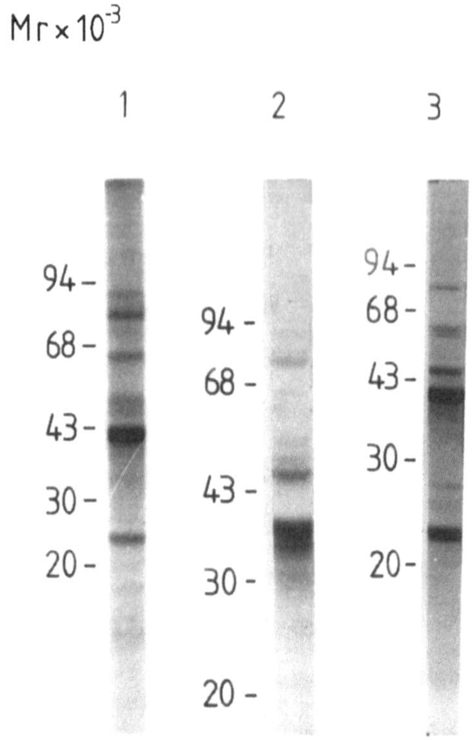

FIGURE 2. Comparison of the protein composition of free mRNP, mRNP from preinitiation complexes, and membrane-associated mRNP. The preparation of mRNP has been described.[32,38] The protein composition was determined by sodium dodecylsulfate polyacrylamide gel electrophoresis.[32] Standards used for molecular weight estimation are listed in the caption of Figure 1B. Lane 1: free mRNP, lane 2: mRNP from preinitiation complexes, lane 3: membrane-associated mRNP.

reports have been published concerning the translatability of nonpolysomal mRNP and the existence of translatable and nontranslatable mRNP has been demonstrated in several systems.[44-48] The apparent contradictory reports may be a consequence of the conditions of isolation or may simply reflect the difference between two functional classes of mRNP isolated from dormant systems, e.g., unfertilized eggs, cryptobiotic embyros, and from developing systems.[11] Contradictory results have also been reported on the translatability of mRNA isolated from cryptobiotic embryos of *Artemia* sp.[49-53] We have been able to dissociate an oligoribonucleotide inhibitor from the repressed mRNA by EDTA treatment suggesting that the conditions of isolation may be responsible for the contradictory results.[52] Our previous results were confirmed when we isolated an inhibitor ribonucleoprotein from repressed nonpolysomal mRNP.[54] mRNP was purified by affinity chromatography on oligo(dT)-cellulose and subsequent gel filtration or sucrose density gradient centrifugation. Each fraction was assayed for its translatability and for the presence of inhibitors of translation in a rabbit reticulocyte lysate.[55] Only a small fraction of mRNP was active in translation and eluted in front of the main poly(A)-containing mRNP (Figure 3). The mRNA of the latter mRNP synthesized a discrete number of proteins. The main proteins have M_r of 50,000, a triplet between 41,500 to 40,000, 30,000, 28,000, and 25000 (Figure 3, Insert). The M_r 50,000

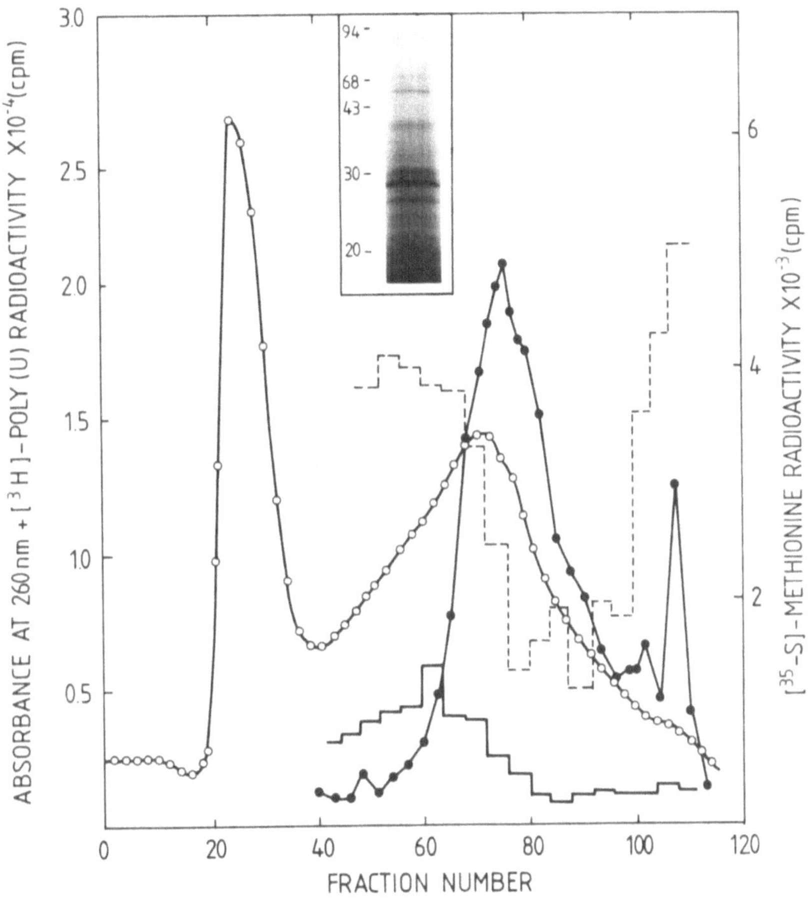

FIGURE 3. Translationally active and repressed poly(A)-containing mRNP. Poly(A)-containing mRNP isolated by affinity chromatography on oligo(dT)-cellulose as described in Figure 1A was further purified by gel filtration on a Sepharose CL-4B column (1.4 cm × 100 cm) in 10 m*M* Hepes pH 7.2, 100 m*M* KCl, 0.1 m*M* DTT, 10% (v/v) glycerol. Absorbance at 260 nm (o——o). Each fraction was assayed for the presence of poly(A)-sequences by hybridization with (^{3}H)-poly(U) (●——●).[30] Fractions were concentrated threefold and assayed for translational activity in a rabbit reticulocyte lysate (——).[54] Insert: analysis by sodium dodecyl sulfate polyacrylamide gel electrophoresis of the proteins synthesized by mRNA extracted from the translationally active mRNP. The inhibitor of translation was assayed by mixing of each fraction with 1 μg globin mRNA before addition to the rabbit reticulocyte lysate (- - - -).[54]

protein may well be eEFTu. More detailed results concerning the synthesized proteins are reported by Daum et al.[56] Inhibitors of translation were detected by addition of 9S globin mRNA to the fractions before translation (Figure 3). A potent inhibitor was localized behind the main mRNP. The latter inhibitor is a small ribonucleoprotein copurified with mRNP on oligo(dT)-cellulose probably by its affinity for RNA.[55] The inhibitor is also associated with the main nontranslatable mRNP (Section III.C). The same results were obtained with mRNP purified by sucrose density gradient centrifugation.[54] The free inhibitor ribonucleoprotein has a sedimentation coefficient of 5 to 6S.

B. Composition of the Inhibitor Ribonucleoprotein

Poly(A)-containing mRNP was purified by sucrose density gradient centrifugation and the inhibitor isolated from the 5 to 6S region by gel filtration on Sephacryl S-200 and ion-exchange chromatography on DEAE-cellulose.[54] The purified inhibitor has an A_{260}/A_{280} ratio of 1.5 and bands at a density of 1.28 g/cm^3 after isopycnic centrifugation in sucrose.[30,54] This density is intermediate between the density of mRNP and ribosomes.[30] From the absorbancy ratio and the density in sucrose an RNA/protein ratio of 1:2.5 is calculated. The elution from Sephacryl S-200 indicates an M_r of 84,000. The composition of the inhibitor ribonucleoprotein was determined by polyacrylamide gel electrophoresis.[54,55] The protein component has an M_r of 64,000 and the RNA has a length of 85 nucleotides. As the inhibition was not removed by phenolization it is a property of the RNA contained in the inhibitor ribonucleoprotein. The RNA has a length similar to other RNA inhibitors reported in the literature (for a review see Reference 57). These inhibitors are isolated from nonpolysomal mRNP, from a ribosomal salt wash, or from the cytosol and have a more complex protein composition than the inhibitor isolated from *Artemia* sp.[47,57-61] The existence of oligo(U) stretches in the 5 to 6S inhibitor ribonucleoprotein was established by binding of the inhibitor RNA to oligo(dA)-cellulose and by hybridization assays with (^{3}H)-poly(A).[62] This property suggests that the association of the inhibitor to mRNP is mediated by the poly(A)-sequence.

C. Association of the Inhibitor Ribonucleoprotein with Nonpolysomal mRNP

The nontranslatable 17S mRNP was separated from the free 5 to 6S inhibitor ribonucleoprotein by sucrose density gradient centrifugation.[54] The inhibitor, presumed to be associated with mRNP, was dissociated from the nonpolysomal poly(A)-containing mRNP by the method of Northemann et al.[47] and further purified by ion-exchange chromatography on DEAE-cellulose as described for the free 5 to 6S inhibitor ribonucleoprotein.[54] First evidence for the association of an inhibitor ribonucleoprotein with nontranslatable mRNP was deduced from the inhibition of 9S globin mRNA by material eluted from the latter DEAE-cellulose in high ionic strength (>500 m*M* KCl). Further evidence came from the analysis of 17S mRNP by polyacrylamide gel electrophoresis.[55] The main small RNA component of the repressed mRNP migrates at the same position as the RNA component of purified inhibitor ribonucleoprotein (iRNP).[55] Also an M_r 64,000 protein of the nontranslatable mRNP is coelectrophoresed with the protein component of the free 5 to 6S inhibitor ribonucleoprotein. More direct evidence for the association of an inhibitor to mRNP is obtained from reconstitution experiments. Initial attempts to reconstitute the mRNP-inhibitor complex by mixing of mRNP and [^{32}P]-pCp-labeled inhibitor RNA were unsuccessful. The failure to reassociate the inhibitor to mRNP may be due to (1) the firm association of the inhibitor to mRNP resulting in a very low exchange with exogeneously added inhibitor RNA, (2) the structure of the poly(A)-protein complex which may require specific alterations before association of the inhibitor to mRNA, and (3) the absence of the iRNA-associated M_r 64,000 protein. Evidence which supported the second hypothesis came from experiments in conditions which temporarily dissociated the poly(A)-binding M_r 38,000 protein from the poly(A)-sequence. The conditions of dissociation are described in Section IV.A.3.a. Dissociation was obtained by incubation of mRNP at 37°C in the presence of 3 m*M* $MgCl_2$ before addition of inhibitor RNA and increase of the salt concentration to 250 m*M* NaCl. In these conditions, association of inhibitor RNA to mRNP was observed (Figure 4). Analysis of the assay mixture by sucrose density gradient centrifugation showed that 52% of the labeled inhibitor RNA sedimented with the 17S mRNP. The remaining inhibitor RNA was located at the top of the gradient. Clearly the mechanism of association of inhibitor RNA to mRNP is very complex and has to be investigated further. So far the association of an inhibitor RNA with mRNP has been demonstrated in embryonic chick muscle, [60,61,63] human placenta,[64] and *Artemia* cryptobiotic embryos.[55]

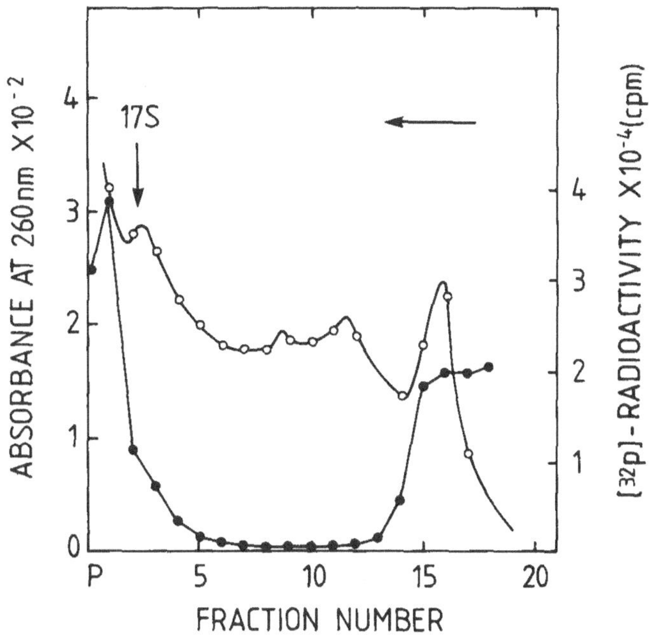

FIGURE 4. In vitro reassociation of the inhibitor RNA with mRNP. Inhibitor RNA was prepared and [^{32}P]-pCp labeled as described in Reference 54. The labeled RNA was further purified by electrophoresis on an 8% (w/v) polyacrylamide gel. RNA was eluted from the gel in 500 m*M* ammoniumsulfate, 0.1 m*M* EDTA, 0.1% (w/v) sodium dodecyl sulfate, and at 37°C precipitated with ethanol and dissolved in distilled water. 17S mRNP was purified by affinity chromatography on oligo(dT)-cellulose and sucrose gradient centrifugation as described in the caption of Figure 1A, concentrated by ultrafiltration, and incubated at 37°C for 5 min in 10 m*M* Hepes pH 7.2, 3 m*M* $MgCl_2$ before addition of purified [^{32}P]-pCp-labeled inhibitor RNA. After cooling to 4°C and increase of the salt concentration to 250 m*M* NaCl iRNA-mRNP mixtures were analyzed by 10 to 30% (w/v) sucrose density gradient centrifugation in 10 m*M* Hepes pH 7.2, 30 m*M* sodium citrate. Centrifugation was in a Beckman SW40 rotor at 154,400 *g* for 19 hr at 4°C. The position of the 17S mRNP and the direction of centrifugation are indicated. Absorbance at 260 nm (○——○). [^{32}P]-pCp-labeled iRNA (●——●). P: pellet.

IV. IDENTIFICATION AND PROPERTIES OF mRNP PROTEINS

A. Poly(A)-Binding Proteins

1. Identification of Poly(A) Binding Proteins

The proteins of nonpolysomal poly(A) containing mRNP were subjected to sodium dodecylsulfate polyacrylamide gel electrophoresis and the poly(A) binding proteins identified by electroblotting on nitrocellulose and incubation of the blot with [5′-^{32}P]-pCp-labeled poly (A). mRNP proteins with M_r 38,000 and 23,500 interacted with polyadenylic acid (Figure 5). This result confirmed our previous identification of the poly(A)-binding proteins by affinity chromatography of mRNP proteins on poly(A)-Sepharose 4B and by ribonuclease degradation of purified mRNP and rechromatography of the poly(A)-protein complex on oligo(dT)-cellulose.[32,38] As already mentioned in Section II.B discrete sedimenting mRNP classes have a different protein composition. This result was further strengthened by electroblotting of proteins from mRNP classes separated by sucrose density gradient centrifugation (Figure 5B). The M_r 38,000 protein is associated with mRNP sedimenting at 17S

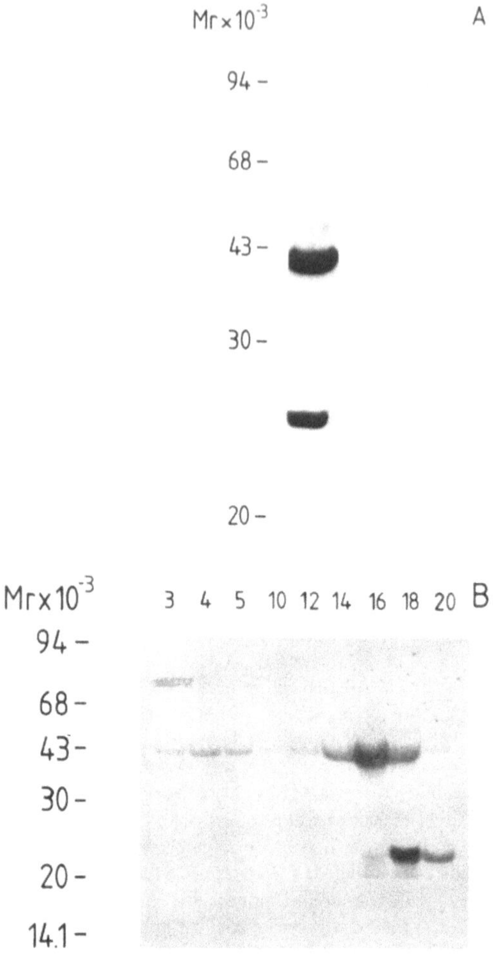

FIGURE 5. Identification of poly(A)-binding proteins. (A) mRNP proteins eluted from oligo(dT)-cellulose-bound RNP by a 1.5 *M* KCl salt wash were separated by sodium dodecylsulfate polyacrylamide gel electrophoresis as described in Reference 32 and electroblotted onto nitrocellulose as described by Richter and Smith.[65] Blotting was at 36 V and 100 mA for 15 hr. Electroblots were washed with 20 m*M* Tris-HCl pH 7.4, 50 m*M* NaCl, 1 m*M* EDTA, 0.02% BSA, 0.02% Ficoll, and incubated for 3 hr with [^{32}P]-pCp-labeled poly(A) (1 to 2 × 10^6 cpm/blot, 12,000 cpm/μg poly A). The blot was washed, dried, and autoradiographed. Fractions of the sucrose density gradient of Figure 2 were used to identify the poly(A)-binding proteins associated with the discrete sedimenting mRNP particles. Proteins were separated by sodium dodecyl sulfate polyacrylamide gel electrophoresis and electroblotted onto nitrocellulose. The identification was as described in A.

and at 35 to 40S. The M_r 23,500 is associated with the poly(A)-sequence of mRNP sedimenting below 15S. The same poly(A)-binding proteins were also identified in mRNP-ribosome preinitiating complexes of cryptobiotic embryos. The M_r 73,000 to 78,000 protein often reported to be associated with the poly (A)-sequence of polysomal mRNP is not detected in nonpolysomal mRNP of *Artemia* sp.[5,66-71] The absence of this protein in cytoplasmic nonpolysomal mRNP has also been observed in Ehrlich ascites tumor cells, rabbit reticulocytes, sea urchin eggs, and duck erythroblasts.[5,71-72]

2. *Purification of the M_r 38,000 Poly(A)-Binding Protein*

The purification of the M_r 38,000 poly(A)-binding protein was a modification of the procedure described previously.[38] Nonpolysomal poly(A)-containing mRNP was bound to oligo(dT)-cellulose and mRNP proteins dissociated from mRNA by a salt wash with 1.5 *M* KC1. mRNP-proteins were further purified by ion-exchange chromatography on phosphocellulose P11. Bound proteins were eluted with a linear salt gradient. The M_r 38,000 protein was recovered at 200 m*M* KC1 and was well separated from the M_r 23,500 poly(A)-binding protein. The poly(A)-binding proteins were subjected to affinity chromatography on poly(A)-Sepharose 4B and the bound proteins eluted discontinuously with 500 m*M* and 1,500 m*M* KC1. The M_r 38,000 poly(A)-binding protein was eluted at 1,500 m*M* KC1 and was concentrated by ultrafiltration. This procedure yielded 2 to 3 mg of purified poly(A)-binding protein from 200 g dry cysts.

3. Properties of the M_r 38,000 Poly(A)-Binding Protein

a. Interaction with RNA

The interaction of the M_r 38,000 poly(A)-binding protein with poly(A) was studied by nitrocellulose filter binding and sucrose density gradient centrifugation. The interaction is very fast and is nearly complete after 5 sec of incubation at 4°C.[73] The protein association is dependent on the pH and the presence of mono- and divalent ions (Figure 6). A gradual increase in the amount of poly(A) retained on nitrocellulose filters was measured when the pH was lowered from pH 10 to pH 5.5 (Figure 6A). The absence of a pH optimum is explained by (1) the existence of several ionic species for the poly(A)-binding protein with a pI ranging from 9.2 to 6.6 (Section IV. 3b) resulting in more protein available for poly(A)-binding at a lower pH, and (2) the helix destabilizing property of the poly(A)-binding protein which, in mild acidic conditions, results in a shift of the formation of poly(A) double stranded helical structures to lower pH values.[74]

The optimal KCl concentration is pH dependent and shifts from 200 m*M* to 50 m*M* KCl upon decrease of the pH from 8.5 to 5.0 respectively (Figure 6B). The decrease in salt concentration compensates for the more positive charge of the protein at low pH. The poly(A)-protein interaction is also dependent on the presence of divalent ions and is strongly decreased above 2.5 m*M* $MgCl_2$ (Figure 6C). The effects of pH and mono- and divalent ions on the protein-poly(A) interaction emphasizes the importance of the salt conditions used in the isolation of mRNP.

In optimal conditions of association (pH 7.4, 100 m*M* KCl) a poly(A)-protein stoichiometry of 10 ± 2 nucleotides per protein was measured by binding on nitrocellulose and 9 ± 2 nucleotides per protein by sucrose density gradient centrifugation.[73] Although the M_r 38,000 protein is located on the poly(A)-sequence of mRNP it is able to interact with other RNA sequences. The RNA-protein stoichiometry is independent of the use of synthetic or natural RNAs. The specificity of the protein for poly(A) was demonstrated by RNA/RNA mixing experiments.[73] Labeled poly(A) of the poly(A)-protein complex was readily exchanged with unlabeled poly(A) after addition of an excess of unlabeled poly(A). Within a few seconds more than 67% of the protein was transfered to unlabeled poly(A). Incubation at 30°C increased the transfer to almost 80%. On the contrary the protein was not exchanged after addition of 18S and 28S rRNA and 80 to 90% of the labeled poly(A) was retained on nitrocellulose filters.

b. Identity Between the Poly(A)-Binding Protein and HD40

An M_r 40,000 protein has been isolated from *Artemia* cryptobiotic embryos by Szer and collaborators.[75,76] The protein was purified from a high salt wash of a ribosomal fraction and is a helix destabilizing protein with an amino acid composition characterized by a high glycine content, the absence of cysteine, and the presence of dimethylarginine. The protein binds to synthetic and natural RNAs in a ratio of 12 to 15 nucleotides per protein and the interaction is strongly reduced in the presence of $MgCl_2$. The helix destabilizing protein is a major component of hnRNP and binds to 30S hnRNP particles in a ratio of 10 to 12 nucleotides per protein.[77] HD40 is similar in molecular weight and amino acid composition

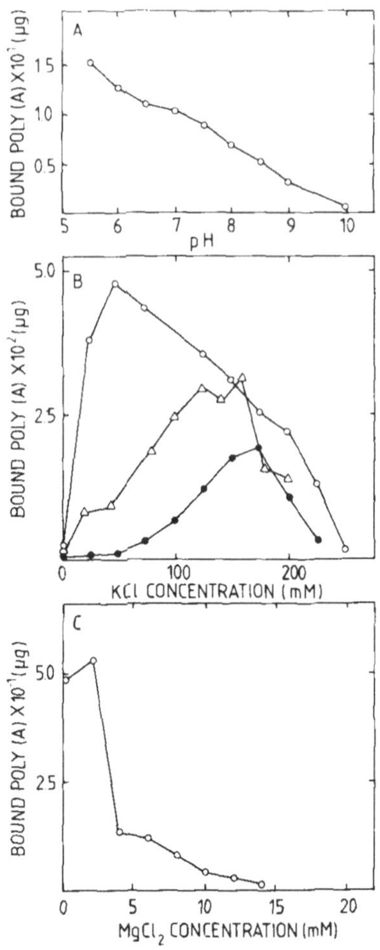

FIGURE 6. Interaction of the M_r 38,000 protein with poly(A). (A) pH dependence of the interaction. Saturating concentrations of the M_r 38000 protein were mixed with poly(A) (specific activity 130,000 cpm/μg) in buffer ranging from pH 5.5 to 10.0 containing 125 m*M* KCl. 20 m*M* Hepes (pH 5.5 to 8.0) and 20 m*M* Tris (pH 7.0 to 10.0) were used as buffers. After incubation at 4°C for 1 min the reaction was stopped by addition of ice-cold buffer of the same pH and the mixture was immediately filtered over nitrocellulose.[38] Control incubations without poly(A)-binding protein were performed at each pH. (B) Dependence of the M_r 38,000 protein-poly(A) interaction on monovalent ions at different pH values. Saturating concentrations of the M_r 38,000 protein were mixed with poly(A) at pH 5.0, 7.5, and 8.5 in 20 m*M* Tris and a varying concentration of KCl. The mixtures were incubated and processed as described in (A). Control incubations without poly(A)-binding protein were performed at 50 m*M* and 250 m*M* KCl. Dependence at pH 5.0 (○——○), at pH 7.5 (△——△), at pH 8.5 (●——●). (C) Dependence of the M_r 38,000 protein-poly(A) interaction on divalent ions. Saturating concentrations of the M_r 38,000 protein were mixed with poly(A) in 20 m*M* Tris-HCl pH 7.5, 125 m*M* KCl and a varying concentration of $MgCl_2$. The mixtures were incubated and processed as described in (A). Controls were performed at each $MgCl_2$ concentration.

to some hnRNP proteins.[78-80] HD40 from cryptobiotic embryos was separated into three isoelectric components with a pI of 8.1, 7.8, and 7.6. Additional components with a pI of 6.8 and 6.6 were detected in developing embryos.[77]

The resemblance in molecular weight and the similarity in interaction with RNA are striking and prompted us to compare the helix destabilizing protein HD40 with the poly(A)-binding protein associated with nonpolysomal mRNP. The poly(A)-binding protein had helix destabilizing properties as was deduced from its effect on the melting of double stranded helical structures formed between poly(A) and poly(U). Increasing concentrations of the protein shifted the melting temperature to lower values.[73]

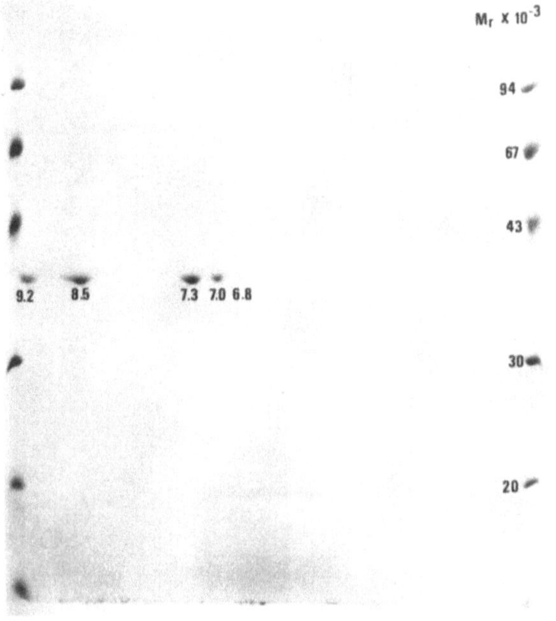

A

FIGURE 7. Analysis of the M_r 38,000 poly(A)-binding protein by two-dimensional gel electrophoresis. The preparation of poly(A)-binding proteins from nonpolysomal mRNP and mRNP-preinitiation complexes is described in Reference 73. Protein samples of 150 μg were analyzed as described by O'Farrell.[82] (A) Poly(A)-binding protein from non-polysomal mRNP. (B) Poly(A)-binding protein from mRNP-preinitiation complexes. (From De Herdt, E., et al., *Eur. J. Biochem.*, 139, 155, 1984. With permission.)

First evidence for the identity between both proteins came from sodium dodecylsulfate polyacrylamide gel electrophoresis. HD40 prepared as described by Marvil et al.[75] and the poly(A)-binding mRNP protein migrated at exactly the same position. Furthermore, analysis of the tryptic digests of the proteins generated the same two main peptides of M_r 21,000 and 18,000 and several identical minor fragments.[38] Further evidence was obtained from the amino acid composition of the proteins. Both compositions were comparable and the M_r 38,000 poly(A)-binding protein also had a high glycine content, contains dimethylarginine, and lacks cysteine.[38] More direct evidence for the identity between the two proteins was obtained from immunodiffusion. Anti-M_r 38,000 protein anti-serum gave rise to a single precipitin line with HD40.[73]

Analysis by two-dimensional gel electrophoresis showed a different composition in ionic species. The preparation of the poly(A)-binding mRNP protein contained seven ionic species with a pI ranging rom 9.2 to 6.6 (Figure 7A). The main components have a pI of 9.2, 8.5, 7.3, and 7.0. Additional minor components of the mRNP associated protein have a pI of 7.6, 6.8, and 6.6. HD40 prepared from the ribosomal fraction only contains the two most basic components with a pI of 8.5 and 9.2 (Figure 7B). The acidic components of the poly(A)-binding protein are generated by phosphorylation of the basic species as is discussed in Section IV.C.2.

Ribosomes purified by differential centrifugation and subsequent sucrose density gradient centrifugation did not contain the M_r 38,000 poly(A)-binding protein. The presence of the

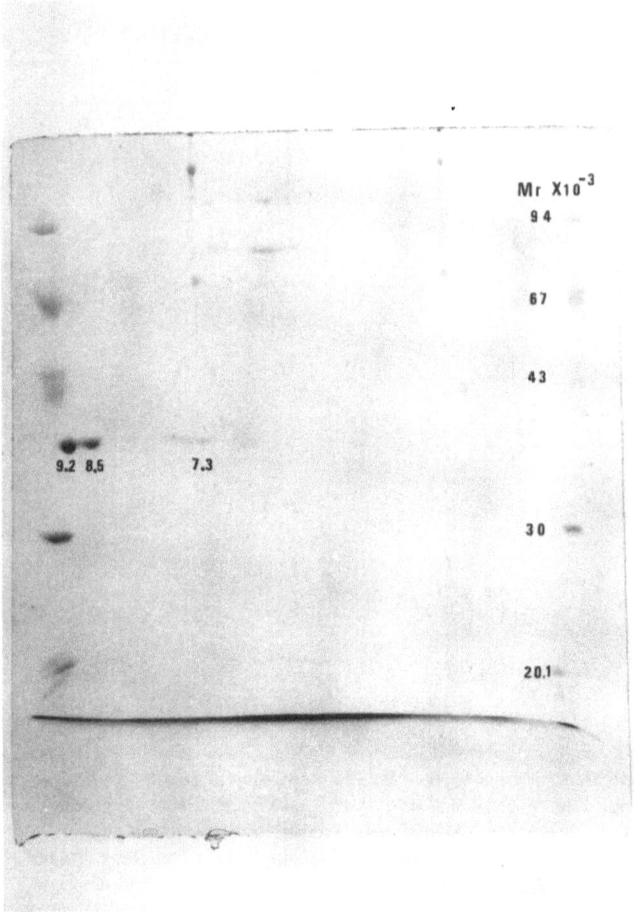

FIGURE 7B.

latter protein in crude ribosomes was a consequence of contamination by mRNP-ribosome preinitiation complexes.[38] This was confirmed by the isolation of poly(A)-containing mRNP from the latter fraction and purification of the poly(A)-binding protein by affinity chromatography on poly(A)-Sepharose 4B (Figure 7B). Two-dimensional gel electrophoresis of the purified poly(A)-binding protein shows the existence of only two components with a pI of 9.3 and 8.5.[73]

The identity between HD40 and the M_r 38,000 poly(A)-binding protein is in agreement with the localization experiments of Thomas and Szer.[81] They reported that in undeveloped embryos only trace amounts of HD40 are present in nuclei and a significant increase is demonstrated after development. HD40 is also present in the cytoplasm as a free protein but the main part is associated with the 10 to 30S poly(A)-containing mRNP.

The existence of several ionic species of the poly(A)-binding protein of which only the basic ones are present in mRNP-ribosome preinitiation complexes and of which none is present in polysomal mRNP suggests that dissociation of the poly(A)-binding protein is correlated with the entry of mRNP into polysomes. As the ionic species are interconverted by phosphorylation-dephosphorylation the latter modification is one of the regulatory mechanisms involved in the activation of stored mRNP.[16]

c. Structure of the Poly(A)-Protein Complex

The poly(A)-sequence is known to undergo a size reduction in the cytoplasm according to a mechanism which favors poly(A)-lengths with regular multiples of 20 to 27 residues.[83-85] This selective poly(A) degradation process may have a function in the translation of mRNA.[86] It is proposed that the periodicity in poly(A)-sequence is due to its interaction with proteins.[83,84]

A periodicity in the poly(A)-length of nonpolysomal mRNP was also observed in *Artemia* sp. The poly(A)-sequences range from 12 to 138 nucleotides as determined by polyacrylamide gel electrophoresis.[38] The distribution is heterogeneous but is resolved in 6 discrete peaks with a length difference of 20 $\pm$ 2. The main poly(A)-sequence has a length of 75 nucleotides in agreement with a length of 65 to 70 nucleotides previously measured by Nilsson and Hultin[87] but much shorter than the length of 126 nucleotides reported by James and Tata.[53] From the protein composition of mRNP and from the protein-RNA ratio it is calculated that approximately 6 to 7 M_r 38,000 proteins are bound to the poly(A)-sequence with a length of 75 nucleotides equivalent to a stoichiometry of 11 to 13 nucleotides per protein. This value is in agreement with the stoichiometry of 10 $\pm$ 2 nucleotides per protein measured from nitrocellulose filter binding and from sucrose density gradient centrifugation (Section IV.A.3.a). The stoichiometry of the M_r 38,000 protein-poly(A) complex and the periodicity in poly(A)-sequences which is twice the stoichiometry value suggests a tandem repetition of the protein on the poly(A)-sequence.

Szer and collaborators studied the structure of the poly(A)-protein complex by mixing of the poly(A)-binding protein with poly(A) of defined sizes.[88] The protein binds to single stranded polynucleotides with a stoichiometry of 1 protein per 10 to 15 nucleotides. In saturating protein concentration protein-protein interactions result in the formation of bead-like structures with a stoichiometry of one protein per eight nucleotides. At a polynucleotide length of 160 nucleotides a disk is formed (3 nm $\times$ 18 nm) containing 20 protein molecules. A single bead sediments at 30S and has a structure similar to native hnRNP. Upon shortening of the poly(A)-sequence, sectors of the disk are formed. This bead formation may be functional in the package of hnRNA necessary for correct processing and subsequent transport from the nucleus to the cytoplasm. The quick transport of the M_r 38,000 poly(A)-binding protein to the nucleus after microinjection into *Xenopus laevis* oocytes supports a function in the nuclear compartment.[89]

The structural studies of Szer and collaborators demonstrate that the concentration of the protein as well as the poly(A)-length determines the structure of the poly(A)-protein complex. As phosphorylation-dephosphorylation affects the affinity of the M_r 38,000 protein for poly(A) (see Section IV.A.3.d), the structure of the complex may be regulated by this protein modification. A small decrease in stoichiometry from 1 protein per 8 nucleotides to 1 protein per 10 to 15 nucleotides, equivalent to a decrease from 10 to 7 proteins per poly(A)-sequence of 75 nucleotides, will have a drastic effect on the structure of the protein-poly(A) complex and may trigger the transition from a stored (condensed form) to a form accessible for the factors involved in the activation of repressed mRNP. After this transition the protein-poly(A) complex is still organized in a tandem-repeat of proteins in order to explain the periodicity in poly(A)-sequence which arise from a specific cytoplasmic degradation process.

d. Possible Functions of the M_r 38,000 Poly(A)-Binding Protein

The identity between the M_r 38,000 poly(A)-binding protein and the helix destabilizing protein HD40 purified from nonpolysomal mRNP and hnRNP respectively already suggests that the poly(A)-binding protein has a cytoplasmic and nuclear function. Since in nauplii, the M_r 38,000 poly(A)-binding protein is not present in nonpolysomal and polysomal mRNP but is only located in hnRNP its function is restricted to the nuclear compartment in developing embryos. This was further supported by microinjecion of labeled poly(A)-binding protein into the cytoplasm of *Xenopus laevis* oocytes.[89] The protein has a half life of approximately 20 hr and is transported to the nucleus where it accumulated to a concentration which is three times higher than calculated from free diffusion. Myoglobin, whose concentration is equilibrated by free diffusion between the cytoplasmic and nuclear compartments, is used as a control. The labeling of the protein by reductive methylation resulted in a reduced binding of the protein to poly(A) and indicated that other interactions must be involved in the accumulation of the protein in the nucleus. hnRNA-associated proteins are known to play a role in the binding of hnRNP to the nuclear matrix and our observations suggest that the M_r 38,000 poly(A)-binding protein may be involved in the latter binding.[89-91]

In crytobiotic embryos the M_r 38,000 poly(A)-binding protein may have a cytoplasmic and nuclear function. A cytoplasmic function was deduced from the effect of the protein on translation and polyadenylation of mRNA and from the effect of phosphorylation/dephosphorylation on the affinity of the protein for the poly(A)-sequence.[73-92] Poly(A)-containing mRNA of *Artemia* sp. and adenylated and deadenylated globin mRNA were translated in a rabbit reticulocyte lysate in the presence of increasing concentrations of poly(A)-binding protein.[73] Independent of the origin and the presence or absence of the poly(A)-binding protein. As the inhibition is counteracted by an increase in mRNA concentration it is not a consequence of an interaction of the protein with translational factors. Apparently the specific location of the protein on the poly(A)-sequence is contradictory to the observed inhibition of poly(A)-lacking mRNA but is explained by the affinity of the protein for other RNA-sequences in the absence of poly(A) (Section IV.A.3.a). The binding of the protein to poly(A)-lacking mRNA may not occur in vivo and the observed inhibition of protein synthesis by the poly(A)-binding protein may have no physiological significance.

A possible function in the polyadenylation by poly(A) polymerase was tested using adenylyl $(3',5')_3$ adenosine ($(A)_4$) as a primer.[73] Poly(A) polymerase is not associated with nonpolysomal mRNP and was isolated from the cytoplasm of cryptobiotic embryos.[93] In vitro assays indicated that the rate of polyadenylation of $(A)_4$ decreased in function of time. The decrease was observed when $(A)_4$ was elongated to approximately eight to ten nucleotides. In the presence of the poly(A)-binding protein the rate of elongation of $(A)_4$ did not decrease as a function of time. After initiation of polyadenylation the rate of polyadenylation remained constant for at least 40 min.[73]

More convincing evidence for a cytoplasmic function came from the effect of phosphorylation/dephosphorylation on the interaction of the M_r 38,000 protein with poly(A). As already discussed in Section IV.A.3.b, striking differences are observed in the phosphorylation of the poly(A)-binding protein associated with nonpolysomal mRNP and with mRNP-ribosome preinitiation complexes. The phosphorylated status of the poly(A)-binding protein is determined by the mRNP-associated protein kinase (Section IV.C) and two cytoplasmic protein phosphatases (Section V.A) and regulates the association of the protein with the poly(A)-sequence. The effect of phosphorylation-dephosphorylation of the protein on its interaction with poly(A) was studied by nitrocellulose filter binding and by electroblotting (Figure 8). Dephosphorylation shifted the acidic ionic components of the protein into basic ones and reduced the binding to poly(A). Conversely, phosphorylation shifted the basic ionic components into acidic ones and increased the interaction with the poly(A)-sequence.

In the latter conditions translatable mRNP was converted into a repressed state (Section VI). These results strongly argue for a function of the poly(A)-binding protein in the repression of mRNP. The M_r 38,000 poly(A)-binding protein seems to be a multifunctional protein which affects several mRNA-related phenomena. All these effects seem to be correlated with the induction of conformational changes in the structure of the poly(A)-protein complex.

4. *Absence of the M_r 38,000 Poly(A)-Binding Protein in Polysomal mRNP*

Polysomal mRNP was prepared from cytoplasmic polysomes of nauplii. Essential in the procedure was the $MgCl_2$ concentration of 2.5 m*M* necessary to prevent the release of

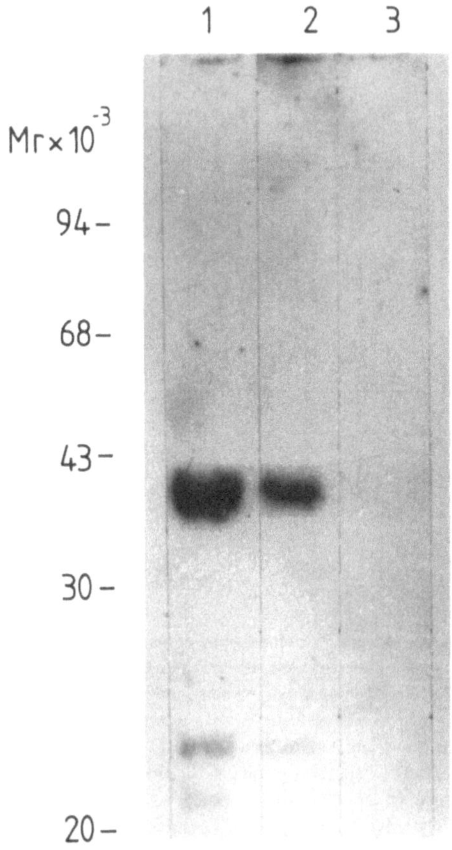

FIGURE 8. Effect of phosphorylation-dephosphorylation on the interaction of the M_r 38,000 protein with poly(A). The M_r 38,000 poly(A)-binding protein was purified from nonpolysomal mRNP as described in Reference 73. 35 μg of the poly(A)-binding protein was phosphorylated by the mRNP-associated protein kinase (Section IV.C) and dephosphorylated by the cytoplasmic protein phosphatases PCS-X and PCS-Y (Section V. A) isolated from *Artemia* as described in Reference 92. The phosphorylation assay contained 10 m*M* Tris-HCl pH 8.0, 12.5 m*M* $MgCl_2$, 50 m*M* KCl, 1 m*M* DTT, 6 μ*M* ATP, and 8 units of protein kinase. Incubation was at 37°C for 2 hr. The reaction was stopped by precipitation with trichloroacetic acid. The dephosphorylation assay contained 17.5 m*M* Tris-HCl pH 7.0, 0.03 m*M* EGTA, 0.03% (v/v) 2-mercaptoethanol, 10 μg BSA, 5 m*M* $MgCl_2$, and 0.01 units of protein phosphatases. Incubation was at 30°C for 2 hr. The reaction was stopped by precipitation with trichloroacetic acid. Protein samples were electrophoresed on dodecylsulfate polyacrylamide gels and electroblotted onto nitrocellulose exactly as described in the caption of Figure 5. Electroblots were incubated with [^{32}P]-pCp-labeled poly(A) (specific activity 12,000 cpm/μg), washed, dried, and autoradiographed. Lane 1: mRNP proteins phosphorylated by the mRNP-associated protein kinase; lane 2: *in situ* mRNP proteins; lane 3: mRNP proteins dephosphorylated by the cytoplasmic protein phosphates. (From Van Hove, L., et al., *Biochem. Biophys. Res. Commun.*, 131, 1241, 1985. With permission.)

poly(A)-binding proteins from mRNP but still sufficient to avoid dissociation of ribosomes into ribosomal subunits. The integrity of the isolated polysomes is shown by sucrose density gradient centrifugation (Figure 9). The majority of the poly(A)-sequences are detected in the 80S monosome peak and are gradually decreasing in heavy polysomes. Polysomes were pooled, dissociated with EDTA and the mRNP purified by affinity chromatography on oligo(dT)-cellulose as described in Section II.B. The main proteins have M_r of 117,000, 77,000, 53,000, 22,000, and 21,000 (Figure 10). Nonpolysomal mRNP was prepared from

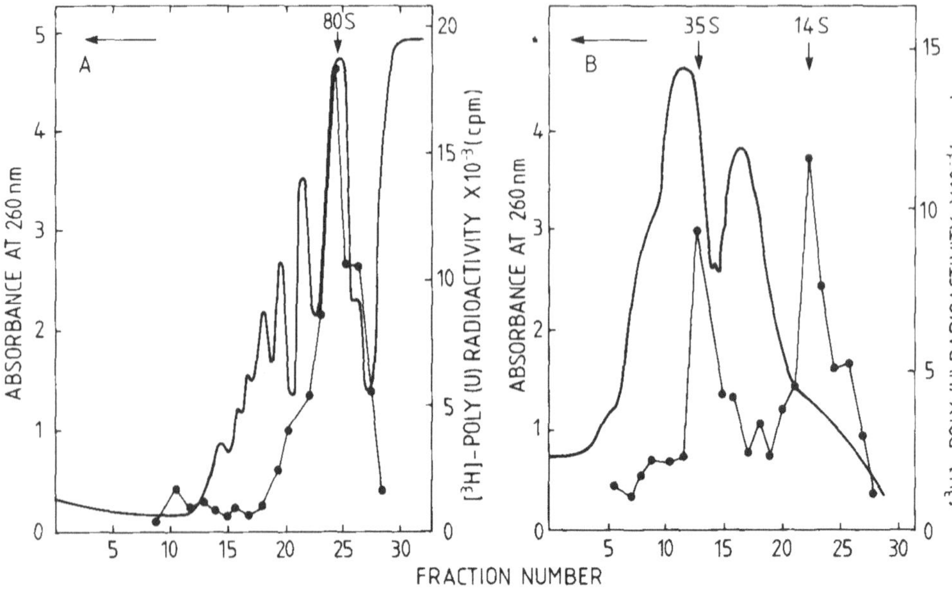

FIGURE 9. Detection of poly(A)-sequences in cytoplasmic polysomes and polysomal mRNP of nauplii. (A) Integrity of polysomes. 1500 A_{260} polysomes were loaded onto a linear 20 to 55% (w/v) sucrose density gradient in 25 m*M* Tris-HCl pH 7.6, 25 m*M* KCl, 2.5 m*M* $MgCl_2$, 5 m*M* 2-mercaptoethanol, 0.1 m*M* PMSF. Centrifugation was in a Beckman Ti14 zonal rotor, at 120,000 *g* for 3 hr. (B) Size distribution of poly(A)-containing polysomal mRNP. 1500 A_{260} polysomes were dissociated by incubation at 50°C for 5 min in 5 m*M* EDTA and centrifuged as described in (A). Centrifugation was at 150,000 *g* for 24 hr. Ribosome subunits sedimenting at 40S and 23S were used as markers.[33] Absorbance at 260 nm (——). Fractions were assayed for the presence of poly(A)-sequences by hybridization with (^{3}H)-poly(U) (●——●).[30] The direction of centrifugation is indicated.

the postribosomal supernatant of nauplii. The latter mRNP is composed of several proteins. The main protein component has an M_r of 53,000 (Figure 10).

The polysomal mRNP proteins of M_r 77,000 and 53,000 and the M_r 53,000 protein of free mRNP have been identified as poly(A)-binding proteins by RNAse degradation and rechromatography on oligo(dT)-cellulose. The M_r 77,000 protein is probably analogous to the highly conserved M_r 78,000 protein often reported to be associated with the poly(A)-sequence of polysomal mRNP.[3] An antigenic relationship has been demonstrated between the latter protein and nuclear poly(A) polymerase from rat hepatoma.[94] It may be noted that poly(A) polymerase of *Artemia* sp. has a comparable molecular weight to the polysomal poly(A)-binding protein but so far the identity between both proteins has not been studied.[93]

As already proposed in Section IV.A.3.d the function of the M_r 38,000 poly(A)-binding protein is restricted to the cytoplasmic and nuclear compartments in cryptobiotic gastrulae and to the nuclear compartments in nauplii which implies that the protein has no function in the translation process in nauplii.

B. Initiation Factor eIF2

1. Initiation Factor eIF2 is Associated with Translatable mRNP

Evidence for the presence of eIF2 on mRNP was obtained from sodium dodecyl sulfate polyacrylamide gel electrophoresis of mRNP proteins and eIF2 (Figure 11). eIF2 is composed of three subunits with M_r 40,000 (α), 42,000 (β), and 52,000 (γ). The α and β subunits migrated at the same position as two minor mRNP proteins. The presence of eIF2 on some mRNP particles was unambiguously demonstrated by electroblotting of mRNP proteins, incubation of the blot with anti-eIF2 antibody and detection with labeled (^{125}I)-protein A (Figure 11).[55]

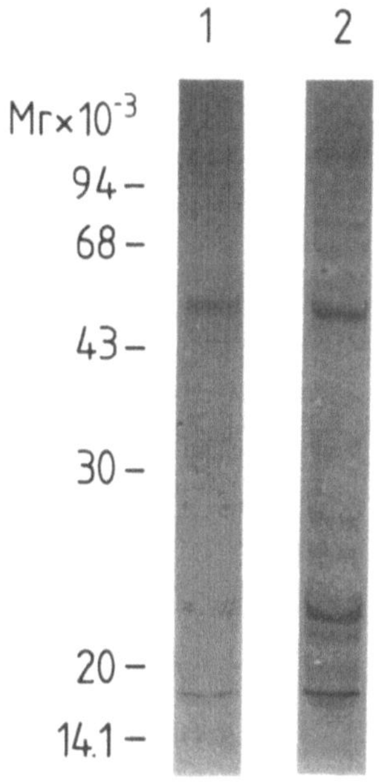

FIGURE 10. Protein composition of poly(A)-containing nonpolysomal and polysomal mRNP of nauplii. The poly(A)-containing nonpolysomal mRNP was prepared from the postribosomal supernatant. Poly(A)-containing polysomal mRNP was purified from polysomes prepared by zonal sucrose density gradient centrifugation of the postmitochondrial supernatant. The centrifugation conditions are described in the caption of Figure 9A. The polysomes containing fractions were pooled, dissociated by EDTA treatment as described in Figure 9B, diluted fivefold with 10 m*M* Hepes pH 7.2, 25 m*M* KCl, 0.1 m*M* DTT, brought to a final concentration of 250 m*M* KCl, and subjected to affinity chromatography on oligo(dT)-cellulose in 10 m*M* Hepes pH 7.2, 250 m*M* KCl, and 0.1 m*M* DTT. Bound mRNP was eluted with 10 m*M* Hepes pH 7.2, 0.1 m*M* DTT at 36°C. The protein composition was determined by sodium dodecylsulfate polyacrylamide gel electrophoresis. Lane 1: nonpolysomal mRNP, lane 2: polysomal mRNP.

The polyclonal rabbit anti eIF2 antibody raised against eIF2 of cryptobiotic embryos (A. J. Wahba, University of Mississippi, Jackson, Miss.) reacted with the α,β, and γ subunits with a different efficiency. The α and β subunits were detected 2 and 3.5 times more efficiently than the γ subunit. Besides the subunits of eIF2 a polypeptide with an M_r of 34,000 was also detected in the eIF2 preparation. The latter protein may be a degradation product of the α subunit.[95] The specificity of the antibody preparation was shown by reaction with proteins of the postmitochondrial supernatant. Reaction was observed with the α and β subunits of eIF2, the M_r 34,000 protein, and two proteins of M_r 71,000 and 26,000 (Figure 11). mRNP proteins of M_r 40,000, 42,000, and 34,000 showed an antigenic reaction with the anti eIF2 antibody. The stoichiometry of the M_r 40,000 and 42,000 proteins was the same as measured for the α and β subunits of eIF2 in the control experiments with post-mitochondrial proteins. The ratio of 1:1.5 and 1:1.8 (α:β) calculated from the autoradiogram and from the stained gel indicated that the β subunit was enriched by its affinity for mRNA necessary for the binding of eIF2 to mRNP or that the α subunit was dissociated from mRNP during mRNP purification or became degraded by proteolytic enzymes.

From the ratio of 1:19 between the β subunit of eIF2 and the M_r 38,000 poly(A)-binding protein of which 6 to 7 molecules are bound to the poly(A)-sequence of mRNP it was calculated that eIF2 is present on 25 to 30% of the mRNP particles.[55]

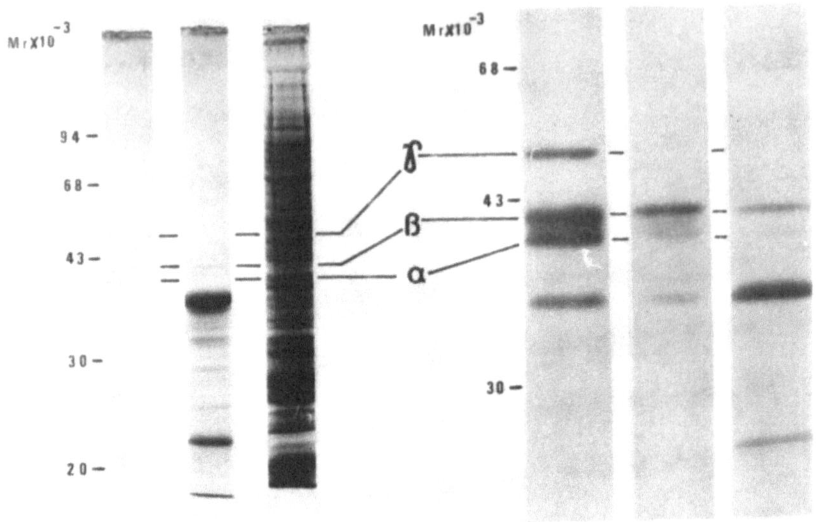

FIGURE 11. Presence of eIF2 among mRNP proteins. eIF2, prepared as described in Reference 104, mRNP proteins prepared from oligo(dT)-cellulose-bound RNP by a 1.5 *M* KCl salt wash and a total protein preparation of the postmitochondrial supernatant were analyzed by sodium dodecylsulfate polyacrylamide gel electrophoresis. Proteins were transferred to nitrocellulose by electroblotting in 25 m*M* Tris-HCl pH 8.3, 192 m*M* glycin, and 20% (v/v) methanol at 36 V for 16 hr. Blots were washed for 45 min in 0.2% (v/v) boric acid pH 8.0, 0.85% (w/v) NaCl, 0.05% (w/v) NaN_3 (boric acid buffer), and 15% (v/v) fetal calf serum and incubated with anti-eIF2 antibody. Blots were subsequently washed with 50 m*M* sodium phosphate pH 7.5, 0.5% (w/v) BSA, 0.5% (v/v) Tween, and 0.05% (w/v) NaN_3 for 10 min, 10 min, 60 min, and 6 hr and with 15% (v/v) calf serum in boric acid buffer. Bound anti-eIF2 antibody was detected by incubation for 60 min with staphylococcal ^{125}I-protein A (60,000 cpm/mℓ) in 15% (v/v) fetal calf serum and boric acid buffer. The blot was washed, dried, and autoradiographed. (Left) stained gel; (Right) autoradiogram. Lane 1: eIF2, lane 2: mRNP-proteins, lane 3: postmitochondrial proteins. (From De Herdt, E., et al., *Eur. J. Biochem.*, 151, 455, 1985. With permission.)

The presence of eIF2 was correlated with the translational activity of mRNP by sucrose density gradient centrifugation (Figure 12). Translationally active mRNP sedimented in front of the 17S mRNP. eIF2 was assayed in the gradient fractions by electroblotting and reaction with anti eIF2 antibody and was detected in fractions sedimenting in front of the 17S mRNP and at the top of the gradient (Figure 12).

So far only a few attempts have been made to correlate mRNA-associated proteins with initiation factors of protein synthesis and no direct relationship has been established although initiation and elongation factors have been identified as RNA-binding proteins.[21,96-98] Our results show that initiation factor eIF2 is associated with the protein moiety of translatable mRNP and is absent in repressed mRNP. The latter mRNP is associated with an inhibitor ribonucleoprotein (Section III. A).

2. Functionality of mRNP-Associated eIF2

Eukaryotic eIF2 is required for the binding of Met-$tRNA_f$ to 40S ribosome subunits in the pathway of initiation of protein synthesis. Met-$tRNA_f$ is not directly bound to the 40S ribosome subunit but first forms a ternary complex with GTP and eIF2.

The ability of the mRNP-associated eIF2 to form a ternary complex with Met-$tRNA_f$ and GTP was shown by assaying each fraction of a sucrose density gradient for Met-$tRNA_f$ binding activity (Figure 12). The Met-$tRNA_f$ binding activity was localized in front of the 17S mRNP and at the top of the gradient and was coincident with eIF2 detected by elec-

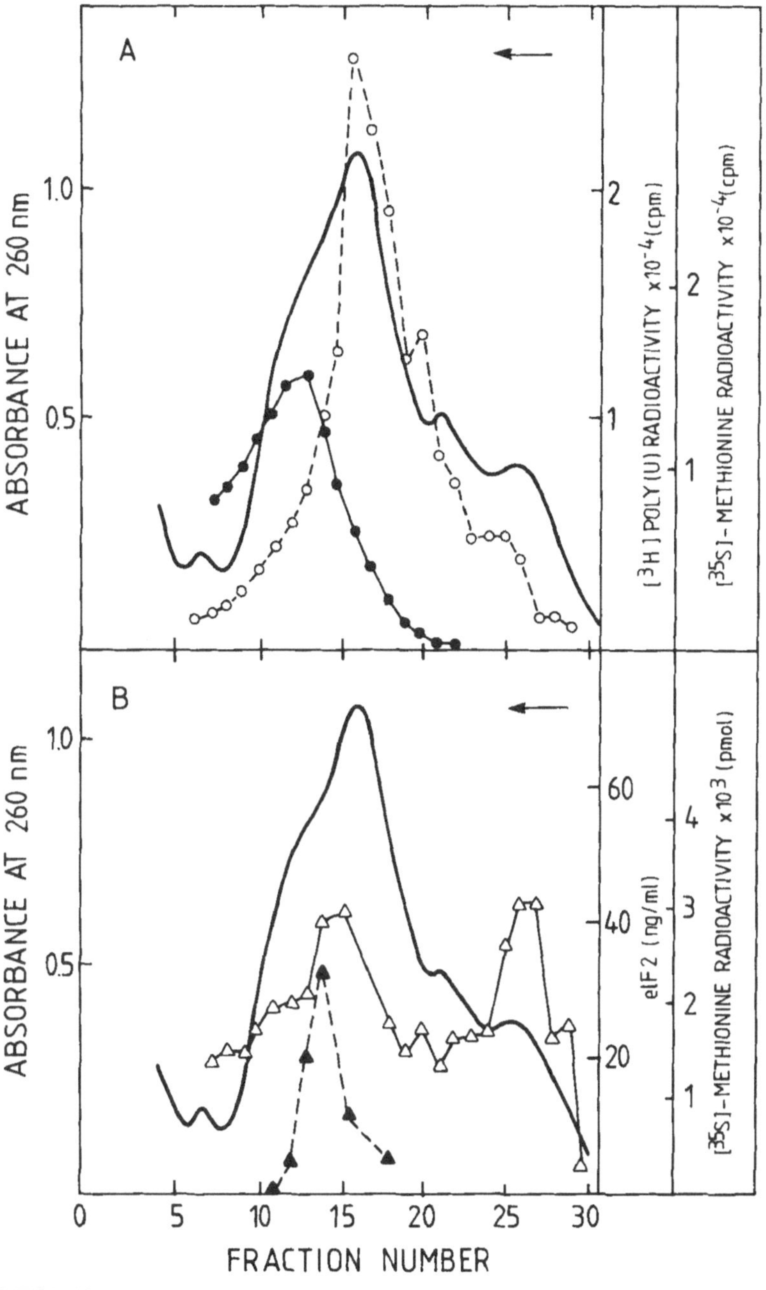

FIGURE 12. Correlation of translation, eIF2 and Met-tRNA$_f$-binding activity. Poly(A)-containing mRNP prepared as described in Figure 1 was centrifuged on a 10 to 30% (w/v) sucrose density gradient in 10 m*M* Hepes pH 7.2, 100 m*M* KCl, 0.1 m*M* DTT in a Beckman SW27 rotor at 96,300 *g* for 17 hr. The direction of centrifugation is indicated. Absorbance at 260 nm (——). (A) Poly(A)-sequences were detected by hybridization with (^{3}H)-poly(U) (○----○).[30] Translational activity of mRNP was measured in rabbit reticulocyte lysate (●——●).[54] (B) Gradient fractions were electrophoresed on sodium dodecylsulfate polyacrylamide gels and electroblotted onto nitrocellulose as described in Figure 11. eIF2 was quantified by anti-eIF2 antibody (▲----▲). Assay for Met-tRNA$_f$-binding activity (△——△).[55] The assay contained 20 m*M* Hepes pH 7.6, 70 m*M* KCl, 1 m*M* DTT, 0.4 m*M* GTP, 0.03 pmol [^{35}S]-Met-tRNA$_f$ (540,000 cpm/pmol). Incubation was at 30°C for 15 min. After addition of 1 mℓ ice-cold buffer containing 20 m*M* Hepes pH 7.5, 70 m*M* KCl, 1 m*M* DTT, the reaction mixture was filtered over cellulose nitrate. The filter was washed, dried, and counted in toluene scintillation liquid.

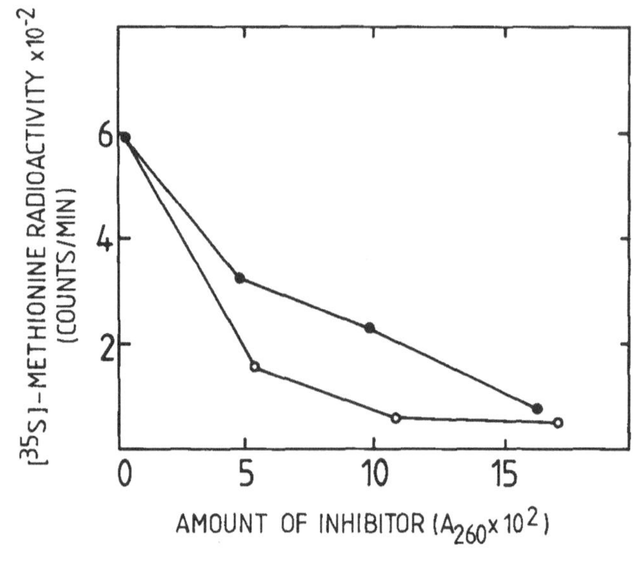

FIGURE 13. Inhibition of eIF2. Met-tRNA$_f$. GTP ternary complex formation by iRNP (O----O) and iRNA (●----●). iRNP was prepared as described in Reference 54. iRNA was extracted from iRNP by phenolization. The assay for ternary complex formation is described in Figure 12.

troblotting. The activity was very low if fractions were assayed without further purification. The mRNP-associated factor had a specific activity of only 2 pmol[^{35}S]-Met tRNA$_f$ per milligram protein in 10 min. The low recovery of the Met-tRNA$_f$-binding activity may be due to the low concentration of the initiation factor,[99] dissociation of eIF2 during mRNP purification, degradation by proteolytic enzymes,[95] and inhibition of ternary complex formation by iRNP dissociated from repressed 17S mRNP (Figure 13).

The initiation factor was purified from salt-washed mRNP proteins by ion-exchange chromatography, phosphocellulose P11 and affinity chromatography on poly(A)-Sepharose 4B. The final preparation was still contaminated with the poly(A)-binding proteins of M_r 38,000 and 23,500 but the specific activity increased more than 40-fold. Due to the loss of activity in additional purification steps we were unable to remove the latter contaminating proteins and subsequent Met-tRNA$_f$-binding experiments were performed with the partially purified eIF2 preparation. The Met-tRNA$_f$ complex formation is GTP dependent (Figure 14A). A saturation of ternary complex formation was obtained at a GTP concentration of 5μ*M*. Approximately the same value was obtained by others for eIF2 prepared from a ribosomal salt wash.[100-102] GDP inhibits the ternary complex formation (Figure 14B). At a ratio of 1:20 (GTP/GDP) complex formation was completely absent. Inhibition of ternary complex formation was also observed at low concentrations of Mg^{2+} (Figure 14C). The effect of Mg^{2+} ions on eIF2 ternary complex formation has led to conflicting results. Although the inhibitory effect of Mg^{2+} ions has been demonstrated many investigators did not observe inhibition by Mg^{2+} ions.[100-103] The observed Mg^{2+} inhibition may be due to the partial purity of the Met-tRNA$_f$- binding factor preparation from nonpolysomal mRNP. The functional significance of the Met-tRNA$_f$-binding activity was further demonstrated by its complex formation with salt-washed 40S ribosome subunits. Assay mixtures containing Met-tRNA$_f$, GTP, factor isolated from mRNP, initiation codon AUG, and salt-washed 40S ribosomes showed that the GTP-dependent complex formed between Met-tRNA$_f$ and mRNP-associated factor can be bound to the 40S ribosome (Figure 15).

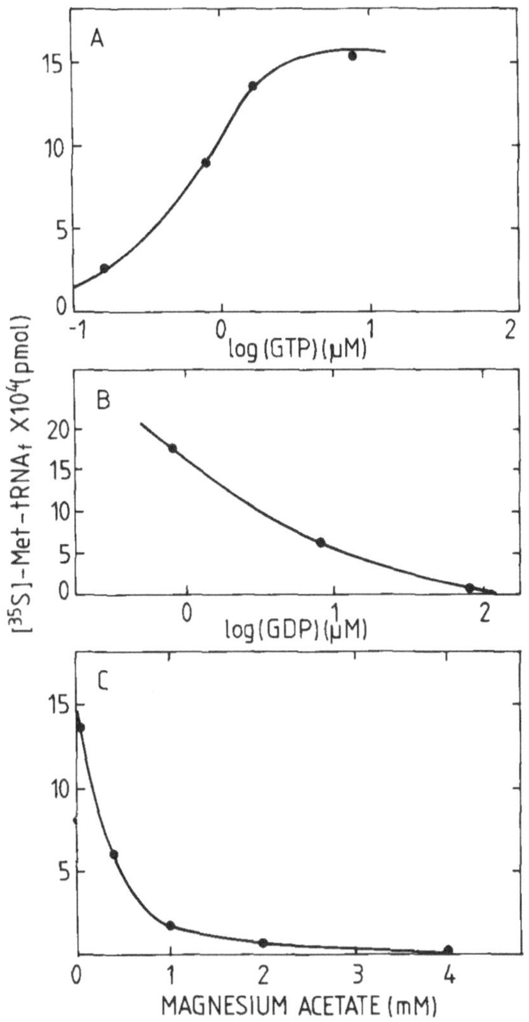

FIGURE 14. Properties of Met-tRNA$_f$-binding activity. The dependence of ternary complex formation on (A) GTP, (B) GDP, and (C) Mg^{2+} ions. The assay contained 20 m*M* Hepes pH 7.6, 70 m*M* KCl, 1 m*M* DTT, and a varying concentration of (A) GTP or (B,C) 3 µ*M* GTP. Incubation was at 30°C for 15 min.

C. mRNP-Associated Protein Kinase

1. Association of a Casein Kinase II with mRNP

Cyclic nucleotide independent protein kinases have been detected in nonpolysomal mRNP of duck erythroblasts, HeLa cells, embryonic chicken muscle, rat liver, mouse plasmacytoma cells, reticulocytes, and *Xenopus* oocytes and have been implicated in the regulation of protein synthesis.[20,22,24,43,105-108] Evidence for the association of a protein kinase with nonpolysomal mRNP of cryptobiotic embryos was obtained from sucrose density gradient centrifugation of mRNP purified by affinity chromatography on oligo(dT)-cellulose. A protein kinase activity is associated with the repressed 17S mRNP (Figure 16). The enzyme was purified to homogeneity by ion-exchange chromatography on phosphocellulose P11 and affinity chromatography on casein-Sepharose 4B and ATP-agarose.[109]

The undegraded enzyme has an M_r of 136,000 and is composed of two subunits with M_r of 36,500 (α) and 28,000 (β). The β subunit was autophosphorylated using ATP and GTP as phosphate donors. Based on the native and subunit molecular weights, autophosphorylation

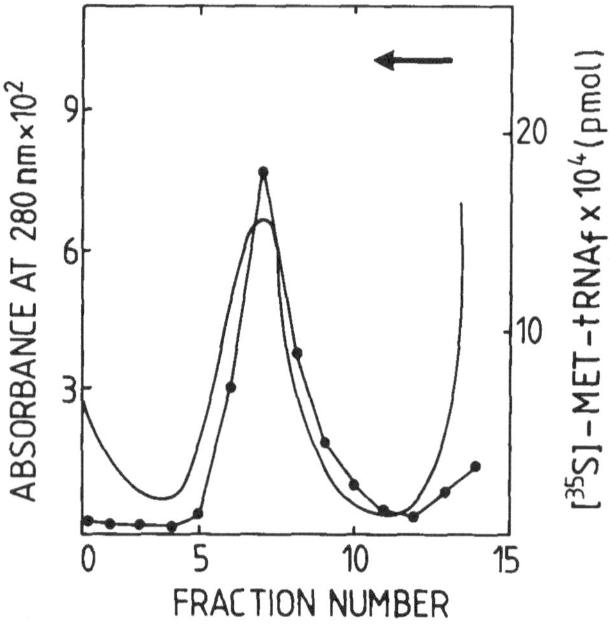

FIGURE 15. Formation of a 40S ribosome preinitiation complex. The assay for the binding of [^{35}S]-Met-tRNA$_f$ to 40S ribosomes contained 20 m*M* Tris-HCl pH 7.4, 100 m*M* KCl, 1 m*M* DTT, 300 μ*M* GTP, 0.1 A_{260} AUG, 0.09 pmol [^{35}S]-Met-tRNA$_f$, 0.9 A_{260} 40S subunits, and eIF2 purified from mRNP proteins by ion-exchange chromatography on DEAE cellulose and phosphocellulose P11 and affinity chromatography on poly(A)-Sepharose 4B. Incubation was at 30°C for 10 min. The reaction mixture was analyzed on a 15 to 50% (w/v) sucrose density gradient in 20 m*M* Hepes pH 7.2, 100 m*M* KCl, 9 m*M* magnesium acetate. Centrifugation was in the Beckman SW65 rotor at 421,000 *g* for 85 min. The direction of centrifugation is indicated. Fractions were collected, diluted, and filtered through cellulose nitrate. [^{35}S]-Met-tRNA$_f$ radioactivity (●—●). Absorbance at 260 nm (—).

of the β subunit, the use of ATP and GTP as phosphate donors, and casein as a substrate, the mRNP-associated protein kinase was classified as a casein kinase II.[109,110] This classification was confirmed by the use of specific inhibitors of the enzyme e.g., heparin, caffeine, and poly(L)-glutamic acid.[111-117]

2. *Substrates of the mRNP-Associated Protein Kinase*

A possible function of the protein kinase in protein synthesis was deduced from the identification of the protein substrates of the enzyme. Exogeneous proteins with a function in mRNP related phenomena have been assayed for their use as substrates of the mRNP-associated kinase (Table 1). The assayed proteins are eIF2,[104] poly(A)-binding proteins of nonpolysomal mRNP,[38] ribosomal proteins,[118] and the RNA-binding 19S protein related to eEF-Ts.[119,120] The main phosphorylated proteins were eIF2 and the poly(A)-binding proteins. In comparison with casein, eIF2 was almost eight times more efficiently phosphorylated than casein. More direct evidence for the in vivo substrates of the enzyme was obtained from the identification of the endogeneous mRNP proteins phosphorylated by the enzyme. The latter proteins were identified by two-dimensional gel electrophoresis of mRNP incubated with [γ-^{32}P]-ATP (Figure 17).[121] Phosphorylated proteins have M_r of 125,000, 65,000, 38,000, 29,000, and 23,500. mRNP-associated eIF2 was not detected among the phosphorylated proteins. The main phosphorylated protein was the M_r 38,000 poly(A)-binding protein

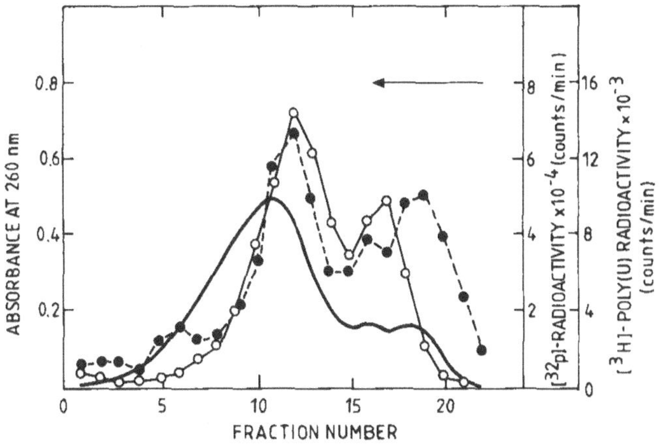

FIGURE 16. Association of a casein kinase II with mRNP. Poly(A)-containing mRNP prepared as described in Figure 1 was analyzed on a 10 to 30% (w/v) sucrose density gradient in 10 m*M* Hepes pH 7.2, 100 m*M* KCl, 7 m*M* 2-mercaptoethanol, and 0.1 m*M* PMSF. Centrifugation was in a Beckman SW27 rotor at 103,000 *g* for 21 hr. The direction of centrifugation is indicated. Absorbance at 260 nm (——). Poly(A)-sequences were assayed by hybridization with (^{3}H)-poly(U) (●----●).[30] Protein kinase activity was measured in standard conditions using casein as a phosphate acceptor(○——○). The assay contained 10 m*M* Tris-HCl pH 8.0, 12.5 m*M* $MgCl_2$, 50 m*M* KCl, 1 m*M* DTT, 7 m*M* 2-mercaptoethanol, 75 μg casein, and 1 μ*M* [γ-^{32}P]-ATP (specific activity 10 Ci/mmol). Incubation was at 37°C for 30 min. The reaction was terminated by addition of trichloroacetic acid to a final concentration of 10% (w/v).

Table 1
PHOSPHORYLATION OF PROTEINS BY THE mRNP-ASSOCIATED PROTEIN KINASE

Acceptor	Incorporation (mol/mol)	Percentage[a]
eIF2	0.301	782
Casein	0.038	100
M_r 23,500 poly(A)-binding protein	0.020	47
M_r 38,000 poly(A)-binding protein	0.016	36
Ribosomal proteins	n.d.	—
19S protein	n.d.	—

Note: eIF-2,the poly(A)-binding proteins, ribosomal proteins, and the 19S protein were prepared from *Artemia* sp. Proteins were phosphorylated with 8 U of mRNP-associated protein kinase in 10 m*M* Tris-HCl pH 8.0, 12.5 m*M* $MgCl_2$, 50 m*M* KCl, 1 m*M* DTT, 100 μ*M* ATP. Incubation was for 2 hr at 37°C. Reactions were stopped by precipitation with trichloroacetic acid.

[a] The radioactivity incorporated into casein and expressed as mol phosphate/mol protein was taken as 100%.

From Thoen, C. et al., *Mol. Biol. Rep.*, 11, 69, 1986. With permission.

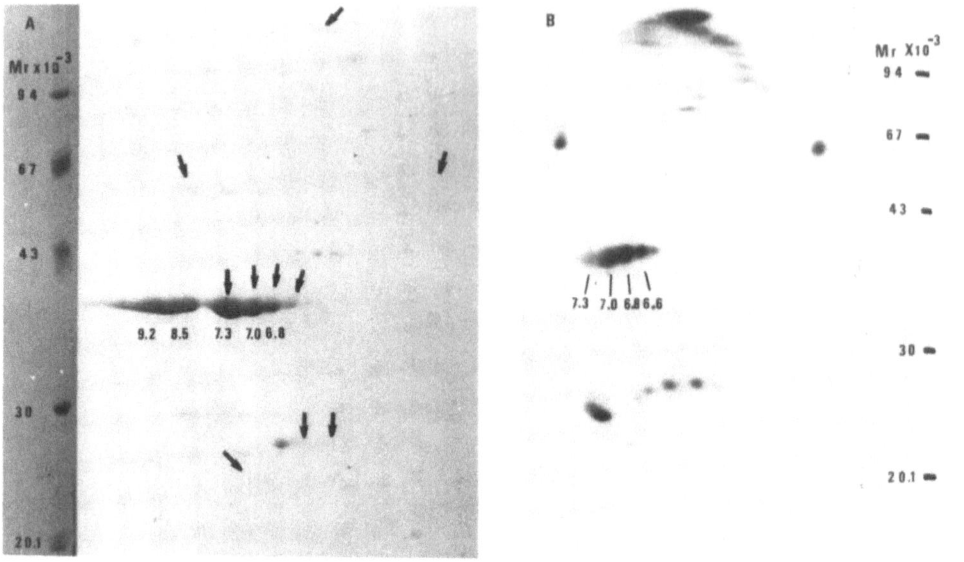

FIGURE 17. Characterization of *in situ* phosphorylated mRNP-proteins. mRNP prepared by affinity chromatography on oligo(dT)-cellulose was incubated in standard protein kinase assay conditions as described in Figure 16. Proteins were analyzed by two-dimensional gel electrophoresis as described by O'Farrell.[82] (A) Stained gel. Arrows indicate the positions of phosphorylated proteins. (B) Autoradiogram. (From Thoen, C., et al., *Mol. Biol. Rep.*, 11, 69, 1986. With permission.)

which existed in several isoforms with a pI of 9.2, 8.5, 7.3, 7.0, 6.8, and 6.6. The components with pI values of 9.2 and 8.5 are not phosphorylated and the acidic isoforms are generated by phosphorylation of the basic ones.[73] As mentioned in Section IV.A.3.d, the latter phosphorylation affects the interaction of the M_r 38,000 protein with the poly(A)-sequence and abolishes the translational activity of nonpolysomal mRNP (Section VI). The phosphorylated proteins with M_r of 125,000, 65,000, and 29,000 have similar molecular weights as the phosphorylated subunits of eIF3.[117] However, due to its molecular weight of 500,000 to 700,000 it is likely that the latter factor was copurified with mRNP.

V. CYTOPLASMIC ENZYMES IMPLICATED IN mRNP ACTIVATION

A. Protein Phosphatases Involved in mRNP Dephosphorylation

1. Characterization of Protein Phosphatases

No protein phosphatase activity was detected in nonpolysomal poly(A)-containing mRNP of cryptobiotic gastrulae using [^{32}P]-phosphorylase a and [^{32}P]-phosphorylase kinase as substrates in conditions where approximately 90% of the phosphate was released from phosphorylase a by protein phosphatases 1 and 2A from rabbit skeletal muscle.[92] As far as we know no protein phosphatase activity has been detected in mRNP of other systems.

Protein phosphatases active in the dephosphorylation of mRNP proteins were detected in the postribosomal supernatant. At least five different protein phosphatases were resolved by ion-exchange chromatography on DEAE-cellulose using phosphorylase a as substrate (Figure 18). Only two polycation-stimulated protein phosphatases designated PCS-X and PCS-Y were active in the dephosphorylation of mRNP proteins. PCS-X and PCS-Y have an M_r of 225,000 and 346,000 respectively as measured by gel filtration on Sephadex G-200. Both protein phosphatases were purified by ion-exchange chromatography on several DEAE matrices, affinity chromatography on poly(L) lysine-agarose, histone-Sepharose 4B, and pro-

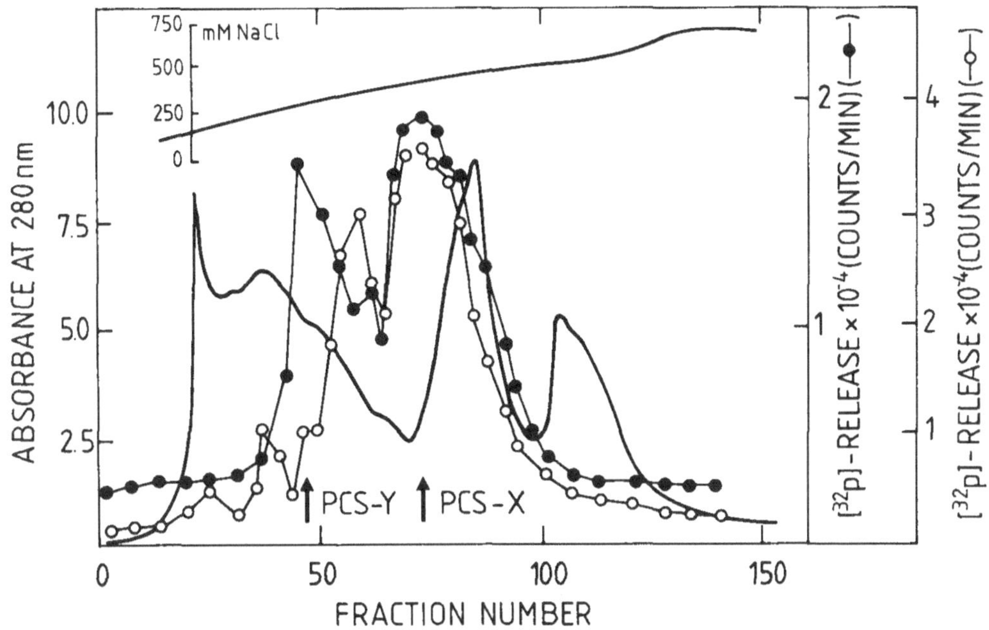

FIGURE 18. Resolution of cytoplasmic protein phosphatases by chromatography on DEAE-cellulose. The postribosomal supernatant of cryptobiotic embryos was prepared from the postmitochondrial supernatant by differential centrifugation at 250,000 *g* for 1 hr in the Beckman R60Ti rotor. The 30 to 60% $(NH_4)_2SO_4$ precipitate of the postribosomal supernatant was dialyzed against 20 m*M* Tris-HCl pH 7.2, 0.1% (v/v) 2-mercaptoenthanol and chromatographed on DEAE-cellulose (60 × 3 cm).[92] Bound proteins were eluted with a linear salt gradient from 50 m*M* to 900 m*M* NaCl (2 × 1ℓ). The flow rate was 140 mℓ/hr and fractions of 10 mℓ were collected. [^{32}P]-labeled phosphorylase *a* was prepared from phosphorylase b using phosphorylase kinase. mRNP proteins were labeled with 6 μ*M* [γ-^{32}P]-ATP (specific activity 110 Ci/mmol) and 8 units of mRNP-assoicated protein kinase. Reactions were stopped by gel filtration on Sephadex G50 to remove ATP. Fractions of the DEAE-cellulose column were assayed for protein phosphatase activity using phosphorylase a (○——○) or mRNP proteins (●——●) as substrates. Assays contained 17.5 m*M* Tris-HCl pH 7.0, 0.03 m*M* EGTA, 0.03% (v/v) 2-mercaptoethanol, 0.35 mg/mℓ BSA, 10 μg [^{32}P]-labeled substrate, and 1 m*M* $MnCl_2$.[92] 5 m*M* caffeine were included when phosporylase a was used as substrate. Fractions were diluted 65 times in the assay for mRNP-protein dephosphorylation. Absorbance at 280 nm (——). The position of PCS-X and PCS-Y is indicated.

tamine-agarose, hydrophobic chromatography on phenyl-Sepharose 4B, and gel filtration on Sephadex G-200.[122] After 12 purification steps PCS-X was purified to near homogeneity and is composed of two subunits with M_r of 75,000 (α) and 40,000 (β).[122] The β subunit was identified as the catalytic subunit by ethanol treatment or freeze-thawing in the presence of 1.7% 2-mercaptoethanol which leads to the dissociation of the catalytic subunit. PCS-Y has a more complex composition and is composed of several proteins with M_r of 100,000, 91,000, 70,000, 60,000, 56,000, and 37,000. The M_r 70,000 and 37,000 have comparable molecular weights to the α and β subunits of PCS-X. Apparently both enzymes are composed of the same core subunits and the other proteins may modulate the activity of the enzymes. Both phosphatases were inhibited by NaF, ATP, and PP_i and are activated by heparin, protamine, poly(L)lysine, and histone H_1.[117] Based on these properties both protein phosphatases were classified as polycation-stimulated protein phosphatases[123] with properties very similar to protein phosphatases 2A from rabbit skeletal muscle.[124,125]

2. *Substrates of the Protein Phosphatases*

Purified PCS-X and PCS-Y dephosphorylated mRNP-proteins approximately 40 times more efficiently than phosphorylase a.[92] Native mRNP was dephosphorylated at a slower rate and less extensively than isolated mRNP proteins. In conditions where PCS-X and PCS-

Y dephosphorylated mRNP proteins to 60 to 80% within 10 min, native mRNP was dephosphorylated only 25% after an incubation of 3 hr. This difference in dephosphorylation indicates that some of the phosphate groups are not accessible to the enzyme and that some phosphorylations have already occurred before mRNP assembly. mRNP substrates of the protein phosphatases were identified by sequential phosphorylation-dephosphorylation and dephosphorylation-phosphorylation of mRNP-proteins.[92] Both enzymes dephosphorylated the poly(A)-binding M_r 38,000 protein. A second substrate with an M_r of 65,000 is detected after dephosphorylation of mRNP proteins prior to phosphorylation.[92] The latter protein has a similar molecular weight as the component of the inhibitor ribonucleoprotein and is one of the major phosphoproteins of mRNP.

B. Polyadenylation of mRNP

1. Absence of Poly(A)-Polymerase in mRNP

Measurements of the poly(A)-polymerase activity in the nuclear, mitochondrial, and postmitochondrial fractions of cryptobiotic embryos indicated that more than 99% of the activity was assayed in the postmitochondrial supernatant using $(A)_4$ as a primer.[93] The enzyme was located in the postribosomal supernatant but is not associated with the nonpolysomal poly(A)-containing mRNP. Previously Sastre and Sebastian located 70 to 80% of the poly(A)-polymerase activity in the cytoplasm and 20 to 30% in the nuclear and membrane fraction using tRNA as a primer.[126,127] Upon development the activities in both compartments equalize. No activity was present in the mitochondrial and microsomal fractions.

Poly(A)-polymerase of cryptobiotic embryos was purified to near homogeneity by ion-exchange chromatography on DEAE matrices and phosphocellulose P11, gel filtration on CL-Sepharose 6B, and affinity chromatography on poly(A)-Sepharose 4B and ATP-agarose.[93] Affinity chromatography on poly(A)-Sepharose 4B separated the preparation into two active fractions.[128] The unbound fraction [poly(A)-polymerase I] contained two proteins with M_r of 94,000 and 70,000. The bound fraction [poly(A)-polymerase II] contained two proteins with M_r of 70,000 and 38,000. The M_r 70,000 protein was identified as poly(A)-polymerase in agreement with the molecular weight measured by Sastre and Sebastian.[129] The Mn^{2+}-dependent enzyme has similar properties as reported for other studied poly(A)-polymerases.[130,131]

2. Primer Specificity of Poly(A)-Polymerase

RNA and RNP were compared for their use as primers for poly(A)-polymerase I and II (Table 2). Both enzymes had a different primer specificity. Oligo(A) primers were elongated by poly(A)-polymerase I to approximately 8 to 10 adenylic acid residues. Poly(A) primers were not at all used by the latter enzyme. Poly(A)-polymerase II elongated oligo(A) to a three- to fourfold greater length than poly(A)-polymerase I. Poly(A)-lacking mRNA was not used by either enzyme. tRNA and rRNAs were not used by poly(A)-polymerase I and were poor primers for poly(A)-polymerase II. No specificity could be deduced from the polyadenylation of RNA primers. A remarkable specificity was observed when ribonucleoproteins were used as primers. Ribosomes, ribosomal subunits, and 22S poly(A)-lacking mRNP were not used as primer. Only the 17S nonpolysomal poly(A)-containing mRNP was efficiently polyadenylated by poly(A)-polymerase II. The difference in primer specificity is due to the presence of a small RNP in the poly(A)-polymerase II fraction.[128] The latter RNP is composed of a small RNA with a length of 83 nucleotides and an M_r 38,000 protein resembling the poly(A)-binding protein. Removal of the small RNP abolished the specificity towards the 17S mRNP.

Table 2
COMPARISON OF THE EFFICIENCY OF PRIMER POLYADENYLATION

	Enzyme fraction I		Enzyme fraction II	
	pmol AMP/pmol primer/μg protein	Percent	pmol AMP/pmol primer/μg protein	Percent
RNA primers				
$(A)_4$	8.0	100.0	28.0	100.0
$(A)_6$	10.2	127.5	n.m.	n.m.
$(A)_8$	10.6	132.5	34.0	121.4
$(A)_{10}$	10.2	127.5	n.m.	n.m.
$(A)_{54}$	0.2	2.5	21.4	76.4
Poly(A)	0.2	2.5	n.d.	n.d.
tRNA	1.8	22.5	7.0	25.0
28S rRNA	n.d.	n.d.	11.2	40.0
18S rRNA	n.d.	n.d.	12.3	44.0
5s rRNA	n.d.	n.d.	0.6	2.1
13S poly$(A)^+$ mRNA	n.d.	n.d.	14.7	52.5
9S poly$(A)^-$ mRNA	n.d.	n.d.	n.d.	n.d.
RNP primers				
80S ribosomes	n.d.	n.d.	n.d.	n.d.
80S ribosomes salt washed with 500 m*M* KCl	n.d.	n.d.	n.d.	n.d.
40S ribosomal subunits	n.d.	n.d.	n.d.	n.d.
60S ribosomal subunits	n.d.	n.d.	n.d.	n.d.
17S poly$(A)^+$ mRNP	n.d.	n.d.	140	500
7S poly$(A)^+$ mRNP	0.2	2.5	n.d.	n.d.
22S poly$(A)^-$ mRNP	n.d.	n.d.	n.d.	n.d.

Note: Poly(A)-polymerase assays (100 μℓ) were in 200 m*M* Tris-HCl pH 8.4, 0.9 m*M* (^{3}H)-ATP (special action 15 cpm/pmol), 20 m*M* NH_4Cl, 0.9 m*M* $MnCl_2$, 1.25 μg poly(A)-polymerase, and primer. Incubation was at 30°C for 3 hr. The polyadenylation was calculated from the slope of concentration-variation curves and was corrected for degradation.[128] Correction factors were calculated from analysis of reaction mixtures by sucrose density gradient centrifugation. Mixtures were centrifuged on a 10 to 30% (w/v) sucrose gradient in 10 m*M* Hepes pH 7.2, 0.1 m*M* DTT, 0.1 m*M* PMSF. Centrifugation was in the Beckman SW40 rotor for 18 hr at 160,000 *g* and at 4°C; n.d. not detectable, n.m. not measured.

From Roggen, E., et al., *Biochim. Biophys. Acta,* 825, 21, 1985. With permission.

VI. REGULATION OF THE REPRESSION OF NONPOLYSOMAL POLY(A)-CONTAINING mRNP

A discrete number of proteins is associated with nonpolysomal mRNP of cryptobiotic gastrulae. Several of these proteins have been identified and their function deduced from in vitro experiments (Section IV). The characterization of the poly(A)-binding proteins, the presence of eIF2 on a fraction of the mRNP, and the association of a protein kinase support the "omnia mecum porto" hypothesis proposed by Spirin which states that the mRNA-associated proteins have a function in mRNA-related phenomena.[132] In particular our identification studies argue for the involvement of mRNP proteins in the regulation of translation of mRNP. Evidence is presented for the existence of several, apparently independent mechanisms involved in the regulation of the activation of repressed mRNP and the subsequent initiation of translation.

Nonpolysomal poly(A)-containing mRNP of cryptobiotic gastrulae mainly exists in a repressed form by association of an iRNP. The RNA component of this particle has the inhibitory property which is enhanced by the M_r 64,000 protein component. The iRNP is one of the regulators of mRNP translation. The association of eIF2 with a small translatable mRNP fraction lacking the iRNP particle demonstrates that the translation of stored mRNP requires the exchange of the inhibitor for eIF2 by a yet unknown mechanism. From iRNP-mRNP reconstitution experiments (Section III. C) some evidence is obtained that at least in vitro the association of iRNP is only possible after a reorganization of the poly(A)-protein structure. Although the RNA component of the inhibitor prevents the ternary complex formation between eIF2, Met-$tRNA_f$, and GTP (Section IV.B.2) no direct evidence is obtained concerning the exchange of iRNP for eIF2.

mRNP translation is also regulated by phosphorylation-dephosphorylation of mRNP proteins by the mRNP-associated protein kinase and two cytoplasmic protein phosphatases. Evidence for a function of the latter protein modification in translation is obtained from a correlation of mRNP phosphorylation and inhibition of translation.[109] The translational activity of mRNP is abolished after incubation of mRNP with ATP or GTP in protein kinase standard assay conditions. Although exogeneous eIF2 is very efficiently phosphorylated by the mRNP-associated protein kinase and although this factor is most likely implicated in translational control, mRNP-associated eIF2 is not phosphorylated by the enzyme. The main phosphorylated proteins are the poly(A)-binding proteins which maintain the mRNP in its repressed form. Conversely, dephosphorylation is required for the activation of mRNP as deduced from the absence of the acidic isospecies of the M_r 38,000 poly(A)-binding protein in mRNP-ribosome preinitiation-complexes (Section IV.A.3.b). Furthermore, the decrease in protein-poly(A) interaction upon dephosphorylation is necessary for the exchange of the M_r 38,000 poly(A)-binding protein with the polysomal M_r 73,000 and 55,000 poly(A)-binding proteins. The latter exchange may result in a structural rearrangement of the poly(A)-protein complex after which mRNA becomes accesible for ribosomes at the onset of the initiation of protein synthesis and before entry into polysomes.

Polyadenylation of mRNA has often been reported to occur in conjunction with the activation of stored mRNP.[133-135] Drastic changes in the polyadenylation of mRNA concomitant with an increase in protein synthesis have been observed in several biological systems after fertilization or resumption of development.[136] *Artemia* is one of these systems.[137] The isolation of a cytoplasmic poly(A)-polymerase very efficient in the elongation of the 17S nonpolysomal poly(A)-containing mRNP makes this enzyme a very attractive candidate to be involved in the activation of repressed mRNP. However, several observations argue against a function of polyadenylation in the activation mechanism. In *Xenopus* oocytes an opposite relationship is observed and maternal histone mRNA is deadenylated when histone synthesis is increased.[138] In several systems the increase in poly(A) length is not always correlated with an increase in protein synthesis.[139] For the efficiently elongated 17S mRNP of *Artemia* we could not observe a change in the repression after elongation. However, this does not rule out the possibility that in vivo elongation of the poly(A)-sequence is required for the binding of factors involved in protein synthesis.

ACKNOWLEDGMENT

This investigation is supported by grants from the Fund for Joint Basic Research of the Belgian National Fund for Scientific Research.

REFERENCES

1. **Spirin, A. S.,** Informosomes, *Eur. J. Biochem.*, 10, 20, 1969.
2. **Preobrazhensky, A. A. and Spirin, A. S.,** Informosomes and their protein components: the present state of knowledge, *Prog. Nucl. Acid Res. Mol. Biol.*, 21, 1, 1978.

3. **Arnstein, H. R. V.**, Messenger ribonucleoprotein complexes in gene expression, *Biochem. Soc. Symp.*, 47, 163, 1983.
4. **Liautard, J. P., Setyono, B., Spindler,E., and Köhler, K.**, Comparison of proteins bound to the different functional classes of messenger RNA, *Biochim. Biophys. Acta*, 425, 373, 1976.
5. **Van Venrooij, W. J., van Eekelen, C. A. G., Janssen, R. T. P., and Princen, J. M. G.**, Specific poly-A-binding protein of 76,000 molecular weight in polyribosomes is not present on poly A of free cytoplasmic mRNP, *Nature (London)* 270, 189, 1977.
6. **Jain, S. K., Roy, R. K., Pluskal, M. G., Croall, D. E., Guha, C., and Sarkar, S.**, A model of translational control involving mRNA-associated proteins in chick embryonic muscle, *Mol. Biol. Rep.*, 5, 79, 1979.
7. **Schweiger, A. and Kostka, G.**, Identification of a 110,000 molecular weight protein associated with heterogeneous nuclear RNA and messenger RNA in rat liver cells, *Exp. Cell Res.*, 125, 211, 1980.
8. **Spirin, A. S.**, Messenger ribonucleoproteins (informosomes) and RNA-binding proteins, *Mol. Biol. Rep.*, 5, 53, 1979.
9. **Brawerman, G.**, The role of the poly(A) sequence in mammalian messenger RNA, *Crit. Rev. Biochem.*, 10, 1, 1981.
10. **Goldenberg, C., Vincent, A., and Scherrer, K.**, Evidence for the protection of specific RNA sequences in globin messenger ribonucleoprotein particles, *Nucl. Acids Res.*, 6, 2787, 1979.
11. **Jenkins, N. A., Kaumeyer, J. F., Young, E. M., and Raff, R. A.**, A test for masked message: the template activity of messenger ribonucleoprotein particles isolated from sea urchin eggs, *Dev. Biol.*, 63, 279, 1978.
12. **Chisick, M. C., Brennessel, B. A., and Biswas, D. K.**, Poly(A)-protein interactions and transport of mRNA in isolated nuclei, *Biochem. Biophys. Res. Commun.*, 91, 1109, 1979.
13. **Moffett, R. B. and Webb, T. E.**, Regulated transport of messenger ribonucleic acid from isolated liver nuclei by nucleic acid binding proteins, *Biochemistry*, 20, 3253, 1981.
14. **Palayoor, T., Schumm, D. E., and Webb, T. E.**, Transport of functional messenger RNA from liver nuclei in a reconstituted cell-free system, *Biochim. Biophys. Acta*, 654, 201, 1981.
15. **Heywood, S. M. and Kennedy, D. S.**, Translational control in embryonic muscle, *Prog. Nucl. Acids Res. Mol. Biol.*, 19, 477, 1976.
16. **Jain, S. and Sarkar, S.**, Poly(riboadendylate)-containing messenger ribonucleoprotein particles of chick embryonic muscles, *Biochemistry*, 18, 745, 1979.
17. **Maundrell, K., Maxwell, E. S., Civelli, O., Vincent, A., Goldenberg, S., Buri, J. F., Imaizumi, M. T., and Scherrer, K.**, Messenger ribonucleoprotein complexes in avian erythroblasts: carriers of posttranscriptional regulation?, *Mol. Biol. Rep.*, 5, 43, 1979.
18. **Bag, J. and Sells, B. H.**, Cytoplasmic nonpolysomal ribonucleoprotein complexes and translational control, *Mol. Cell. Biochem.*, 40, 129, 1981.
19. **Vincent, A., Goldenberg, S., Standart, N., Civelli, O., Imaizumi-Scherrer, T., Maundrell, K., and Scherrer, K.**, Potential role of mRNP proteins in cytoplasmic control of gene expression in duck erythroblasts, *Mol. Biol. Rep.*, 7, 71, 1981.
20. **Auerbach, S. and Pederson, T.**, Phosphorylation of messenger RNA-bound proteins in Hela cells, *Biochem. Biophys. Res. Commun.*, 63, 149, 1975.
21. **Ovchinnikov, L. P., Spirin, A. S., Erni, B., and Staehelin, T.**, RNA-binding proteins of rabbit reticulocytes contain the two elongation factors and some of the initiation factors of translation, *FEBS Lett.*, 88, 21, 1978.
22. **Bag, J. and Sells, B. H.**, The presence of protein kinase activity and acceptors of phosphate groups in nonpolysomal cytoplasmic messenger ribonucleoprotein complexes of embryonic chicken muscle, *J. Biol. Chem.*, 254, 3137, 1979.
23. **Blanchard, J. M., Brunel, C., and Jeanteur, P.**, Characterization of an endogeneous protein kinase activity in ribonucleoprotein structures containing heterogeneous nuclear RNA in Hela cell nuclei, *Eur. J. Biochem.*, 79, 117, 1977.
24. **Cardelli, J. and Pitot, H. C.**, Characterization of protein kinase activity associated with rat liver polysomal messenger ribonucleoprotein particles, *Biochemistry*, 19, 3164, 1980.
25. **Lorberboum, H., Galski, H., Scharf, C., Weinstein, D., de Groot, N., and Hochberg, A. A.**, Alkaline phosphatase and protein kinase(s) activities in free cytoplasmic mRNPs from human term placenta, *Mol. Biol. Rep.*, 11, 29, 1986.
26. **Dutrieu, J.**, Observations biochimiques et physiologiques sur le dèveloppement *d'Artemia salina* leach, *Arch. Zool. Exp. Gen.*, 99, 1, 1960.
27. **Finamore, F. J. and Clegg, J. S.**, Biochemical aspects of morphogenesis in the brine shrimp *Artemia salina*, in *The Cell Cycle*, Padilla, G. M., Whitson, G. L., and Cameron, I. L., Eds., Academic Press New York, 1969, chap. 12.
28. **Clegg, J. S. and Golub, A. L.**, Protein synthesis in *Artemia salina* embryos. II. Resumption of RNA and protein synthesis upon cessation of dormancy in the encysted gastrula, *Dev. Biol.*, 19, 178, 1969.

29. **Nilsson, M. O. and Hultin, T.,** Characteristics and intracellular distribution of messenger like RNA in encysted embryos of *Artemia salina, Dev. Biol.*, 38, 138, 1974.
30. **Slegers, H. and Kondo, M.,** Messenger ribonucleoprotein complexes of cryptobiotic embryos of *Artemia salina, Nucl. Acids Res.*, 4, 625, 1977.
31. **Simons, J., De Herdt, E., Kondo, M., and Slegers, H.,** Membrane associated poly(A)-containing ribonucleoproteins in *Artemia salina* gastrula embryos, *FEBS Lett.*, 91, 53, 1978.
32. **Slegers, H., De Herdt, E., and Kondo, M.,** Non-polysomal poly(A)-containing messenger ribonucleoproteins of cryptobiotic gastrulae of *Artemia salina, Eur. J. Biochem.*, 117, 111, 1981.
33. **Blobel, G.,** Isolation of a 5S-RNA protein complex from mammalian ribosomes, *Proc. Natl. Acad. Sci. U. S. A.*, 68, 1881, 1971.
34. **Hoerz, W. and McCarthy, K. S.,** Initiation of protein synthesis in a rabbit reticulocyte lysate system, *Biochim. Biophys. Acta,* 228, 526, 1971.
35. **Svardal, A. M. and Pryme, I. F.,** Aspects of the role of the endoplasmic reticulum in protein synthesis, *Subcell. Biochem.*, 7, 117, 1980.
36. **Slegers, H., De Herdt, E., and Kondo, M.,** Characterization of protein components of poly(A)-containing messenger ribonucleoproteins from cryptobiotic gastrulae of *Artemia salina, Mol. Biol. Rep.*, 5, 65, 1979.
37. **Bag, J. and Sells, B. H.,** Heterogeneity of the nonpolysomal cytoplasmic (free) mRNA protein complexes of embryonic chicken muscle, *Eur. J. Biochem.*, 99, 507, 1979.
38. **De Herdt, E., Piot, E., Kondo, M., and Slegers, H.,** The 38000-Mr poly(A)-binding protein of non-polysomal messenger ribonucleoproteins of cryptobiotic gastrulae of *Artemia salina, Eur. J. Biochem.*, 122, 453, 1982.
39. **Greenberg, J. R.,** Messenger RNA metabolism of animal cells. Possible involvement of untranslated sequences and mRNA-associated proteins, *J. Cell Biol.*, 64, 269, 1975.
40. **Jeffery, W. R.,** Characterization of polypeptides associated with messenger RNA and its polyadenylate segment in Ehrlich Ascites messenger ribonucleoprotein, *J. Biol. Chem.*, 252, 3525, 1977.
41. **Bag, J. and Sells, B. H.,** Isolation and characterization of the nonpolysomal cytoplasmic messenger ribonucleoprotein complexes of rat liver, *Can. J. Biochem.*, 57, 1051, 1979.
42. **Kloetzel, P. M., Johnson, R. M., and Sommerville, J.,** Interaction of the hnRNA of amphibian oocytes with fibril-forming proteins, *Eur. J. Biochem.*, 127, 301, 1982.
43. **Dearsly, A. L., Johnson, R. M., Barrett, P., and Sommerville, J.,** Identification of a 60-kDa phosphoprotein that binds stored messenger RNA of *Xenopus* oocytes, *Eur. J. Biochem.*, 150, 95, 1985.
44. **Civelli, O., Vincent, A., Maundrell, K., Buri, J. F., and Scherrer, K.,** The translational repression of globin mRNA in free cytoplasmic ribonucleoprotein complexes, *Eur. J. Biochem.*, 107, 577, 1980.
45. **Jackle, H.,** In vitro translatability of RNP particles from insect eggs and embryos (Smittia spec, Chironomidae, *Diptera), J. Exp. Zool.*, 212, 177, 1980.
46. **Liautard, J. P. and Egly, J. M.,** In vitro translation studies of the cytoplasmic nonpolysomal particles containing messenger RNA, *Nucl. Acids Res.*, 8, 1793, 1980.
47. **Northemann, W., Schmelzer, E., and Heinrich, P. C.,** Characterization of 20 S and 40 S non-polysomal cytoplasmic messenger ribonucleoprotein particles from rat liver, *Eur. J. Biochem.*, 112, 451, 1980.
48. **Geoghean, T., Cereghini, S., and Brawerman, G.,** Inactive mRNA-protein complexes from mouse sarcoma-180 ascites cells, *Proc. Natl. Acad. Sci. U.S.A.*, 76, 5587, 1979.
49. **Grosfeld, H. and Littauer, U. Z.,** Cryptic form of mRNA in dormant *Artemia* salina cysts, *Biochem. Biophys. Res. Commun.*, 67, 176, 1975.
50. **Felicetti, L., Amaldi, P. P., Moretti, S., Campioni, N. G., and Urbani, C.,** Intracellular distribution, sedimentation values and template activity of polyadenylic acid-containing RNA stored in *Artemia* salina cysts, *Cell Differentiation,* 4, 339, 1975.
51. **Sierra, J. M., Filipowicz, W., and Ochoa, S.,** Messenger RNA in undeveloped and developing *Artemia salina* embryos, *Biochem. Biophys. Res. Commun.*, 69, 181, 1976.
52. **Slegers, H., Mettrie, R., and Kondo, M.,** Evidence for a cytoplasmic translational inhibitor RNA in *Artemia salina* gastrula embryos, *FEBS Lett.*, 80, 390, 1977.
53. **James, T. C. and Tata, J. R.,** Messenger RNA during early embryogenesis in *Artemia salina.* Altered translatability and sequence complexity, *Differentiation,* 16, 9, 1980.
54. **Piot, E., Backhovens, H., and Slegers, H.,** The inhibitor ribonucleoprotein of poly(A)-containing non-polysomal messenger ribonucleoproteins of *A. salina* cryptobiotic embryos, *FEBS Lett.*, 175, 16, 1984.
55. **De Herdt, E., Piot, E., Wahba, A., and Slegers, H.,** Initiation factor eIF2 associated with non-polysomal poly(A)-containing messenger ribonucleoproteins of cryptobiotic gastrulae of *Artemia salina, Eur. J. Biochem.*, 151, 455, 1985.
56. **Daum, H. A. III, Bragg, P. W., Sittman, D. B., Dholakia, J. N., Woodley, C. L., and Wahba, A. J.,** The expression of a gene for eukaryotic elongation factor Tu in *Artemia* during development, *J. Biol. Chem.*, 260, 16347, 1985.
57. **Pluskal, M. G. and Sarkar, S.,** Cytoplasmic low molecular weight ribonucleic acid species of chick embryonic muscles, a potent inhibitor of messenger ribonucleic acid translation in vitro, *Biochemistry,* 20, 2048, 1981.

58. **Vincent, A., Civelli, O., Maundrell, K., and Scherrer, K.**, Identification and characterization of the translationally repressed cytoplasmic globin messenger-ribonucleoprotein particles from duck erythroblasts, *Eur. J. Biochem.*, 112, 617, 1980.
59. **Kühn, B., Villringer, A., Falk, H., and Heinrich, P. C.**, Inhibition of cell-free protein synthesis by low-molecular-weight RNAs from free cytoplasmic ribonucleoprotein particles, *Eur. J. Biochem.*, 126, 181, 1982.
60. **McCarthy, T. L., Siegel, E., Mroczkowski, B., and Heywood, S. M.**, Characterization of translational-control ribonucleic acid isolated from embryonic chick muscle, *Biochemistry*, 22, 935, 1983.
61. **Bag, J., Hubley, M., and Sells, B. H.**, A cytoplasmic ribonucleoprotein complex containing a small RNA inhibitor of protein synthesis, *J. Biol. Chem.*, 255, 7055, 1980.
62. **De Herdt, E., Kondo, M., and Slegers, H.**, The use of Taka-diastase in a [^{3}H]poly(A) hybridization assay of oligo(U) sequences in RNA, *J. Biochem. Biophys. Methods*, 5, 51, 1981.
63. **Bag, J., Maundrell, K., and Sells, B. H.**, Low Mr RNAs of chicken muscle cells and their interaction with messenger and ribosomal RNAs, *FEBS Lett.*, 143, 163, 1982.
64. **Lorberboum, H., Digweed, M., Erdman, V. A., Servadio, Y., Weinstein, D., de Groot, N., and Hochberg, A. A.**, Small cytoplasmic RNAs from human placental free mRNPs. Structure and their effect on in vitro protein synthesis, *Eur. J. Biochem.*, 155, 279, 1986.
65. **Richter, J. D. and Smith, L. D.**, Developmentally regulated RNA binding proteins during oogenesis in *Xenopus laevis*, *J. Biol. Chem.*, 258, 4864, 1983
66. **Tomcsanyi, T., Komaromy, L., and Tigyi, A.**, Structural characterization of polysomal poly(A)-protein particles in rat liver, *Eur. J. Biochem.*, 114, 421, 1981.
67. **Lindberg, U. and Sundquist, B.**, Isolation of messenger ribonucleoproteins from mammalian cells, *J. Mol. Biol.*, 86, 451, 1974.
68. **Barrieux, A., Ingraham, H. A., David, D. N., and Rosenfeld, M. G.**, Isolation of messenger-like ribonucleoproteins, *Biochemistry*, 14, 1815, 1975.
69. **Kish, V. M. and Pederson, T.**, Poly(A)-rich ribonucleoprotein complexes from HeLa cell messenger RNA, *J. Biol. Chem.*, 251, 5888, 1976.
70. **Schwartz, H. and Darnell, J. E.**, The association of a protein with the polyadenylic acid of HeLa cell messenger RNA: evidence for a transport role of a 75,000 molecular weight polypeptide, *J. Mol. Biol.*, 104, 833, 1976.
71. **Vincent, A., Goldenberg, S., and Scherrer, K.**, Comparison of proteins associated with duck-globin mRNA and its polyadenylated segment in polyribosomal and repressed free messenger ribonucleoprotein complexes, *Eur. J. Biochem.*, 114, 179, 1981.
72. **Peters, C. and Jeffery, W. R.**, Postfertilization poly(A)-protein complex formation on sea urchin maternal messenger RNA, *Differentiation*, 12, 91, 1978.
73. **De Herdt, E., Thoen, C., Van Hove, L., Roggen, E., and Slegers, H.**, Identification and properties of the 38,000-Mr poly(A)-binding protein of non-polysomal messenger ribonucleoproteins of cryptobiotic gastrulae of *Artemia salina*, *Eur. J. Biochem.*, 139, 155, 1984.
74. **Massoulié, J.**, Associations de poly(A) et poly(U) en milieu acide. Phènomènes irrèversibles, *Eur. J. Biochem.*, 3, 439, 1968.
75. **Marvil, D. K., Nowak, L., and Szer, W.**, A single-stranded nucleic acid-binding protein from *Artemia salina*. I. Purification and characterization, *J. Biol. Chem.*, 255, 6466, 1980.
76. **Nowak, L., Marvil, D. K., Thomas, J. O., Boublik, M., and Szer, W.**, A single-stranded nucleic acid-binding protein from *Artemia salina*. II. Interaction with nucleic acids, *J. Biol. Chem.*, 255, 6473, 1980.
77. **Thomas, J. O., Raziuddin, Sobota, A., Boublik, M., and Szer, W.**, An RNA helix-destabilizing protein is a major component of *Artemia salina* nuclear ribonucleoproteins, *Proc. Natl. Acad. Sci. U.S.A.*, 78, 2888, 1981.
78. **Christensen,M. E., Beyer, A. L. Walker, B., and Lestourgeon, W. M.**, Identification of N^6,N^6-dimethylarginine in a nuclear protein from the lower eukaryote physarum polycephalum homologous to the major proteins of mammalian 40S ribonucleoprotein particles, *Biochem. Biophys. Res. Commun.*, 74, 621, 1977.
79. **Fuchs, J. P., Judes, C., and Jacob, M.**, Characterization of glycine-rich proteins from the ribonucleoproteins containing heterogeneous nuclear ribonucleic acid, *Biochemistry*, 19, 1087, 1980.
80. **Wilk, H. E., Werr, H., Friedrich, D., Kiltz, H. H., and Schafer, K. P.**, The core proteins of 35 S hnRNP complexes. Characterization of nine different species, *Eur. J. Biochem.*, 146, 71, 1985.
81. **Thomas, J. O. and Szer, W.**, RNA-helix-destabilizing proteins, *Prog. Nucl. Acids Res. Mol. Biol.*, 27, 157, 1982.
82. **O'Farrell, P. H.**, High resolution two-dimensional electrophoresis of proteins, *J. Biol. Chem.*, 250, 4007, 1975.
83. **Baer, B. W. and Kornberg, R. D.**, Repeating structure of cytoplasmic poly(A)-ribonucleoprotein, *Proc. Natl. Acad. Sci. U.S.A.*, 77, 1890, 1980.
84. **Kelly, J. M. and Cox, R. A.**, Periodicity in the length of 3′-poly(A) tails from native globin mRNA of rabbit, *Nucl. Acids Res.*, 10, 4173, 1982.

85. **Krowczynska, A. and Brawerman, G.,** Structural features in the 3′-terminal region of polyribosome-bound rabbit globin messenger RNAs, *J. Biol. Chem.*, 261, 397, 1986.
86. **Palatnik, C. M., Wilkins, C., and Jacobson, A.,** Translational control during early Dictyostelium development: possible involvement of poly(A) sequences, *Cell,* 36, 1017, 1984.
87. **Nilsson, M. O. and Hultin, T.,** Poly(A)-containing cytoplasmic RNA in dormant cysts of *Artemia salina, FEBS Lett.*, 52, 269, 1975.
88. **Thomas, J. O., Glowacka, S. K., and Szer, W.,** Structure of complexes between a major protein of heterogeneous nuclear ribonucleoprotein particles and polyribonucleotides, *J. Mol. Biol.*, 171, 439, 1983.
89. **De Herdt, E., Marbaix, G., Tencer, R., and Slegers, H.,** Subcellular distribution in *Xenopus laevis* oocytes of a microinjected poly(A)-binding protein from *Artemia salina* gastrulae, *Eur. J. Biochem.*, 132, 623, 1983.
90. **Miller, T. E., Huang, C. Y., and Pogo, A. O.,** Rat liver nuclear skeleton and ribonucleoprotein complexes containing hnRNA, *J. Cell Biol.*, 76, 675, 1978.
91. **Van Eekelen, C. A. G. and Van Venrooij, W. J.,** HnRNA and its attachment to the nuclear protein matrix, *J. Cell Biol.*, 88, 554, 1981.
92. **Van Hove, L., Thoen, C., Cohen, P., and Slegers, H.,** Dephosphorylation of cytoplasmic non-polysomal messenger ribonucleoproteins from cryptobiotic gastrulae of *Artemia salina, Biochem. Biophys. Res. Commun.*, 131, 1241, 1985.
93. **Roggen, E. and Slegers, H.,** Isolation and characterization of cytoplasmic poly(A) polymerase from cryptobiotic gastrulae of *Artemia salina, Eur. J. Biochem.*, 147, 225, 1985.
94. **Rose, K. M., Jacob, S. T., and Kumar, A.,** Poly(A) polymerase and poly(A)-specific mRNA binding protein are antigenically related, *Nature (London)*, 279, 260, 1979.
95. **Zardeneta, G., Kramer, G., and Hardesty, B.,** Structure and function of peptide initiation factor 2: differential loss of activities during proteolysis and generation of a terminal fragment containing the phosphorylation sites of the α subunit, *Proc. Natl. Acad. Sci. U.S.A.*, 79, 3158, 1982.
96. **Hellerman, J. G. and Shafritz, D. A.,** Interaction of poly(A) and mRNA with eukaryotic initiator Met-$tRNA_f$ binding factor: identification of this activity on reticulocyte ribonucleic acid protein particles, *Proc. Natl. Acad. Sci, U.S.A.*, 72, 1021, 1975.
97. **Barrieux, A. and Rosenfeld, M. G.,** Comparison of mRNA binding by Met-$tRNA_f$ binding protein and mRNA-associated proteins, *J. Biol. Chem.*, 252, 392, 1977.
98. **Vlasik, T. N., Ovchinnikov, L. P., Radjabov, Kh. M., and Spirin, A. S.,** Translation factors of the wheat embryo extract are RNA-binding proteins, *FEBS Lett.*, 88, 21, 1978.
99. **Benne, R., Amesz, H., Hershey, J. W. B., and Voorma, H. O.,** The activity of eukaryotic initiation factor eIF2 in ternary complex formation with GTP and Met-$tRNA_f$, *J. Biol. Chem.*, 254, 3201, 1979.
100. **Walton, G. M. and Gill, G. N.,** Nucleotide regulation of a eukaryotic protein synthesis initiation complex, *Biochim. Biophys. Acta,* 390, 231, 1975.
101. **Ochiai-Yanagi, S. and Mazumder, R.,** Role of GTP in eukaryotic polypeptide-chain initiation. Purification and properties of a factor from *Artemia salina* embryos which interacts with initiator transfer RNA and guanine nucleotides, *Eur. J. Biochem.*, 68, 395, 1976.
102. **Benne, R., Wong, C., Luedi, M., and Hershey, J. W. B.,** Purification and characterization of initiation factor IF-E2 from rabbit reticulocytes, *J. Biol. Chem.*, 251, 7675, 1976.
103. **Benne, R., Kasperaitis, M., Voorma, H. O., Ceglarz, E., and Legocki, A. B.,** Initiation factor eIF-2 from wheat germ. Purification, functional comparison of eIF-2 from rabbit reticulocytes and phosphorylation of its subunits, *Eur. J. Biochem.*, 104, 109, 1980.
104. **MacRae, T. H., Roychowdhury, M., Houston, K. J., Woodley, C. L., and Wahba, A. J.,** Protein synthesis in brine shrimp embryos. Dormant and developing embryos of *Artemia salina* contain equivalent amounts of chain initiation factor 2, *Eur. J. Biochem.*, 100, 67, 1979.
105. **Gander, E. S., Stewart, A. G., Morel, C. M., and Scherrer, K.,** Isolation and characterization of ribosome-free cytoplasmic messenger-ribonucleoprotein complexes of avian erythroblasts, *Eur. J. Biochem.*, 38, 443, 1973.
106. **Egly, J. M., Schmitt, M., Elkaim, R., and Kempf, J.,** Protein kinases and their protein substrates in free messenger ribonucleoprotein particles and polysomes from mouse plasmacytoma cells, *Eur. J. Biochem.*, 118, 379, 1981.
107. **Rittschof, D. and Traugh, J. A.,** Identification of casein kinase II and phosphorylated proteins associated with messenger ribonucleoproteins particles from reticulocytes, *Eur. J. Biochem.*, 123, 333, 1982.
108. **Egly, J. M., Elkaim, R., and Pierre, M.,** mRNP proteins, initiation factors and phosphorylation, *Mol. Biol. Rep.*, 5, 91, 1979.
109. **Thoen, C., Van Hove, L., Piot, E., and Slegers, H.,** Purification and characterization of the messenger ribonucleoprotein-associated casein kinase II of *Artemia salina* cryptobiotic gastrulae, *Biochim. Biophys. Acta,* 783, 105, 1984.
110. **Hathaway, G. M. and Traugh, J. A.,** Casein kinases. Multipotential protein kinases, *Curr. Topics Cell Reg.*, 21, 101, 1982.

111. **Meggio, F., Pinna, L. A., Marchiori, F., and Borin, G.,** Poly glutamyl peptides: a new class of inhibitors of type-2 casein kinases, *FEBS Lett.*, 162, 235, 1983.
112. **Feige, J. J., Pirollet, F., Cochet, C., and Chambaz, E. M.,** Selective inhibition of a cyclic nucleotide-independent protein kinase (G-type casein kinase) by naturally occurring glycosaminoglycans, *FEBS Lett.*, 121, 139, 1980.
113. **Hathaway, G. M., Lubben, T. H., and Traugh, J. A.,** Inhibition of casein kinase II by heparin, *J. Biol. Chem.*, 255, 8038, 1980.
114. **Maenpaä, P. H.,** Effects of polyamines and polyanions on a cyclic nucleotide-independent and a cyclic AMP-dependent protein kinase, *Biochim. Biophys. Acta*, 498, 294, 1977.
115. **Meggio, F., Deana, A. D., Brunati, A. M., and Pinna, L. A.,** Inhibition of rat cytosol casein kinases by heparin, *FEBS Lett.*, 141, 257, 1982.
116. **Le Comte, M. C., Galaud, C., and Boivin, P.,** Inhibition of human erythrocyte casein kinase by methylxanthines, *FEBS Lett.*, 116, 45, 1980.
117. **Plana, M., Guasch, M. D., and Itarte, E.,** Modulators of rat liver cytosol casein kinases 1 and 2, *Biochem. Biophys. Res. Commun.*, 109, 1284, 1982.
118. **Reichert, G. and Issinger, O. G.,** Characterization of native 40S particles from Krebs II mouse ascites tumor cells. Resolution, nomenclature and molecular weights of the nonribosomal proteins, *Biochim. Biophys. Acta*, 654, 268, 1981.
119. **De Herdt, E., Slegers, H., and Kondo, M.,** Identification and characterization of a 19-S complex containing a 27000-Mr protein in *Artemia salina, Eur. J. Biochem.*, 96, 423, 1979.
120. **De Herdt, E., De Voeght, F., Clauwaert, J., Kondo, M., and Slegers, H.,** A cryptobiosis-specific 19S protein complex of *Artemia salina* gastrulae, *Biochem. J.*, 194, 9, 1981.
121. **Thoen, C., Van Hove, L., and Slegers, H.,** Identification of the substrates of the casein kinase II associated with non-polysomal messenger ribonucleoproteins of *A. salina* cryptobiotic embryos, *Mol. Biol. Rep.*, 11, 69, 1986.
122. **Thoen, C., Van Hove, L., and Slegers, H.,** The function of phosphorylation-dephosphorylation in messenger ribonucleoprotein translation, *Adv. Protein Phosphatases*, 3, 121, 1986.
123. **Merlevede, W.,** Protein phosphates and the protein phosphatases. Landmarks in an eventful century, *Adv. Protein Phosphatases*, 1, 1, 1985.
124. **Pelech, S. and Cohen, P.,** The protein phosphatases involved in cellular regulation. I. Modulation of protein phosphatases-1 and 2A by histone H1, protamine, polylysine and heparin, *Eur. J. Biochem.*, 148, 245, 1985.
125. **Tung, H. Y. L., Alemany, S., and Cohen, P.,** The protein phosphatases involved in cellular regulation. II. Purification, subunit structure and properties of protein phosphatases-$2A_0$, $2A_1$ and $2A_2$ from rabbit skeletal muscle, *Eur. J. Biochem.*, 148, 253, 1985.
126. **Sastre, L. and Sebastian, J.,** Poly(A) polymerase activity during the early development of *Artemia*, in *The Brine Shrimp Artemia*, Vol. 2, Persoone, G., Sorgeloos, P., Roels, O., and Jaspers, E., Eds., Universa Press, Wetteren, Belgium, 1980, 325.
127. **Sastre, L. and Sebastian, J.,** Developmental changes in poly(A) polymerase activity in *Artemia, Eur. J. Biochem.*, 135, 69, 1983.
128. **Roggen, E., Dams, E., and Slegers, H.,** The primer specificity of cytoplasmic poly(A) polymerase from cryptobiotic gastrulae of *Artemia salina, Biochim. Biophys. Acta*, 825, 21, 1985.
129. **Sastre, L. and Sebastian, J.,** Purification and properties of a polyadenylate polymerase from *Artemia* dormant embryos, *Biochim. Biophys. Acta*, 661, 54, 1981.
130. **Jacob, S. T. and Rose, K. M.,** RNA polymerases and poly(A) polymerase from neoplastic tissues and cells, *Methods Cancer Res.*, 14, 191, 1978.
131. **Edmonds, M.,** Poly(A) adding enzymes, in *The Enzymes*, Vol. 15, Boyer, P. D., Ed., Academic Press, New York, 1982, 217.
132. **Spirin, A. S.,** Eukaryotic messenger RNA and informosomes. Omnia mea mecum porto, *FEBS Lett.*, 88, 15, 1978.
133. **Slater, I. and Slater, D. W.,** Polyadenylation and transcription following fertilization, *Proc. Natl. Acad. Sci. U.S.A.*, 71, 1103, 1974.
134. **Egrie, J. C. and Wilt, F. H.,** Changes in poly (adenylic acid) polymerase activity during sea urchin embryogenesis, *Biochemistry*, 18, 269, 1979.
135. **Sastre, L. and Sebastian, J.,** In vitro polyadenylation of messenger ribonucleoprotein particles stored in *Artemia* dormant embryos, *Biochem. Biophys. Res. Commun.* 108, 1609, 1982.
136. **Rosenthal, E. T., Tansey, T. R., and Ruderman, J. V.,** Sequence-specific adenylations and deadenylations accompany changes in the translation of maternal messenger RNA after fertilization of *Spisula* oocytes, *J. Mol. Biol.*, 166, 309, 1983.

137. **James, T. C., Hentschel, C. C., and Tata, J. R.,** Gene expression during resumption of development of encysted *Artemia* gastrulae, in *Biochemistry of Artemia Development,* Bagshaw, J. C. and Warner, A. H., Eds., University Microfilms Int., Ann Arbor, Mich., 1979, 167.
138. **Ruderman, J. V., Woodland, H. R., and Sturgess, E. A.,** Modulations of histone messenger RNA during the early development of *Xenopus laevis, Dev. Biol.,* 71, 71, 1979.
139. **Slater, D. W., Slater, I., and Gillespie, D.,** Post-fertilization synthesis of polyadenylic acid in sea urchin embryos, *Nature (London),* 240, 333, 1972.

Chapter 3

MOLECULAR ASPECTS OF DEVELOPMENT IN THE BRINE SHRIMP *ARTEMIA*: STRUCTURAL AND FUNCTIONAL STUDIES ON THE EUKARYOTIC POLYPEPTIDE INITIATION FACTOR 2 AND EXPRESSION OF THE EUKARYOTIC ELONGATION FACTOR Tu DURING DEVELOPMENT

Albert J. Wahba and Charles L. Woodley

TABLE OF CONTENTS

I. INTRODUCTION

During embryogenesis of *Artemia*, gastrulae may develop into adults or may encyst and enter a dormant state. The encysted embryos are released into the environment where they may become dessicated. Cell-free extracts from these dormant embryos contain mRNA,[1-5] large quantities of 80S ribosomes,[6,7] elongation factors,[8-9] termination factors,[10] and aminoacyl tRNA synthetases;[11] however, they are inactive in the translation of natural mRNA.[12,13] The absence of protein synthesis in these extracts was attributed partially to a deficiency of chain initiation factors. The GTP-dependent Met-tRNA$_f$ binding protein, the eukaryotic initiation factor 2 (eIF-2), was reported to increase over 20-fold following resumption of embryo development.[12,13] According to these results, embryo development following encystment would depend upon the replacement of depleted or inactive initiation factors. Furthermore, the lack of active chain initiation factors in cell-free extracts of dormant embryos may reflect intracellular conditions prior to encystment which may trigger cessation of protein synthesis and eventual cyst formation.

One of the major proteins synthesized in vivo by *Artemia* is a 50,000 M_r polypeptide which was identified as the eukaryotic elongation factor Tu (eEF-Tu, eEF-lα.)[8,9] This factor, which is responsible for binding aminoacyl-tRNA to the ribosome,[9] has been shown to be present in abundance in the cytoplasm of both prokaryotic and eukaryotic cells. The importance of eEF-Tu in protein synthesis makes the regulation of its synthesis a central problem in the study of development.

The molecular aspects of development we have chosen to examine include (1) the amount and activity of polypeptide chain initiation factors in dormant and developing embryos; (2) the nature of the stored message; (3) the selection of transcripts for translation; and (4) changes in specific mRNAs during the different stages of development.

The 8 to 18 hr period of development following hydration, but prior to emergence of the embryos, is characterized by relatively synchronous development, the synthesis of polypeptides on newly formed polysomes, and the absence of DNA synthesis and cell division.[6,14-17] Since embryonic development is divorced from cell division throughout pre-emergence, *Artemia* offers a unique system for examining the relationship between protein synthesis, its regulation, and differentiation.

Most of our understanding of polypeptide chain initiation in eukaryotes is derived from studies with extracts of rabbit reticulocytes. The first two steps in the initiation process are the recognition of the initiator tRNA by chain initiation factor 2 (eIF-2)[18-20] and the selection of a message in the formation of the 40S initiation complex. Several ancillary factors stimulate eIF-2 activity in the formation[21] and binding of the ternary complex (eIF-2 · GTP · Met-$tRNA_f$) to 40S ribosomal subunits (for reviews see References 21 and 23). After formation of the 43S preinitiation complex, the message with its cap-associated proteins binds to the ribosomal subunit to form the 48S initiation complex.[23] Recruitment of a message by the mRNA-binding proteins is influenced by the nature of the message, which may include capping, methylation, secondary structure, or proteins associated with the RNA.[24-32] Other factors may also play a role in positioning the message on the ribosomal subunit.[13,20,33-34] Hydrolysis of ATP is required for message binding, although the role of the hydrolysis, as well as the movement of the ribosome from the 5′ cap structure to the initiation codon, are not as yet clearly understood.[23,35-39] In the presence of eIF-5, the 60S ribosomal subunit attaches to the 48S initiation complex, the eIF-2-bound GTP is hydrolyzed, and eIF-2 is released.[40-43] At this stage, the chain-initiation factor is presumably released as an eIF-2 · GDP complex.[43] A protein found in the cytosol facilitates the exchange of bound GDP for GTP.[44-48] The presence of such a factor would allow eIF-2 to enter another round of initiation. A pool of active 40S ribosomal subunits, necessary for the initiation process, may be maintained by two other factors, eIF-3 and eIF-6. These factors prevent the reassociation of ribosomal subunits.[50,51] After formation of the 80S initiation complex, protein synthesis continues through the elongation and termination steps.

We chose eIF-2 as the starting point of our studies with *Artemia* since this factor catalyzes the first step of the initiation process and may serve as a control point for polypeptide chain initiation.

II. ISOLATION AND CHARACTERIZATION OF eIF-2 FROM *ARTEMIA* EMBRYOS

A. eIF-2 in Dormant and Developing *Artemia* Embryos

1. Assay of eIF-2 Activity

eIF-2 activity is assayed either by its ability to form a ternary complex containing GTP and the initiator tRNA (eIF-2 · GTP · [^{35}S] Met-$tRNA_f$)[21,52] or a binary complex with GDP

Table 1
ACTIVITY OF eIF-2 FROM DIFFERENT PREPARATIONS OF *ARTEMIA* EMBRYOS

Type of embryo	eIF-2 activity[a]
Dormant	
1-hr hydration	3.0
3-hr hydration	4.0
24-hr hydration	4.2
Developing	
11-hr	3.5
Emerged	
Nauplii (24 hr)	3.1

[a] Activity of eIF-2 in a 35/60% saturated ammonium sulfate fraction of the ribosomal salt wash and is expressed as pmol [^{35}S]Met-tRNA$_f$ bound in a (eIF-2 · GTP Met-tRNA$_f$) ternary complex.

(eIF-2 · [^{3}H] GDP).[53] The formation of each of these complexes is quantitated by a filtration assay in which the binary or ternary complex is adsorbed to a nitrocellulose membrane, washed with buffer to remove excess radioactive label ([^{35}S] Met-tRNA$_f$ or [^{3}H]GDP), dried, and counted for radioactivity in a liquid scintillation counter.

2. Purification of eIF-2

eIF-2 may be purified from the 0.5 *M* NH_4Cl wash of dormant or developing *Artemia* ribosomes.[52,54] The activity is in the 35 to 60% saturated ammonium sulfate fraction of the ribosomal salt wash. This fraction is first chromatographed on heparin-Sepharose (applied in 100 m*M* KC1 and eluted between 200 and 400 m*M* KC1), and then on phosphocellulose (applied in 400 m*M* KC1 and eluted with 650 m*M* KC1) The protein was successively applied to a Mono S and a Mono Q column (Pharmacia, Piscataway, N.J.) in buffer containing 100 m*M* KC1 and eluted with a linear gradient as previously reported for the reticulocyte factor.[55] Activity was found in fractions eluting from Mono S and Mono Q at 270 and 235 m*M* KC1, respectively. The factor may also be purified by the same procedure from the postribosomal supernatant fraction.

3. Activity and Distribution of eIF-2 in Dormant and Developing Artemia Embryos

When the ammonium sulfate fractions of extracts from dormant embryos were assayed for eIF-2 activity, 53 and 47% of the activity was found in the ribosomal salt wash and postribosomal supernatant, respectively.[52] Similarly, in developing embryos, 64% of the activity was associated with the ribosomal salt wash and 36% with the postribosomal supernatant. The total amount of eIF-2 activity, in contrast to earlier reports,[16] was very similar in dormant and in developing embryos.[52] Per 100 g net weight of embryos, 116 and 110 units of eIF-2 were obtained from dormant and 11-hr developing embryos, respectively.

Factor activity was determined in ammonium sulfate fractions of ribosomal salt washes of embryos that were allowed to hydrate at 4°C or to develop for different times (Table 1). Neither the time of hydration nor the time of development up to emergence (18 to 24 hr) had any significant effect on the amount of eIF-2 activity that was recovered. The overall level of eIF-2 activity did not change during *Artemia* development.[52]

Preparations of eIF-2 from dormant and developing embryos were also tested for their ability to stimulate in vitro polypeptide synthesis in a cell-free system that contained *Artemia*

Table 2
EFFECT OF eIF-2 FROM DORMANT AND DEVELOPING EMBRYOS ON POLYPEPTIDE SYNTHESIS

Source of eIF-2[a]	eIF-2 added (μg)	Leucine incorporated (pmol)	Stimulation (%)
None	0	70	0
Dormant embryos	0.8	89	27
	1.9	105	50
	3.8	121	72
Developing embryos	0.8	86	23
	1.9	116	66
	3.8	146	108

[a] Each reaction mixture contained 360 μg of reticulocyte unfractionated salt wash, 0.5 A_{260} unit of *Artemia* ribosomes, and 0.15 A_{260} unit of poly(A)-rich mRNA.

From MacRae, T. H. et al., *Eur. J. Biochem.*, 100, 67, 1979. With permission.

80S ribosomes, globin mRNA, and a limiting level of reticulocyte ribosomal salt wash (Table 2). Both preparations were similarly active in stimulating polypeptide synthesis at low levels of the factor, although eIF-2 from dormant embyros was approximately 30% less active at the highest levels tested.[52] In the absence of any cofactors, eIF-2 preparations from *Artemia* embyros typically bound approximately 2400 pmol Met-$tRNA_f$ per milligram factor.

B. Comparison of eIF-2 from *Artemia*, Wheat Germ, Yeast, and Rabbit Reticulocytes

1. Subunit Composition of eIF-2 Preparations

On dodecylsulfate polyacrylamide gel electrophoresis, eIF-2 preparations from dormant and developing embyros appeared to be identical, each having three polypeptides with M_r's of 52,000, 43,000, and 42,000 (Figure 1, panel A).[52] eIF-2 preparations from wheat germ embryos, yeast, and rabbit reticulocytes were also analyzed by dodecylsulfate polyacrylamide gel electrophoresis.[56] Although the eIF-2 from each preparation is composed of a complex containing three nonidentical subunits in a 1:1:1 ratio, the M_r's of these subunits are different (Figure 1).[21,57] These differences are tabulated in Table 3. Yeast and rabbit eIF-2 preparations are composed of a "lower" and two "upper" subunits, whereas those from wheat germ and *Artemia* contain two "lower" and a single "upper" subunit.

2. Identification of eIF-2 Subunits by Phosphorylation with Casein Kinase II (CK II) and the Rabbit Reticulocyte Heme-Controlled Repressor (HCR)

The specificity of reticulocyte HCR for phosphorylating the α subunit and casein kinase II for phosphorylating the β subunit of eIF-2 preparations has been used as a means to identify the respective subunits.[21,58-61] The rabbit reticulocyte HCR was prepared as previously described.[58] CK II from rabbit muscle was a gift from Erwin Reimann (Department of Biochemistry, Medical College of Ohio, Toledo). The *Artemia* kinase was isolated from dormant *Artemia* embryos[57] and appears to be similar to CK II.

Our data demonstrate that eIF-2 preparations diverse eukaryotic sources contain a subunit recognized by reticulocyte HCR and casein kinase II isolated from either *Artemia* embryos or rabbit muscle. After incubation of the kinase and eIF-2 preparation with [γ-^{32}P]ATP, the reaction products were analyzed by electrophoresis in dodecylsulfate polyacrylamide gels

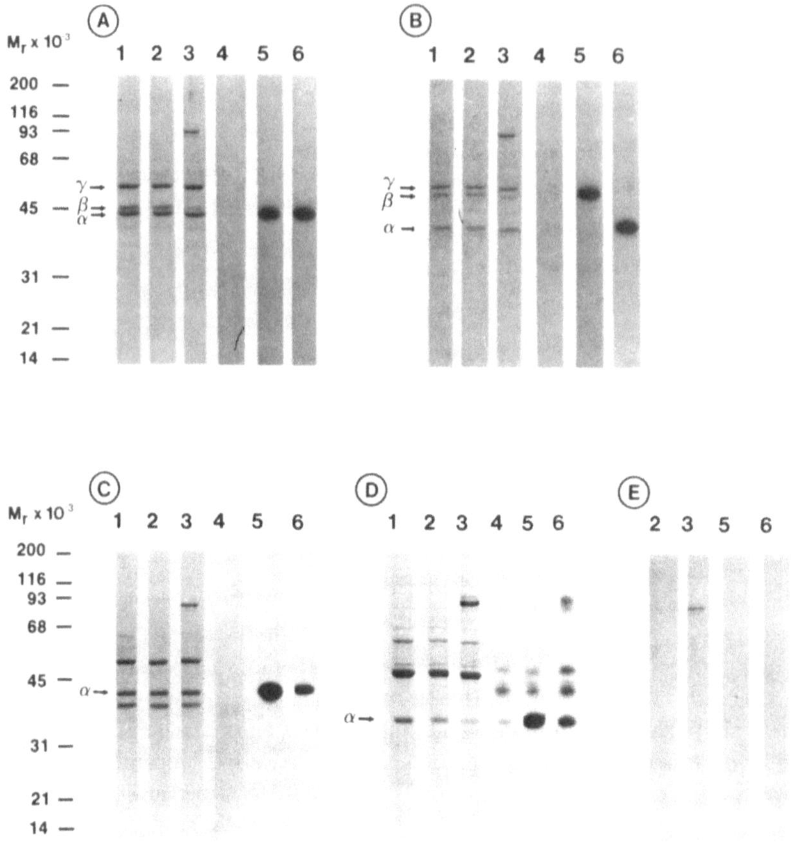

FIGURE 1. M_r of eIF-2 subunits from *Artemia*, rabbit reticulocytes, wheat germ, and yeast and their phosphorylation by HCR and CK II. Phosphorylation was carried out with [γ-^{32}P]ATP as described, and the products were eletrophoresed in SDS polyacrylamide (10%) gels and stained with Coomassie Brilliant Blue R-250 (lanes 1 to 3) and autoradiographed (lanes 4 to 6). Two μg of *Artemia* (A), rabbit reticulocyte (B), wheat germ embryos (C), or yeast (D) eIF-2 were used. The samples in panel E were without eIF-2. Lanes 2 and 5 included 0.2 μg of the *Artemia* kinase while reaction mixtures for lanes 3 and 6 contained 1 μg of HCR. The phosphorylation was terminated with an equal volume of SDS buffer. The molecular weight of each protein standard is indicated at the right. Not shown is the phosphorylation of the different eIF-2 preparations by 0.2 μg of the muscle casein kinase II. This reaction yielded identical results as the *Artemia* kinase (lanes 2 and 5). (From Mehta, H. B. et al., *J. Biol. Chem.*, 261, 6705, 1986. With permission.)

(Figure 1). Reticulocyte HCR efficiently phosphorylated the smallest (α) subunit of eIF-2 from *Artemia*, yeast, and reticulocytes. In wheat germ, the middle subunit is the one recognized by HCR. With CK II from rabbit muscle or *Artemia*, the β subunit of reticulocyte eIF-2 becomes phosphorylated. This is in agreement with earlier reports that CK II specifically phosphorylates the β subunit of reticulocyte eIF-2.[60,61] However, with eIF-2 preparations from yeast, *Artemia*, and wheat germ, CK II preparations phosphorylated the same subunit that was recognized by reticulocyte HCR. Figure 1 shows the results obtained with the *Artemia* casein kinase.

The phosphorylation of *Artemia* eIF-2 by HCR or CK II did not affect its ability to form binary (eIF-2 · GDP) or ternary (eIF-2 · GTP Met_f) complexes.[57]

Table 3
THE M_R eIF-2 SUBUNITS FROM VARIOUS SOURCES

Source[a]	Subunit M_r[b] (kdalton)		
	1	2	3
Artemia	41	43	52
Wheat Germ	36	40	55
Yeast	38	54	55
Rabbit reticulocyte	37	52	55

[a] The wheat germ eIF-2 was generously provided by Joanne Ravel (Department of Chemistry, University of Texas, Austin) and yeast eIF-2 was from Naba Gupta (Department of Chemistry, University of Nebraska, Lincoln).

[b] The M_r of the eIF-2 subunits as determined in a 10% polyacrylamide gel containing 0.1% dodecylsulfate.

3. Peptide Mapping of eIF-2 Subunits from Artemia Embryos and Rabbit Reticulocytes

a. Phosphopeptide Maps

We compared the sites on *Artemia* and reticulocyte eIF-2 (α) that are phosphorylated by HCR as well as those on the α subunit of the *Artemia* factor that are recognized by HCR and CK II. eIF-2 preparations from *Artemia* embyros and rabbit reticulocytes were phosphorylated by HCR and reisolated by chromatography on phosphocellulose.[21] The sites phosphorylated on the reticulocyte and *Artemia* eIF-2 preparations were compared by both limited and extensive trypsin digestion.

i. Phosphorylation of *Artemia* and Reticulocyte eIF-2 (α) by HCR

The tryptic digestion of eIF-2 was terminated at different times by the addition of the soybean trypsin inhibitor and the products were analyzed by dodecylsulfate polyacrylamide gel electrophoresis[56] and subsequent autoradiography. As the trypsin digestion proceeded the α subunit of *Artemia* eIF-2 began to disappear and the cleavage was complete by 5 to 10 min (Figure 2).[57] The limited trypsin digestion appeared to cleave the α subunit (M_r 41,000) into two products, a Coomassie Brilliant Blue R-250 stained fragment (M_r 38,000) (Figure 2A) and a peptide estimated to have an M_r of approximately 4,000. This peptide was detected only upon autoradiography (Figure 2B). This establishes that the site of phosphorylation on the α subunit of *Artemia* eIF-2 resides in a 4000 M_r peptide. This is similar to the results obtained by Zardeneta et al.[62] for rabbit reticulocyte eIF-2. Recently, the α subunit of eIF-2 was phosphorylated by HCR and the sequence of the first 52 amino acids was determined.[63] Overlapping tryptic and chymotryptic peptides indicated that phosphorylation was at serine 51 in the sequence Leu-Leu-Ser^{48}-Glu-Leu-Ser^{51}-Arg-Arg.

Generation of analogous phosphopeptides during limited tryptic digestion of the α subunits of *Artemia* and reticulocyte eIF-2 prompted us to determine if similar phosphopeptides were obtained upon prolonged trypsin digestion and subsequent two-dimensional peptide analyses. For this purpose the [^{32}P]-labeled α subunit of eIF-2 was excised from a dodecylsulfate polyacrylamide gel and digested with DPCC-treated trypsin. Labeled peptides with nearly identical mobilities were obtained after digestion for 24 hr of both the *Artemia* and reticulocyte eIF-2(α-P).[57] This is illustrated in Figure 3A and D. Further digestion of [^{32}P]-labeled peptides resulted in the progressive loss of two of the phosphopeptides with all of the remaining radioactivity in a single peptide (Figure 3C and F). The results for the reticulocyte factor were similar to those of Gross et al.[64]

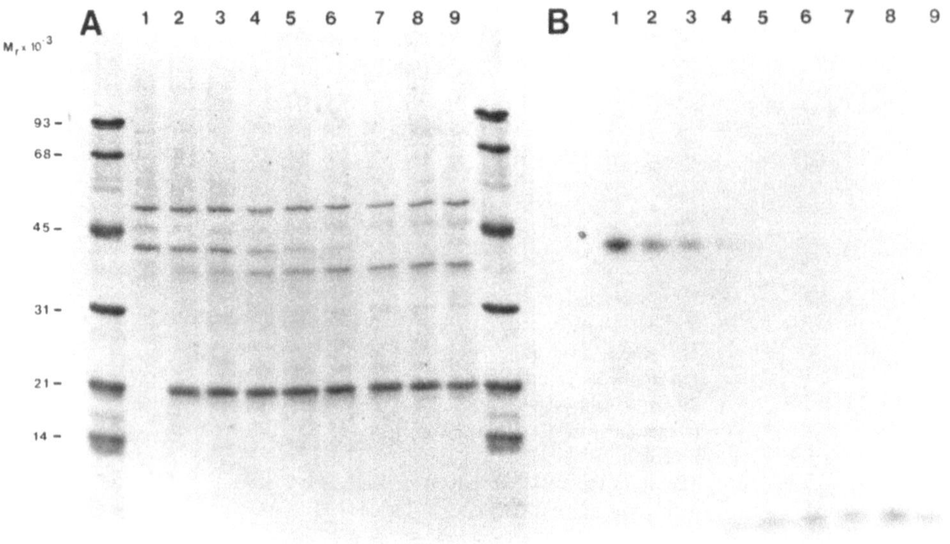

FIGURE 2. Time course for limited trypsin digestion of *Artemia* eIF-2 phosphorylated by HCR. Phosphorylated *Artemia* eIF-2 was digested as described [57] and aliquots containing approximately 4 μg of eIF-2 were removed at the indicated times and fractionated by electrophoresis in SDS polyacrylamide (12.5%) gels. Panel A, Coomassie Brilliant Blue R-250 stained gel; panel B, the autoradiogram of the gel. Lanes 1 to 9 represent aliquots taken at 0, 0.5, 1, 3, 6, 9, 16, 20, and 30 min, respectively. (From Mehta, H. B. et al., *J. Biol. Chem.*, 261, 6705, 1986. With permission.)

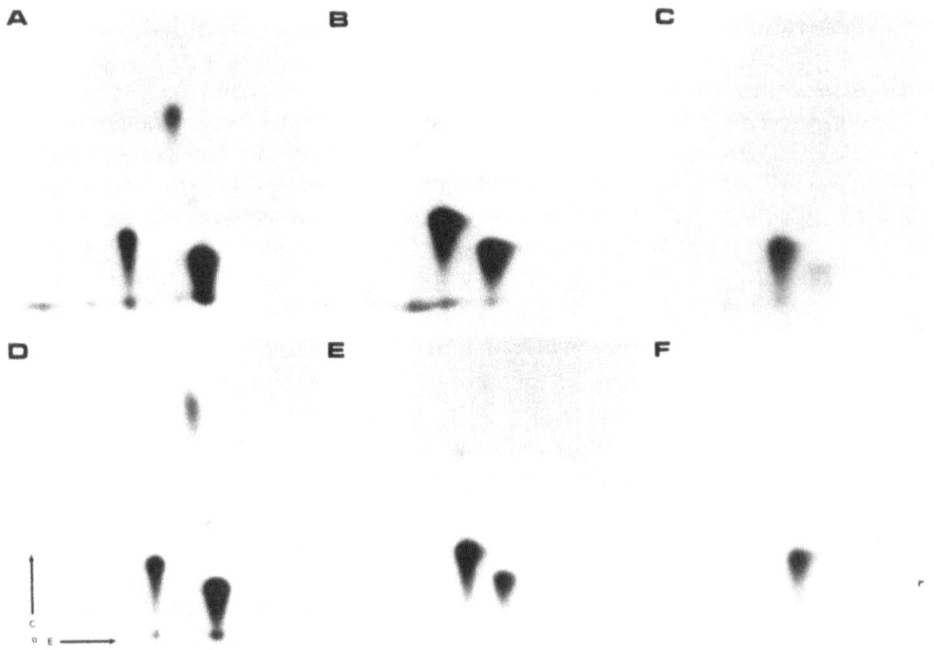

FIGURE 3. Extensive trypsin digestion of *Artemia* and reticulocyte eIF-2 phosphorylated by HCR. Autoradiograms of two-dimensional thin-layer chromatograms of ^{32}P-labeled *Artemia* (panels A to C) and reticulocyte eIF-2 (panels D to F) after trypsin digestion for 1, 3, and 5 days. Trypsin digestion and electrophoresis were as described.[57] Approximately 5000 cpm of ^{32}P-labeled phosphopeptide were applied to each chromatogram, and exposure and development times were reduced to prevent overexposure of the films for panels C and F. o, Origin; C, direction of the first dimension; E, electrophoresis in the second dimension. (From Mehta, H. B. et al., *J. Biol. Chem.*, 261, 6705, 1986. With permission.)

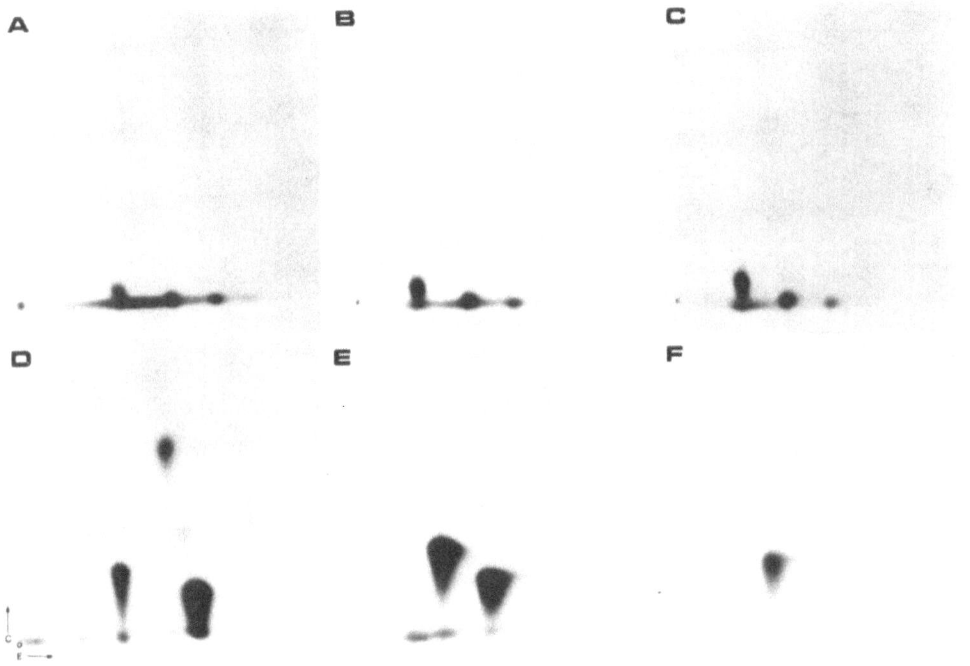

FIGURE 4. Two-dimensional tryptic maps of phosphopeptides derived from *Artemia* elf-2 phosphorylated by reticulocyte HCR or *Artemia* casein kinase II. The *Artemia* eIF-2 was phosphorylated by either reticulocyte HCR (upper panel) or by *Artemia* casein kinase II (lower panel), followed by trypsin digestion and two-dimensional mapping as described.[57] Approximately 5000 cpm of each ^{32}P-labeled phosphopeptide was applied to each chromatogram. O, origin; C, direction of the first dimension; and E, electrophoresis in the second dimension. (From Mehta, H.B. et al., *J. Biol. Chem.*, 261, 6705, 1986. With permission.)

Phosphopeptides 1, 2, and 3 seem to be related by a precursor/product relationship, the sequence apparently being 1 to 2 to 3 (Figure 3). The complete digestion to generate peptide 3 required unusually large amounts of trypsin and long periods of digestion (up to 5 days).[64] This may be due to either (1) Arg-Arg, Arg-Lys, or Lys-Lys sequences in the peptide which may produce random and incomplete trypsin cleavage at these sites or (2) the presence of trace amounts of another protease in the trypsin preparation. Recent sequence data indicate that the amino acid sequence,[53-55] just to the -COOH side of the phosphorylation site[63] is Arg-Arg-Arg.[65] This would also explain the number of iodopeptides observed in Figure 5.

ii. Sites on *Artemia* eIF-2(α) Phosphorylated by *Artemia* CK II

As discussed above, CK II and HCR phosphorylated the α subunit of *Artemia* eIF-2, but the *Artemia* kinase did not phosphorylate the α subunit of reticulocyte eIF-2. In order to check the phosphorylation site specificity of the two kinases, *Artemia* eIF-2 was phosphorylated with each kinase.[57,66] After phosphorylation of *Artemia* eIF-2 (α) by casein kinase, short term (30-min) trypsin digestion chased the ^{32}P label from the 41,000 M_r α subunit into 35,000 to 38,000 M_r phosphopeptides (data not shown). This is in contrast to the 4000 M_r phosphopeptide that resulted when the α subunit was phosphorylated by HCR and digested with trypsin (Figure 2). After extensive trypsin digestion of the phosphorylated α subunits, two-dimensional peptide mapping further demonstrated that different sites were phosphorylated by the two kinase preparations (Figure 4). The amino acid phosphorylated by these two kinases was determined by chromatography on thin-layer cellulose plates of the acid hydrolysate of ^{32}P-labeled eIF-2(α).[21] In each case the results indicated that serine was the phosphate acceptor. Presumably, the site phosphorylated by HCR is homologous to that of

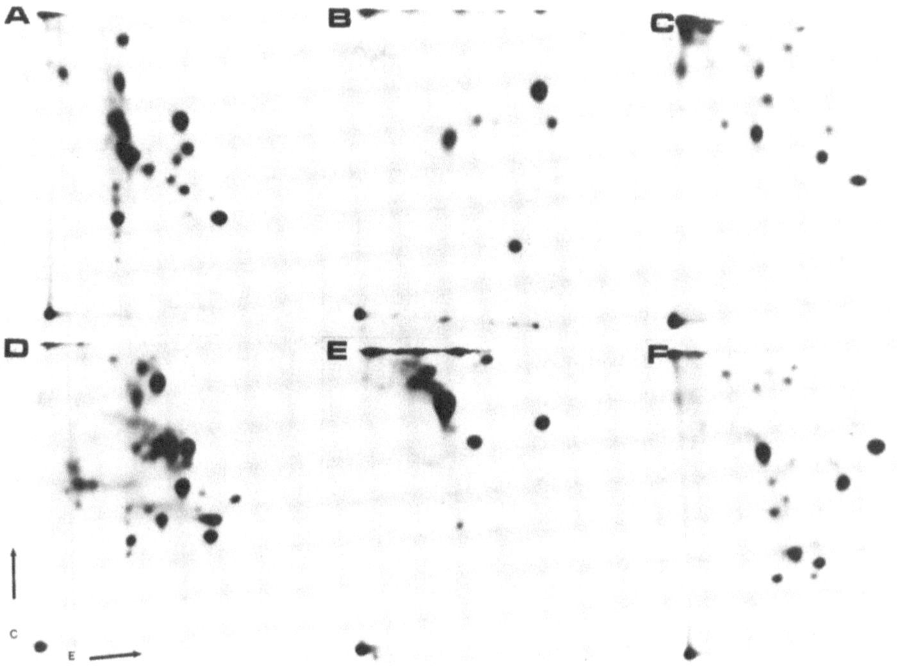

FIGURE 5. Fingerprint analyses of 5-day digested ^{125}I-labeled tryptic peptides of individual *Artemia* and reticulocyte eIF-2 subunits. Individual subunits were digested with trypsin for 5 days as described [57] and analyzed by two-dimensional thin-layer chromatography. Panels A to C, *Artemia* eIF 2(α), (β), and (γ) subunits, respectively; panels D to F, the reticulocyte eIF-2(α), (β), and (γ) subunits, respectively. (From Mehta, H. B. et al., *J. Biol. Chem.*, 261, 6705, 1986. With permission.)

the reticulocyte factor, but nothing is known about the site on *Artemia* eIF-2(α) that is recognized by CK II.

b. Iodopeptide Maps of Subunits from Artemia and Reticulocyte eIF-2

As a result of our observations on the structural similarities between eIF-2(α) from *Artemia* and rabbit reticulocytes, we extended these studies to determine if other common structural features exist between these eIF-2 preparations. Fingerprint analyses of ^{125}I-labeled peptides were then performed, and the results are shown in Figures 5 and 6.[57,66] In contrast to the ^{32}P-labeled eIF-2(α) phosphopeptides (Figures 2 and 3), the iodopeptide patterns of the α, β, and γ subunits of *Artemia* eIF-2 (Figure 5, panels A to C) did not appear to be similar to those from the corresponding subunits of the reticulocyte factor (panels D to F). These results did not change even after 5 days of continued digestion with DPCC-trypsin (Figure 5). Although a few peptides may migrate to similar positions on these two-dimensional maps, the overall maps do not appear to be highly analogous.

Structural similarities observed from phosphopeptide maps cannot be extrapolated to iodopeptide maps of the α, β, and γ subunits of the two eIF-2 preparations. The results of the iodopeptide mapping indicate the lack of obvious similarities between any of the subunits from these two sources. However, the technique utilized would not detect small stretches of conserved amino acid sequences. The phosphopeptide generated after extensive trypsin digestion (Figure 3) could not be precisely matched with any of the iodopeptides (Figure 5) presumably because the phosphopeptide either lacks tyrosine or migrates differently due to the presence of the phosphate.

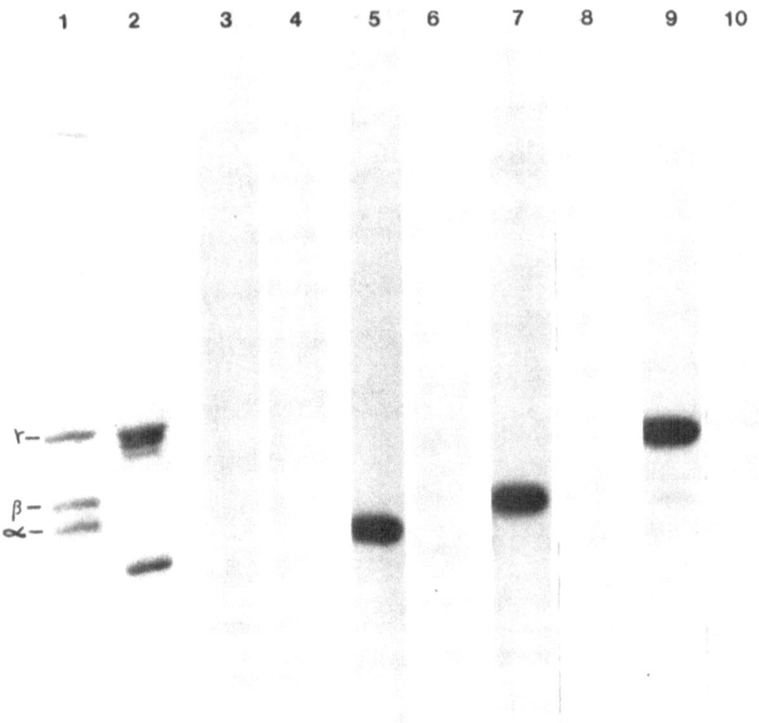

FIGURE 6. Immunological reactivity of *Artemia* and reticulocyte eIF-2 to antibodies raised against *Artemia* eIF-2 subunits. Lanes 1 and 2, Coomassie Brilliant Blue R-250 stained dodecyl sulfate polyacrylamide gel of *Artemia* and reticulocyte eIF-2, respectively. Lanes 3 to 10, the autoradiogram of the ^{125}I-labeled immunoblot showing *Artemia* and reticulocyte eIF-2 reactivity with antibodies raised against individual subunits of *Artemia* eIF-2. Lanes 3 and 4, *Artemia* and reticulocyte eIF-2 reactivity with preimmune serum; lanes 5 and 6, with anti-*Artemia* eIF-2(α); lanes 7 and 8, anti-*Artemia* eIF-2(β); and lanes P and 10, with anti-*Artemia* eIF-2 (γ), respectively. (From Mehta H. B. et al., *J. Biol. Chem.*, 261, 6705, 1986. With permission.)

C. Immunological Relationship Between eIF-2 Preparations from *Artemia* and Reticulocytes

To examine further the structural similarities between the *Artemia* and rabbit reticulocyte eIF-2 preparations, antibodies were prepared in rabbits against homogeneous α, β, and γ subunits of the brine shrimp factor.[57,66] The DEAE-cellulose purified IgG fraction from each serum was used to study the shared antigenic determinants between the two eIF-2 preparations. In the simple agar gel immunodiffusion test, each of the three antibodies showed a single prominent precipitin line with native *Artemia* eIF-2, but had no cross reactivity with reticulocyte eIF-2. A more sensitive technique, Western blotting, demonstrated that anti-*Artemia* eIF-2(α), eIF-2(β), and eIF-2(γ) IgG specifically reacted with the respective subunits of *Artemia* eIF-2, but did not exhibit consistent cross reactivity with the corresponding subunits of reticulocyte eIF-2 (Figure 6). Because the antibodies to *Artemia* eIF-2 were prepared in rabbits, this technique would reveal differences rather than similarities between *Artemia* and reticulocyte eIF-2 preparations.

The immunological techniques used in this work revealed that there are antigenic determinants present in the *Artemia* eIF-2 subunits that are absent or modified in the subunits of the reticulocyte factor. Brown-Luedi et al.[67] compared the immunological relatedness of eIF-2 preparations from two mammalian systems, rabbits and humans. They concluded on

eIF-2

α subunit

Artemia	NH_2 –	Pro	—	Leu	Ser	Pro	Arg	Tyr	Tyr	Ala	–
Rabbit	NH_2 –	*	Gly	*	*	Cys	*	Phe	*	Gln	–

Artemia	Glu	Lys	Phe	Pro	Glu	Thr	Glu	Asp	Val	Val
Rabbit	His	*	*	*	*	Val	*	*	*	*

FIGURE 7. The NH_2- terminii of the α subunit of *Artemia* and rabbit reticulocyte eIF-2.

the basis of Western immunoblotting that the antigenic determinants for the α subunit were only moderately conserved and those of the β subunit the least conserved. Recently, Altmann et al.[68] found that antibodies raised against either yeast or sheep reticulocyte eIF-4E did not cross react with the other factor.

D. The NH_2-Terminal Sequence of the α Subunit of eIF-2

The NH_2-terminii of the α subunits from *Artemia* and reticulocyte eIF-2 preparations contained sufficient homology to be recognized by the heme-controlled repressor and to yield similar phosphopeptides upon extensive trypsin cleavage. We have started to sequence the NH_2-terminii of these α subunits in order to examine further the homology between the factors. This is illustrated in Figure 7. Of the first 18 amino acids of the *Artemia* subunit, 14 amino acids in the reticulocyte factor are either the same or represent a conservative change (Tyr-6 in *Artemia* to Phe-7 in rabbit). Between these two sequences there are four nonconservative changes and 1 amino acid (Gly) inserted after the initial proline in the rabbit sequence. Further work is necessary to determine how homologous the remaining sequence is, especially in the regions where the factors are phosphorylated. cDNA clones of the individual eIF-2 subunits would be important, not only for the comparison of phosphorylation sites, but for locating amino acid sequences which are homologous to known guanine nucleotide binding sites.

III. COMPARISON OF *ARTEMIA* AND RETICULOCYTE eIF-2 ACTIVITIES

A. Effect of Mg^{2+} on Binary (eIF-2 · GDP) and Ternary (eIF-2 · GTP · Met-$tRNA_f$) Complex Formation

1. Binary Complex Formation

Reticulocyte and *Artemia* eIF-2 preparations form a binary complex with GDP or GTP.[21,69] With *Artemia* eIF-2, the K_D for GDP is 3×10^{-7} *M* and that for GTP approximately 1×10^{-5} *M*.[49] Similar values were obtained for reticulocyte eIF-2.[69] We also examined the extent of GDP binding to eIF-2 preparations from *Artemia* and rabbit reticulocytes, and the effect of Mg^{2+} on binary (eIF-2 · GDP) complex formation.[53] In the absence of Mg^{2+}, *Artemia* and reticulocyte eIF-2 bind GDP. Addition of Mg^{2+} increases the binding of [^3H]GDP to *Artemia* eIF-2, whereas binding to reticulocyte eIF-2 is reduced. GDP was found to be

Table 4
EFFECT OF Mg^{2+} ON TERNARY COMPLEX FORMATION[a]

	$[^{35}S]$Met-tRNA$_f$ bound (cpm)	
Factors added	$-Mg^{2+}$	$+Mg^{2+}$
Artemia eIF-2	6,000	5,700
Reticulocyte eIF-2	12,000	3,800

[a] Standard nitrocellulose membrane filtration assay conditions were used. Where indicated, reaction mixtures contained 1 m*M* magnesium acetate. Concentrations of the factors used were reticulocyte eIF-2, 2 μg; *Artemia* eIF-2, 2 μg. A filtration background of 150 cpm for $[^{35}S]$Met-tRNA$_f$ has been subtracted from each value. The specific activity of the $[^{35}S]$Met-tRNA$_f$ was 15,000 cpm/pmol.

Table 5
EFFECT OF Mg^{2+} ON GUANINE NUCLEOTIDE EXCHANGE

	$[^3H]$GDP retained in binary complex (pmol)	
Additions	$-Mg^{2+}$	$+Mg^{2+}$
Artemia		
eIF-2	0.49	0.93
eIF-2 + GTP	0.02	0.10
Reticulocyte		
eIF-2	0.90	0.97
eIF-2 + GTP	0.06	0.97

Note: The two-stage assay mixture in 75 μℓ contained 100 m*M* KCl, 20 m*M* Tris-Cl (pH 7.8), 0.2 m*M* GTP, and, as indicated, 1 m*M* Mg^{2+}, 1.0 pmol of reticulocyte eIF-2 · GDP, or 1.2 pmol *Artemia* eIF-2 · GDP. Stage one was incubated 10 min at 30°C and either water or GTP was added to 0.15 m*M* and the reaction continued for 10 min.

associated with purified eIF-2 preparations from both *Artemia* and reticulocytes (0.5 to 0.6 pmol/pmol factor). In the presence of Mg^{2+} this GDP bound to reticulocyte eIF-2 was apparently stabilized, thus reducing the subsequent binding of $[^3H]$GDP.[53] Although purified *Artemia* eIF-2 also contained GDP, Mg^{2+} did not stabilize the bound GDP sufficiently to reduce further $[^3H]$GDP binding.

2. Ternary Complex Formation

The effect of Mg^{2+} on ternary complex formation by *Artemia* and reticulocyte eIF-2 was determined.[53] Addition of Mg^{2+} to ternary complex formation assays with the *Artemia* factor had no consistent effect on the binding of Met-tRNA$_f$ (Table 4). However, the addition of 1 m*M* Mg^{2+} to reactions containing reticulocyte eIF-2 produced a 60 to 70% reduction in ternary complex formation. These results are consistent with the hypothesis that the GDP isolated with the reticulocyte factor cannot be released in the presence of Mg^{2+}, and that this bound GDP prevents the factor from participating in the subsequent reactions.

B. 40S Initiation Complex Formation with *Artemia* eIF-2

Three types of eIF-2 preparations were assayed for AUG-dependent binding of Met-tRNA$_f$ on 40S ribosomal subunits: (1) eIF-2, (2) eIF-2 phosphorylated by the reticulocyte HCR, and (3) eIF-2 reisolated after phosphorylation. The nonphosphorylated as well as the phosphorylated and reisolated factor are active in 40S initiation complex formation (Table 5). However, the specific activity of the nonphosphorylated eIF-2 is approximately 30% higher. If the factor, after treatment with HCR, is not reisolated by chromatography on phosphocellulose, it is not active in binding Met-tRNA$_f$ to 40S ribosomal subunits. This suggests that some component in the HCR preparation, not phosphorylation of the eIF-2 per se, is responsible for the inhibition of formation of the 40S initiation complex.

C. Stimulation of Cell-Free Protein Synthesis by eIF-2

The ability of purified *Artemia* eIF-2 to stimulate protein synthesis was assayed in two different translation systems. Purified *Artemia* eIF-2 stimulates amino acid incorporation in

Table 6
FORMATION OF BINARY (eIF-2 · GDP) AND TERNARY (eIF-2 GTP MET-tRNA) COMPLEX WITH PHOSPHORYLATED [eIF-2(α-P)] AND NONPHOSPHORYLATED eIF-2 FROM RABBIT RETICULOCYTES AND *ARTEMIA*[a]

	[³H]GDP bound (cpm)		[³⁵S]Met-tRNA$_f$ bound (cpm)	
Additions	$-Mg^{2+}$	+1 m*M* Mg^{2+}	$-Mg^{2+}$	+1 m*M* Mg^{2+}
Reticulocyte				
eIF-2	11,200	3,300	8,700	1,700
eIF-2 + factor	10,300	11,000	9,900	7,700
eIF-2 (α-P)	9,500	1,300	6,900	1,600
eIF-2 (α-P) + factor	9,000	1,400	8,800	1,600
Artemia				
eIF-2	5,300	8,800	4,200	6,200
eIF-2 + factor	8,200	8,500	7,500	8,400
eIF-2(α-P)	4,900	8,200	4,800	6,400
eIF-2(α-P) + factor	6,600	8,500	6,300	7,100

[a] Reaction mixtures contained, where indicated, 0.5 μg (3.8 pmol) of reticulocyte eIF-2 or eIF-2(α-P); 1 μg (7.1 pmol) of *Artemia* eIF-2 or eIF-2(α-P); nucleotide exchange factor, 0.2 μg (approximately 0.4 pmol assuming 70% purity); and 0.2 m*M* GTP. Filtration backgrounds for [³H]GDP, [³⁵S]Met-tRNA$_f$, and ³²P-labeled eIF-2 have been subtracted.

a partially fractionated system containing salt-washed ribosomes from dormant *Artemia* embryos, rabbit reticulocyte globin poly(A)-rich mRNA, and a limiting level of the 0.5 *M* KCl ribosomal wash fraction from rabbit reticulocytes (Table 6). The *Artemia* eIF-2 was phosphorylated by reticulocyte eIF-2 and reisolated by chromatography on phosphocellulose.[21] With 1.5 μg of mock phosphorylated or phosphorylated eIF-2, protein synthesis is stimulated approximately 1.8-fold. The increase in amino acid incorporation with eIF-2(α-P) is linear during the entire incubation period. [21] During this time, only a small amount of radioactivity is lost from the [³²P]eIF-2(α-P), indicating that the stimulation does not result from dephosphorylation of the factor. Similar results were obtained when *Artemia* eIF-2(α-P) was used in a reticulocyte lysate protein-synthesis system.[21]

D. The Interaction of the Guanine Nucleotide Exchange Factor and eIF-2 and the Roles of Mg^{2+} and Phosphorylation

As noted above, binary and ternary complex formation with reticulocyte eIF-2 is reduced in the presence of Mg^{2+}. The addition of Mg^{2+} to the reticulocyte eIF-2 stabilizes the binding of GDP associated with the purified factor. GDP binding to reticulocyte eIF-2, but not the *Artemia* factor, is stabilized by the addition of Mg^{2+} and this GDP prevents GTP binding and subsequent ternary complex formation (Table 7). A factor was isolated which reduced the Mg^{2+} sensitivity of eIF-2 and facilitated the replacement of eIF-2-bound GTP.[44-48,53,55] This problem is known as the guanine nucleotide exchange factor (GEF, RF, Co-eIF-2B) and is composed of 5 subunits of M_r 82,000, 67,000, 55,000, 39,000 and 29,000. This protein can be isolated free or in a complex with eIF-2.[55]

1. Kinetics of Nucleotide Exchange and Ternary Complex Formation

The kinetics of GDP release from a preformed eIF-2 · GDP binary complex, and subsequent ternary complex formation with *Artemia* and reticulocyte eIF-2 are illustrated in Figure 8.[53] These reactions are carried out in the presence of 1 m*M* Mg^{2+}. As reported earlier [21,53] and

Table 7
ACTIVITY OF *ARTEMIA* eIF-2(α-P) IN 40S INITIATION COMPLEX FORMATION[a]

Expt. no.	Factor added (μg)		[^{35}S]Met-tRNA$_f$ bound cpm	pmol
I			900	0.10
II	eIF-2	1.0	2100	0.24
		2.5	6500	0.72
III	eIF-2(α-P)	1.4	2600	0.28
		3.0	4800	0.53
IV	eIF-2(α-P)	1.9	1000	0.10
		3.8	1000	0.11

[a] No eIF-2 was present in Expt. I. In Expt. II, eIF-2 was mock phosphorylated by omitting ATP, then re-isolated by phosphocellulose chromatography. In Expt. III, after phosphorylation by HCR, eIF-2 was chromatographed on phosphocellulose. This chromatography was not done in Expt. IV

From Woodley, C. L. et al., *Eur. J. Biochem.*, 117, 543, 1981. With permission.

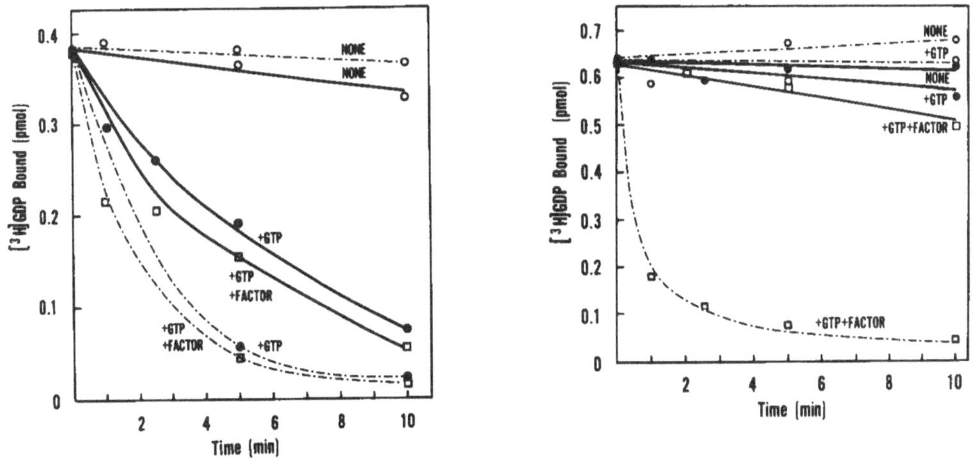

FIGURE 8. Kinetics of GDP release from eIf-2 · GDP or eIF-2(α-P) · GDP. The assay mixture in 75 μℓ contained 100 m*M* KCl, 20 m*M* Tris-Cl (pH 7.8), 1 m*M* Mg^{2+}, 0.2 m*M* GTP, and, as indicated, 42 ng of the guanine nucleotide exchange factor,[55] 0.6 pmol of reticulocyte eIF 2 · GDP, or eIF-2(α-P) · GDP (right panel) or 0.4 pmol *Artemia* eIF-2 · GDP or eIF-2(α-P) · GDP (left panel). No addition (○); GTP (●); GTP and nucleotide-exchange factor (□); eIF-2 · GDP (— · —) and eIF-2(α-P) · GDP (——). (From Wahba, A. J. and Woodley, C. L., *Prog. Nucl. Acids Res. Mol. Biol.*, 31, 484, 1984. With permission.)

shown in Figure 8A, *Artemia* eIF-2 readily exchanges GTP for prebound GDP in the presence of 1 m*M* Mg^{2+}. Addition of the reticulocyte nucleotide exchange factor has no effect on the rate or extent of nucleotide exchange. Under the same reaction conditions, reticuloctye eIF-2 requires the addition of GEF for GDP release and subsequent ternary complex formation (Figure 8B). The addition of Met-tRNA$_f$ does not influence the rate of nucleotide exchange

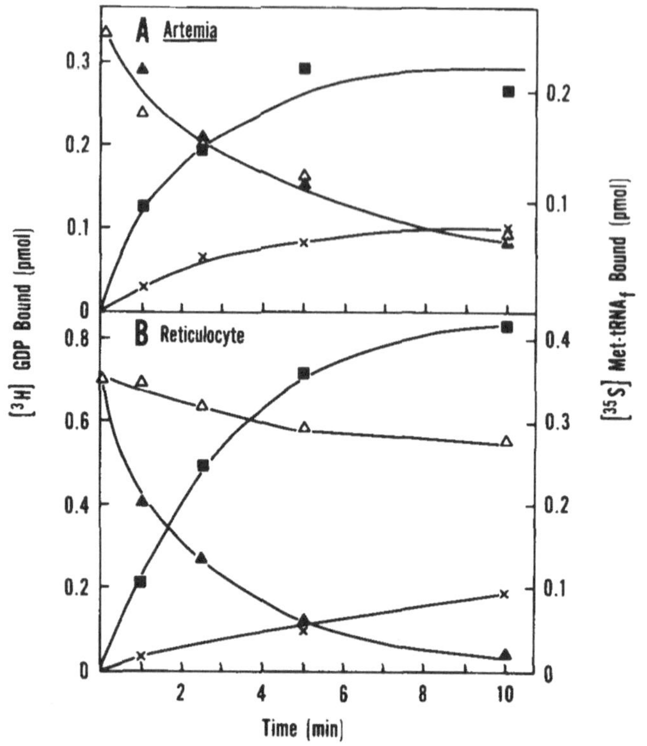

FIGURE 9. Kinetics of GDP release and Met-tRNA$_f$ binding during ternary complex formation with *Artemia* and reticulocyte binary complexes. The components of the assay mixture were as described in Figure 6. Either *Artemia* eIF-2 · GDP (0.35 pmol) (panel A), or reticulocyte eIF-2 · GDP (0.7 pmol) (panel B) was used. GDP retained with factor(▲—▲) and without factor (△—△); Met-tRNA$_f$ bound with (■—■) or without (X—X) factor. (From Mehta, H. B. et al., *J. Biol. Chem.*, 258, 3438, 1983. With permission.)

with eIF-2 from either *Artemia* or reticulocytes (compare Figure 8 and 9), and the amount of Met-tRNA$_f$ bound increases stoichiometrically with the release of GDP from the binary complex.

The purification of *Artemia* eIF-2 includes chromatography on Mono S, which should remove any nucleotide exchange factor activity.[55] We examined the possibility that catalytic levels of an *Artemia* nucleotide exchange factor might be present in *Artemia* eIF-2 preparations. Addition of *Artemia* eIF-2 to a reticulocyte eIF-2 · GDP binary complex did not promote nucleotide exchange.[96] Although our in vitro assays suggest that *Artemia* eIF-2 does not require the guanine nucleotide exchange factor, we cannot eliminate the possibility that under certain assay conditions *Artemia* eIF-2 would require GEF, or that a factor similar to GEF exists in *Artemia*.

2. *Effect of the Reticulocyte HCR On eIF-2 Activities and Interaction with GEF*

The regulation of eIF-2 activity in rabbit reticulocytes has been extensively studied. The primary point of regulation may occur at the GDP/GTP nucleotide exchange step and involves two proteins, GEF[45,55,70-72] and HCR.[73,74] The phosphorylation of the α subunit of mammalian eIF-2 may be an important mechanism for regulating the activity of the factor. In the absence of heme or the presence of ds-RNA[75-77] a cAMP-independent protein kinase isolated from rabbit reticulocytes phosphorylates a serine residue near the NH_2-terminus of the α subunit

Table 8
STIMULATION OF POLYPEPTIDE SYNTHESIS BY PHOSPHORYLATED eIF-2α IN A FRACTIONATED CELL-FREE PROTEIN SYNTHESIS SYSTEM[a]

Additions	[³H]leucine incorporated	Total radioactivity (cpm) in [³²P]eIF-2(α-P) at	
		0 min	60 min
None	1,500	—	—
Salt wash	55,500	—	—
Salt wash + eIF-2	102,500	—	—
Salt wash + eIF-2(α-P)	100,500	20,800	15,300

[a] Each 50-μℓ reaction mixture contained 0.26 A_{260} unit of salt-washed *Artemia* ribosomes, 0.063 A_{260} unit of rabbit hemoglobin poly(A)-rich mRNA, 36 μg of rabbit reticulocyte ribosomal salt wash, 10 μCi [³H]leucine (50,000 cpm/pmol), and 1.5 μg *Artemia* eIF-2 or 1.4 μg of eIF-2(α-P). Under these assay conditions, amino acid incorporation and loss of radioactivity were linear for the assay period. The ³²P-labeled eIF-2(α-P) had 19,280 cpm/μg of protein.

From Woodley, C. L. et al., *Eur. J. Biochem.* 117, 543, 1981. With permission.

of reticulocyte eIF-2. [57,63,64] This phosphorylation blocks the nucleotide exchange factor from catalyzing the GTP-dependent release of GDP from an eIF-2(α-P) · GDP binary complex (Table 8) and thus prevents recycling of the factor.[45,70-72] At one time this regulation was thought to be common to all eukaryotes.[34] We have phosphorylated *Artemia* and reticulocyte eIF-2 preparations with HCR and determined if their activity were altered.

a. Binary and Ternary Complex Formation with Phosphorylated eIF-2

Preparations of eIF-2 from *Artemia* and reticulocytes were phosphorylated with [³²P]ATP and reticulocyte HCR and reisolated by chromatography on phosphocellulose.[21] The ³²P-labeled eIF-2(α-P) was subjected to isoelectric focusing[78] and autoradiographed. In each case only one labeled polypeptide was detected.

The effect of the nucleotide exchange factor and Mg^{2+} on GDP binding and ternary complex formation with phosphorylated eIF-2 from *Artemia* and reticulocytes was examined. In the absence of Mg^{2+}, isolated phosphorylated and nonphosphorylated eIF-2 from either source form stable binary or ternary complexes (Table 8). However, Mg^{2+} inhibits both binary and ternary complex formation with reticulocyte eIF-2[53] and eIF-2 (α-P), but not with either *Artemia* eIF-2 or eIF-2(α-P) (Table 8). Addition of the nucleotide exchange factor to binary or ternary complex reactions readily reverses the Mg^{2+} inhibition with reticulocyte eIF-2, but not with eIF-2(α-P). On the other hand, there is an increase in GDP or Met-$tRNA_f$ binding with *Artemia* eIF-2 or eIF-2 (α-P) upon addition of Mg^{2+}. Maximum binding of Met-$tRNA_f$ with either eIF-2 or eIF-2(α-P) occurs in the presence of both Mg^{2+} and the nucleotide exchange factor. The ability of the nucleotide exchange factor to promote GDP/GTP exchange with these eIF-2 preparations was also determined.

b. Nucleotide Exchange with Phosphorylated eIF-2

Preformed, isolated eIF-2·GDP binary complexes were used to examine the kinetics of nucleotide exchange with *Artemia* and reticulocyte eIF-2(α-P). With either reticulocyte eIF-

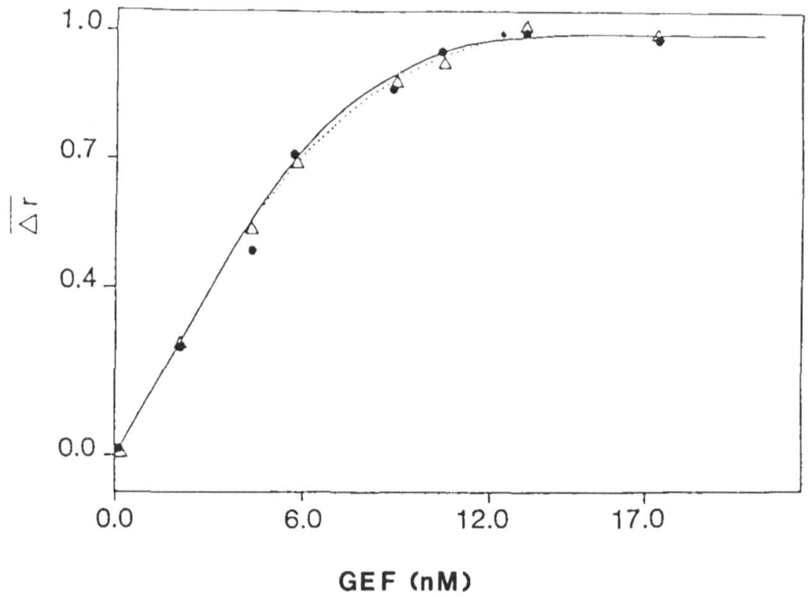

FIGURE 10. Titration of eIF-2 with the nucleotide exchange factor GEF. The y axis is the the normalized anisotropy change: $(r_{obs} - r_{eIF\text{-}2})/(r_{eIF\text{-}2\ GEF} - r_{eIF\text{-}2})$. The triangles represent the data points for the binding of eIF-2 · GDP (protein = 8 n*M*, GDP total = 0.1 μ*M*) to GEF. The anisotropy range was 0.13 to 0.20. The range varied slightly with different preparations, probably due to the difficulty in removing free dye. Circles are the observations for 8 n*M* eIF-2 and 20 μ*M* GDP titrated with GEF. In this case, the anisotropy range was 0.16 to 0.22. The solid and dotted lines represent the calculated equilibrium curves for K_{eq} = 0.15 and 0.20 n*M*, respectively. The temperature was 20°C. (From Goss, D. J. et al., *J. Biol. Chem.*, 259, 7374, 1984. With permission.)

2 · GPD or eIF-2(α-P) · GDP nucleotide release does not occur either in the presence or the absence of GTP (Figure 8A). Only upon addition of both GTP and the guanine nucleotide exchange factor is GDP released from the reticulocyte eIF-2 · GDP complex; GEF does not promote nucleotide release from the eIF-2(α-P) GDP binary complex.

The *Artemia* binary complex is stable in the absence of free nucleotide, however, in contrast to the reticulocyte binary complex, the eIF-2 · GDP preparations readily release GDP in the presence of added GTP (Figure 8B).[21,53] Addition of the nucleotide exchange factor does not produce any significant effect on the extent of nucleotide exchange with *Artemia* eIF-2 · GDP or eIF-2(α-P) · GDP. *Artemia* eIF-2(α-P) exchanges GTP for GDP at approximately 50% the rate of the nonphosphorylated preparation.

3. Measurement of the Interaction of eIF-2 with the Guanine Nucleotide Exchange Factor by Fluorescence Techniques

Labeling of eIF-2 with dansyl chloride permitted direct measurement of the interaction of phosphorylated and nonphosphorylated *Artemia* and reticulocyte eIF-2 with GEF.[72] Dansyl-eIF-2 shows a large change in fluorescence anisotropy when GEF was added to the solution. Figure 10 shows a titration curve for GEF binding to fluorescently labeled eIF-2. Equilibrium constants for GEF binding to eIF-2 and eIF-2(α-P) are given in Table 9. The apparent equilibrium constant for the reaction eIF-2 · GDP · GEF ⇌ eIF-2 · GDP + GEF did not change significantly with increasing GDP concentrations from 0.01 to 0.20 μ*M*. The equilibrium constant corresponding to eIF-2 + GDP + GEF is for GEF binding to eIF-2 where GDP is added to a final concentration of 20 μ*M* before addition of GEF. The addition of 1 m*M* Mg^{2+} and stoichiometric or higher amounts of GDP causes a small

Table 9
THE EQUILIBRIUM CONSTANTS FOR GEF INTERACTION WITH DANSYL-LABELED PHOSPHORYLATED AND NONPHOSPHORYLATED eIF-2 FROM *ARTEMIA* AND RABBIT RETICULOCYTES

Additions	Experimental conditions[a]	K_D (nmolar)[b]
Reticulocyte		
eIF-2 + GEF	A	0.05
eIF-2 + GEF + GDP	B	0.15
eIF-2(α-P) + GEF + GDP	C	0.31
Artemia		
eIF-2 + GEF	A	0.55
eIF-2 + GEF + GDP	B	0.60
eIF-2(α-P) + GEF + GDP	C	0.15

[a] Experimental conditions were A: [eIF-2] = 5 to 9 n*M*. No GDP was added for the titration, however some GDP is usually present in eIF-2 preparations. These concentrations were low enough that less than 10% of the eIF-2 was complexed with GDP. B: [eIF-2] = 8 to 20 n*M*, GDP was 20 μ*M* before addition of GEF. All titrations were performed in buffer A (see text). The temperature was 20°C.

[b] K_D is the equilibrium constant for the dissociation of the GEF · eIF-2 or GEF · eIF-2(α-P) complex.

From Goss, D. J. et al., *J. Biol. Chem.*, 259, 7374, 1984. With permission.

anisotropy change for eIF-2, from an initial value of approximately 0.09 to about 0.11 to 0.12. For reticulocyte eIF-2, GEF binding is about three to four times stronger than in the presence of GDP. The *Artemia* factor has less than 1/10 the affinity for GEF as does reticulocyte eIF-2, and this binding was not effected by the presence of GDP. Upon phosphorylation of reticulocyte eIF-2, GEF has approximately 1/2 the affinity for eIF-2(α-P) + GDP as it does for the nonphosphorylated factor. In contrast to the response of the reticulocyte factor, phosphorylation of *Artemia* eIF-2(α) by HCR increases the affinity of the factor for GEF approximately three- to fourfold.

For nonphosphorylated reticulocyte eIF-2, the addition of GEF caused a fluorescence anisotropy change from about 0.11 to 0.19 indicating that eIF-2 and GEF are forming a complex. At low levels of GDP (stoichiometric, 5 to 15 n*M* to about 500 n*M*), the addition of GTP caused a decrease in anisotropy, possibly from dissociation of the eIF-2 GEF complex. Addition of GTP to a concentration of 10 to 15 μ*M* lowered the anisotropy from 0.19 to about 0.08. The phosphorylated reticulocyte eIF-2 behaves quite differently. In this case, 1 to 10 μ*M* of GTP does not cause a decrease in fluorescence anisotropy. These results demonstrate that although the *Artemia* eIF-2 does bind to the guanine nucleotide exchange factor, its pattern of interaction with GEF is significantly different from that of reticulocyte eIF-2.

IV. ISOLATION OF eIF-2 COFACTORS FROM *ARTEMIA* EMBRYOS

Several factors that enhance eIF-2 activity have been isolated from extracts of rabbit reticulocytes and wheat germ.[21,23,48,79,80] We have characterized three distinct factors from the wash of *Artemia* ribosomes that stimulate ternary complex formation and Met-$tRNA_f$

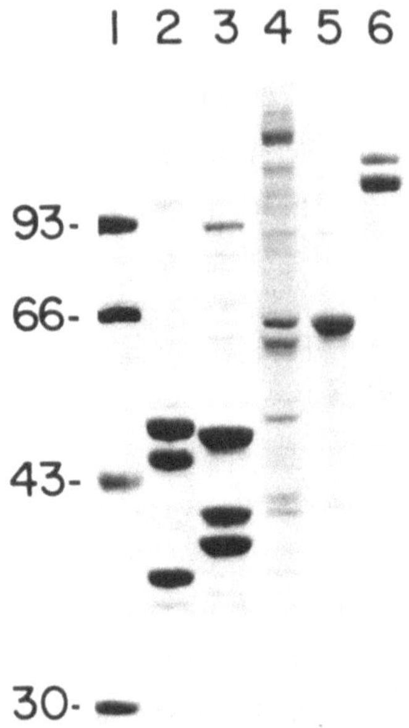

FIGURE 11. SDS polyacrylamide gel electrophoresis of eIF-2, Co-eIF-2(A), Co-eIF-2(B). The electrophoresis was in 10% gels containing dodecyl sulfate. Lane 1, M_r standards; lane 2, 10 μg of reticulocyte eIF-2; lane 3, 10 μg *Artemia* eIF-2; lane 4, 12 μg of Co-eIF-2(A) hydroxyl apatite fraction; 5, 6.5 μg Co-eIF-2 (A) affinity purified; lane 6, 3.8 μg of Co-eIF-2(B). (From Woodley, C. L. et al., *Eur. J. Biochem.*, 117, 543, 1981. With permission.)

binding to 40S ribosomal subunits.[21,66] Although it is not possible to correlate any of the *Artemia* factors with specific reticulocyte or wheat germ counterparts, some of the similarities are discussed in this section. We have designated the three *Artemia* factors as Co-eIF-2(A), Co-eIF-2(B), and Co-eIF-2(C).

A. Isolation of Cofactors for eIF-2

Co-eIF-2(A) and Co-eIF-2(B) are isolated from the 40 to 60% saturated ammonium sulfate fraction of the ribosomal salt wash by chromatography on heparin-Sepharose, DEAE-cellulose, phosphocellulose, and hydroxyapatite.[21] Neither factor can be assayed in the initial steps because of the presence of eIF-2. Co-eIF-2(A)[21] preparations are at least 90% pure and contain a single polypeptide of M_r 65,000 (Figure 11, lane 2). Analysis of the Co-eIF-2(B) preparation in SDS polyacrylamide gels shows two polypeptides of M_r 105,000 and 112,000 in a 1:1 ratio (Figure 11, lane 3).

Co-eIF-2(C), unlike Co-eIF-2(A) and Co-eIF-2(B), is isolated from the 0 to 40% saturated ammonium sulfate fraction of the ribosomal high salt wash of 12-hr developing embryos.[66] Fractionation of the phosphocellulose fraction of Co-eIF-2(C) by sucrose gradient centrifugation gives two peaks of activity sedimenting at 12 and 15S. The 12S fraction contains three polypeptides of M_r 115,000, 71,000, and 49,000 as determined by electrophoresis in SDS polyacrylamide gels. In contrast, the 15S fraction contains 9 to 11 polypeptides, 3 of which are identical to those of the 12S fraction. In all assays the 15S fraction behaves similarly to the 12S fraction.

Table 10
EFFECT OF AURINTRICARBOXYLIC ACID ON CO-eIF-2(A) AND CO-eIF-2(B) ACTIVITIES[a]

	[^{35}S]Met-tRNA$_f$ bound		
Additions	−ATA	+ATA (4.5 μ*M*)	+ATA (9.0 μ*M*)
eIF-2	625	95	75
eIF-2 + Co-eIF-2(A)	2765	550	215
eIF-2 + Co-eIF-2(B)	3050	2475	2215

[a] Incubation mixtures for ternary complex formation contained 100 m*M* KCl, 1 m*M* Mg^{2+}, 0.16 m*M* GTP, 2 pmol [^{35}S]Met-tRNA$_f$, 0.74 μg of eIF-2, 3.9 μg of Co-eIF-2(A), or 3.7 μg of C0-eIF-2(B). After 5 min at 30°C, aurintricarboxylic acid was added; the reaction was continued for an additional 5 min and terminated by addition of cold buffer.

From Woodley, C. L. et al., *Eur. J. Biochem.*, 117, 543, 1981. With permission.

B. Characterization of the eIF-2 Cofactors

1. Co-eIF-2(A) and Co-eIF-2(B)

a. Ternary Complex Formation

Stimulation of ternary complex formation is used as the assay for Co-eIF-2(A) and Co-eIF-2(B).[21] At levels of eIF-2 below 2 to 3 μg/75 μℓ of reaction mixture, the eIF-2 is most sensitive to stimulation by Co-eIF-2(A) and Co-eIF-2(B).[21] Each factor stimulates eIF-2 activity to a similar extent, and this effect is observed for both the rate and extent of ternary complex formation. Either protein produces maximum stimulation and no additive effect is observed in the presence of satuaration levels of either cofactor.

Artemia eIF-2, like its reticulocyte counterpart, is strongly inhibited by low concentrations (4 to 10 μ*M*) of aurintricarboxylic acid (ATA) (Table 10).[21,23,66] In reactions containing 0.74 μg of the initiation factor, ternary complex formation is inhibited 85 to 90% by ATA. The presence of 3.9 μg of Co-eIF-2(A) does not protect eIF-2 from inhibition, whereas a comparable amount of Co- eIF-2(B) provides 70 to 80% protection.

b. Formation of 40S Initiation Complex

Binding of the initiator tRNA (Met-tRNA$_f$) to 40S ribosomal subunits is dependent upon the addition of eIF-2 and GTP,[21,52] and this binding is stimulatd three- to fourfold by the addition of the template AUG (Table 11). Maximum binding of Met-tRNA$_f$ to *Artemia* 40S ribosomal subunits requires the addition of approximately 4 μg of Co-eIF-2(B) per 75 μℓ of reaction mixture.[21] Addition of either Co-eIF-2(A) or Co-eIF-2(B) results in approximately a twofold increase in the extent of 40S initiation complex formation, and no additive effect is observed in the presence of saturating levels of either cofactor.

2. *Co-eIF-2(C)*

a. Formation of Ternary Complex

Stimulation of ternary complex formation by the phosphocellulose fraction of Co-eIF-2(C) was assayed with different levels of eIF-2 (Table 12).[66] With 0.8 μg of eIF-2, Co-eIF-2(C) increases Met-tRNA$_f$ binding 3.3-fold. As observed with Co-eIF-2(A) and Co-eIF-2(B), the degree of stimulation by Co-eIF-2(C) is maximum at low levels of the eIF-2. At high levels of eIF-2 there is no stimulation of Met-tRNA binding.

b. Reversal of the Inhibition by mRNA of Ternary Complex Formation

The binding of Met-tRNA$_f$ is decreased by 75% by the addition of 1μg of mRNA to the

Table 11
AUG-DEPENDENT BINDING OF [^{35}S]MET-tRNA$_f$ TO 40S RIBOSOMES: STIMULATION BY Co-eIF-2(A) AND Co-eIF-2(B)[a]

Additions	[^{35}S]Met-tRNA$_f$ bound (cpm)	Stimulation (fold)
eIF-2	1000	1.0
eIF-2 + Co-eIF-2(A)	2030	1.8
eIF-2 + Co-eIF-2(B)	2725	2.3

[a] The 40S initiation complex assay contained 0.5 A_{260} unit of 40S ribosomal subunits, 1.1 μg of eIF-2, and where indicated, 3.9 μg of Co-eIF-2(A) or 3.7 μg of Co-eIF-2(B). A blank of 300 cpm for Met-tRNA$_f$ binding in the absence of AUG was subtracted from each value.

From Woodley, C. L. et al., *Eur. J. Biochem.*, 117, 543, 1981. With permission.

Table 12
EFFECT OF Co-eIF-2(C) ON TERNARY COMPLEX FORMATION WITH INCREASING AMOUNTS OF eIF-2[a]

	[^{35}S]Met-tRNA$_f$ bound		
eIF-2 (μg)	−Co-eIF-2(C) (cpm)	+Co-eIF-2(C) (cpm)	Stimulation (fold)
0.8	2,500	8,400	3.3
2.1	7,900	15,000	1.9
5.0	25,000	27,000	1.1
9.9	35,000	36,000	1.0

[a] Blanks in the absence of eIF-2 (110 cpm) and with Co-eIF-2(C) alone (330 cpm) were subtracted from the appropriate values. eIF-2 activity was assayed by the nitrocellulose membrane filtration technique.

ternary complex assay (Table 13). This inhibition is substantially relieved by Co-eIF-2(C), and the addition of 1 m*M* Mg^{2+} aids in reversing this inhibition.

c. *Formation of 40S Initiation Complex.*

The eIF-2-dependent binding of Met-tRNA$_f$ to 40S ribosomal subunits in the presence of AUG is assayed by sucrose density gradient centrifugation (Table 14).[21] In the presence of eIF-2, Met-tRNA$_f$ binding to 40S ribosomal subunits is stimulated twofold by addition of AUG. Co-eIF-2(C) stimulates Met-tRNA$_f$ binding six- to eightfold in the presence or the absence of AUG. In the absence of eIF-2, Co-eIF-2(C) has no significant activity, either alone or in combination, in binding Met-tRNA$_f$ to 40S ribosomal subunits.

Co-eIF-2(C) also stimulates 40S initiation complex formation with natural message as template. The binding of Met-tRNA$_f$ and ascites mRNA to 40S ribosomal subunits is determined by pelleting the initiation complex through a sucrose gradient (Table 15). In the absence of ancillary factors, no message and very little Met-tRNA$_f$ is bound. In the absence of eIF-2, Co-eIF-2(C) has no effect on Met-tRNA$_f$ binding, but produces significant message binding. However, in the presence of eIF-2, the addition of Co-eIF-2(C) significantly increases Met-tRNA$_f$ binding but not message binding. It should be noted that the fraction

Table 13
EFFECT OF mRNA ON TERNARY COMPLEX FORMATION IN THE PRESENCE OF Co-eIF-2(C)[a]

Factor	[^{35}S]Met-tRNA$_f$ bound (cpm)			
	None	Mg^{2+}	mRNA	Mg^{2+} + mRNA
eIF-2	3100	3100	800	1200
+ Co-eIF-2(C)	6100	6700	4600	5700
+ Co-eIF-2(C) (12S)	5900	5900	2200	3600
+ Co-eIF-2(C) (15S)	5700	5600	2700	3800

[a] Ternary complex formation was assayed in a one-stage reaction. Each incubation mixture (75 μℓ) contained 10 μg of bovine serum albumin, 0.92 μg of *Artemia* eIF-2, 3 pmol of [^{35}S]Met-tRNA$_f$, 0.2 m*M* GTP and, where indicated, 1 m*M* Mg^{2+}, 1 μg of reticulocyte poly(A)-containing RNA, and 11 μg of the phosphocellulose 15S or 12S fraction of Co-eIF-2(C). A blank in the absence of eIF-2 (70 cpm) has been subtracted from each value.

From Wahba, A. J. and Woodley, C. L., *Prog. Nucl. Acids Res. Mol. Biol.*, 31, 484, 1984. With permission.

Table 14
EFFECT OF Co-eIF-2(C) ON AUG-DEPENDENT MET-tRNA$_f$ BINDING TO 40-RIBOSOMAL SUBUNITS[a]

Additions	[^{35}S]Met-tRNA$_f$ bound (cpm)	
	−AUG	+AUG
eIF-2	400	900
+ Co-eIF-2(C)	3300	5000
+Co-EIF-2(C) (12S)	3000	3800
Co-eIF-2(C)	100	300

[a] Reaction mixtures contained, where indicated, 1.6 μg of *Artemia* eIF-2, 11 μg of phosphocellulose fraction Co-eIF-2(C), 12 μg of mRNA binding factor, and 8 μg of the sucrose gradient 12S Co-eIF-2(C) fraction. A value of 350 cpm, obtained in the absence of eIF-2 has been subtracted from each value. Each reaction contained 3 pmol of [^{35}S]Met-tRNA$_f$(19,500 cpm/pmol).

which elutes from phosphocellulose after Co-eIF-2(C) binds mRNA but not Met-tRNA$_f$ to 40S ribosomal subunits in the absence of eIF-2, and that addition of eIF-2 further increases this mRNA binding. At this point there is no obvious correlation between the *Artemia* Co-eIF-2(C) or the mRNA-binding fraction and the message binding factors isolated from rabbit reticulocytes. Further work is required to determine if this RNA binding fraction is specifically retained on a m^{7}GTP-Sepharose column and contains the equivalent of the cap binding protein complex. Maximum stimulation of message and Met-tRNA$_f$ binding requires the addition of eIF-2, Co-eIF-2(C), and the mRNA binding fraction. No significant Met-tRNA$_f$ binding is obtained when either Co-eIF-2(C) or the mRNA binding fraction is added alone. This is consistent with results obtained in the AUG-dependent assay (Table 14). Although Co-eIF-2(C) stimulates ternary complex formation, its primary function appears to be in stabilizing the 40S initiation complex.

Table 15
40S INITIATION COMPLEX FORMATION WITH ^{3}H-LABELED ASCITES mRNA[a]

Additions	mRNA bound (cpm)		Met-tRNA$_f$ bound (cpm)	
	−eIF-2	+eIF-2	−eIF-2	+eIF-2
None	0	0	0	300
Co-eIF-2(C)	2400	300	80	2300
mRNA binding fraction	1500	3300	90	300
Co-eIF-2(C) + mRNA binding fraction	2900	4200	70	3400

[a] Each reaction contained 3 pmol of [^{35}S]Met-tRNA$_f$ (15,700 cpm/pmol), 0.4 μg of ^{3}H-labeled ascites mRNA (70,000 cpm/pmol) and, where indicated, 1.6 μg of eIF-2, 11 μg of phosphocellulose fraction Co-eIF-2(C), and 12 μg mRNA binding fraction. A blank of 540 cpm for ^{3}H-labeled mRNA and 375 cpm for [^{35}S]Met-tRNA$_f$ has been subtracted from each value.

From Wahba, A. J. and Woodley, C. L., *Prog. Nucl. Acids Res. Mol. Biol.*, 31, 484, 1984. With permission

V. EXPRESSION OF eEF-Tu DURING DEVELOPMENT OF *ARTEMIA* EMBRYOS

The unusual developmental cycle of *Artemia* presents a unique opportunity for investigating the regulation of cellular processes. Upon hydration dormant embryos rapidly revive and initiate protein and RNA synthesis, although DNA synthesis and cell division do not begin until after emergence of the prenauplius. Since development during this period is relatively synchronous a population of closely staged individuals is available. This is of clear value in the study of gene expression during development.

A. Changes in the Steady-State Level of eEF-Tu in Developing *Artemia* Embryos

We began by examining changes in the in vivo proteins at different stages of development.[81,82] Two-dimensional gel electrophoresis[78] followed by silver staining[88] reveals that during development a variety of proteins change with respect to their relative proportions (Figure 12). Among the proteins identified by their co-migration with purified standards were tubulin, actin, and eEF-Tu, all of which demonstrated stage-specific quantitative changes. Of these, eEF-Tu increased the most dramatically (Figures 12A to C).

B. Steady-State Levels of Message Coding for eEF-Tu in Developing Embryos

1. Changes in the Level of Translatable Poly(A)$^+$ Message

In order to examine the steady-state level of message specific for eEF-Tu and other proteins during development, poly(A)$^+$ RNA was isolated [84-87] from various stages of development and assayed for mRNA activity in a nuclease-treated rabbit reticulocyte cell-free system[49] (Figure 13). RNA isolated from each stage directed protein synthesis at a level equivalent to that of globin mRNA. [81,82] The translation products of these reactions were subjected to electrophoresis in one-dimensional SDS polyacrylamide gels. The overall distribution of [^{35}S]Met-labeled polypeptides did not vary significantly with the RNAs prepared from the different stages and most of the polypeptides seen in a given stage were also present in the other stages (Figure 14). However, prominent changes are evident in the quantity of a polypeptide(s) of approximate M_r 28,000, as well as for proteins co-migrating with tubulin, actin, and eEF-Tu. Scanning densitometry shows that the products migrating at the M_r of eEF-Tu comprise only 7% of total in vitro translation products when the mRNA is from dormant embryos, whereas this percentage rises to approximately 20% when the translation

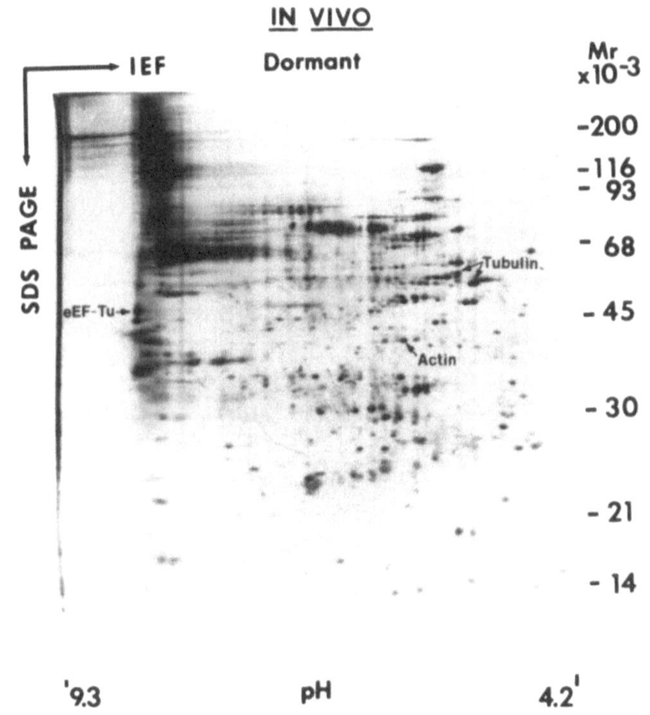

A

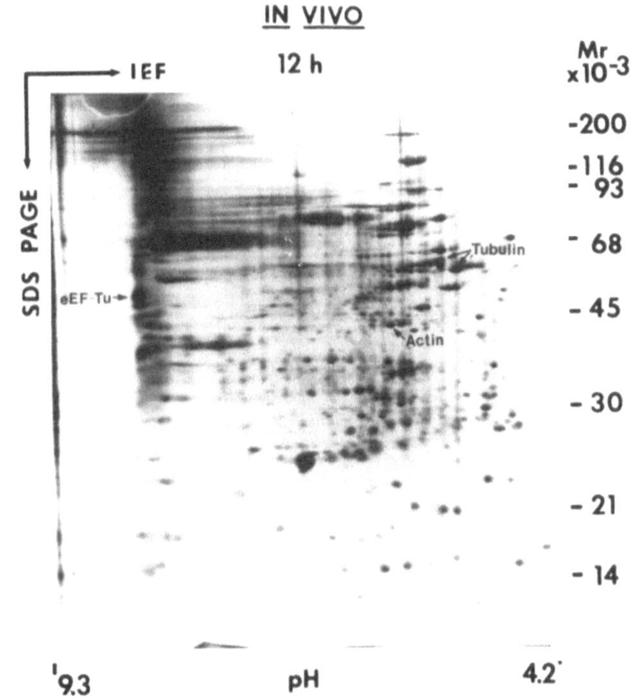

B

FIGURE 12. Two-dimensional gel electrophoresis of proteins extracted from *Artemia*. Proteins were isolated from dormant (Figure 12A), 12-hr developing embryos (Figure 12B), and nauplii (Figure 12C) and subjected to two-dimensional electrophoresis.[78] Each gel contained approximately 20 μg of protein, which was visualized by silver staining.[83] (From Daum, III, H. D. et al., *J. Biol. Chem.*, 260, 16347, 1985. With permission.)

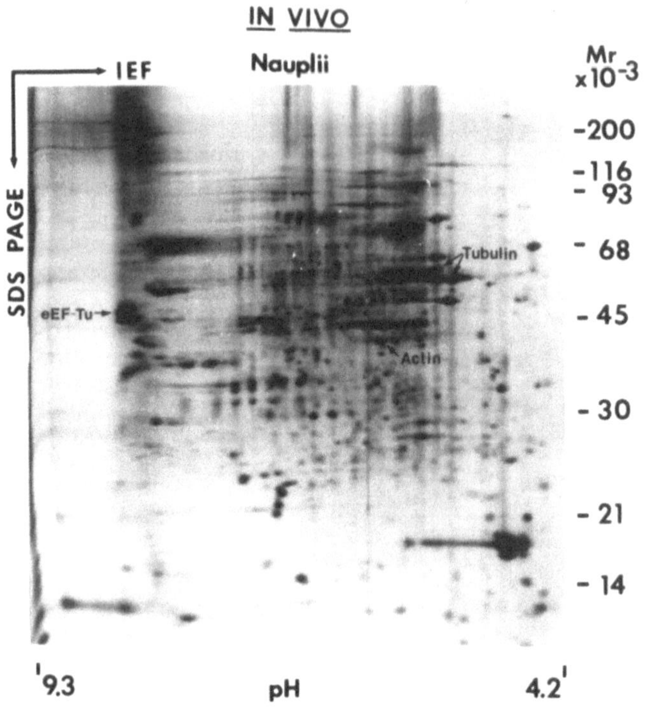

FIGURE 12C.

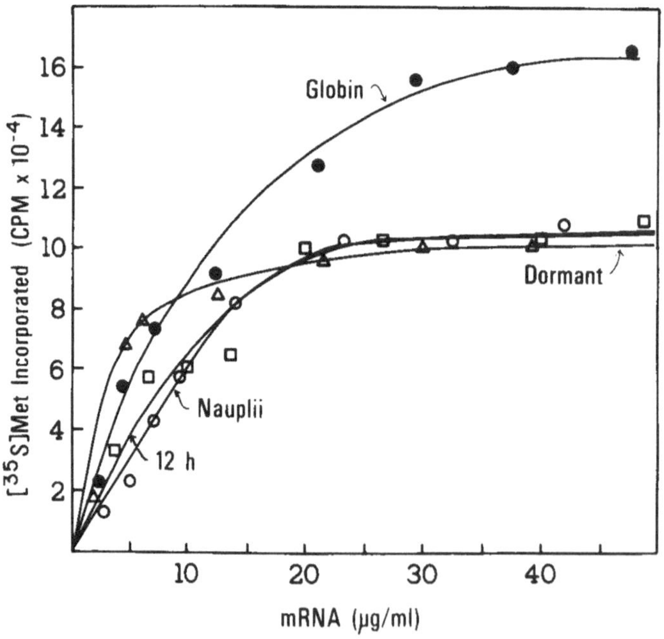

FIGURE 13. Translation of *Artemia* and reticulocyte mRNA in a nuclease-treated reticulocyte lysate. Poly(A)-containing RNA was prepared from rabbit reticulocytes, from dormant and 12-hr developing *Artemia* embryos, and from nauplii. [^{35}S]Methionine incorporation into hot trichloroacetic acid-insoluble peptides was assayed in a nuclease-treated reticulocyte lysate (11 μℓ) containing mRNA as indicated. Globin (●); dormant *Artemia* embryos (△); 12-hr developing embryos (□); and nauplii (○). (From Daum, III, H. D. et al., *J. Biol. Chem.*, 260, 16347, 1985. With permission.)

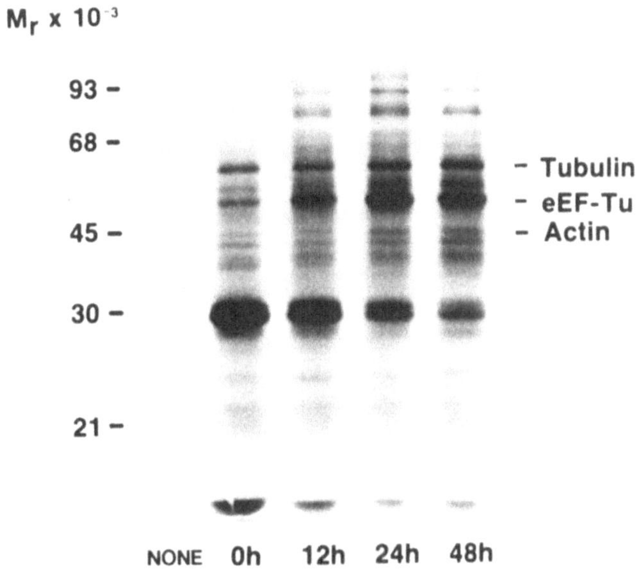

FIGURE 14. Autoradiograph of *Artemia* and globin mRNA translation products separated in SDS polyacrylamide gels. The translation products as produced in Figure 13 were analyzed by electrophoresis in 10% polyacrylamide gels containing SDS. After staining with Coomassie Blue R-250, the gels were treated with EN^{3}HANCE (NEN) and subjected to autoradiography. Each lane represents the radioactive translation products obtained with the indicated message. The markers are eEF-Tu; tubulin; actin; and the 27,000 M_r major translation product of dormant embryos. (From Daum, III, H. D. et al., *J. Biol. Chem.*, 260, 16347, 1987. With permission.)

is directed by mRNA from 24-hr embryos (Figure 14, scans not shown). Although other proteins co-migrate with eEF-Tu in this one-dimensional system, these results are consistent with the changes observed in the two-dimensional gels of the in vivo proteins (Figure 12).

These translation products were also analyzed by NEPHGE/SDS/polyacrylamide gel electrophoresis[56,78] (Figure 15). In this system, eEF-Tu is separated from all other proteins, and may be more easily analyzed than in conventional IEF/SDS gels (Figure 12). The autoradiogram for the NEPHGE/SDS gel of the products resulting from the translation of dormant message is shown in Figure 15A. The change in the level of the labeled product corresponding to eEF-Tu (Figure 15B) clearly demonstrates that synthesis of eEF-Tu directed by mRNA from embryos at each stage of development increases dramatically following hydration and resumption of development. Scanning densitometry of the eEF-Tu area (Figure 15B) shows that the amount translatable eEF-Tu mRNA increases by more than threefold between the dormant and 12-hr stages (Table 16, column 3).[81,82]

2. Changes in the Relative Abundance of eEF-Tu-Specific RNA During Development

The expression of eEF-Tu mRNA at various developmental stages was examined with a cDNA probe, prepared by primer-extending eEF-Tu mRNA with a mixed-sequence 14-base oligodeoxyribonucleotide. This sequence was based on amino acids 259-263.[81,82] By comparison with the published amino acid sequence, the 900-base primary cDNA product was shown by sequencing to be specific for eEF-Tu.[81,82] Futhermore, the nucleotide sequence was in almost complete agreement with that recently published by van Hemert et al.,[90] differing only at the wobble position of five codons. These changes would not alter the amino acid sequence. The shorter cDNA products were premature termination products.

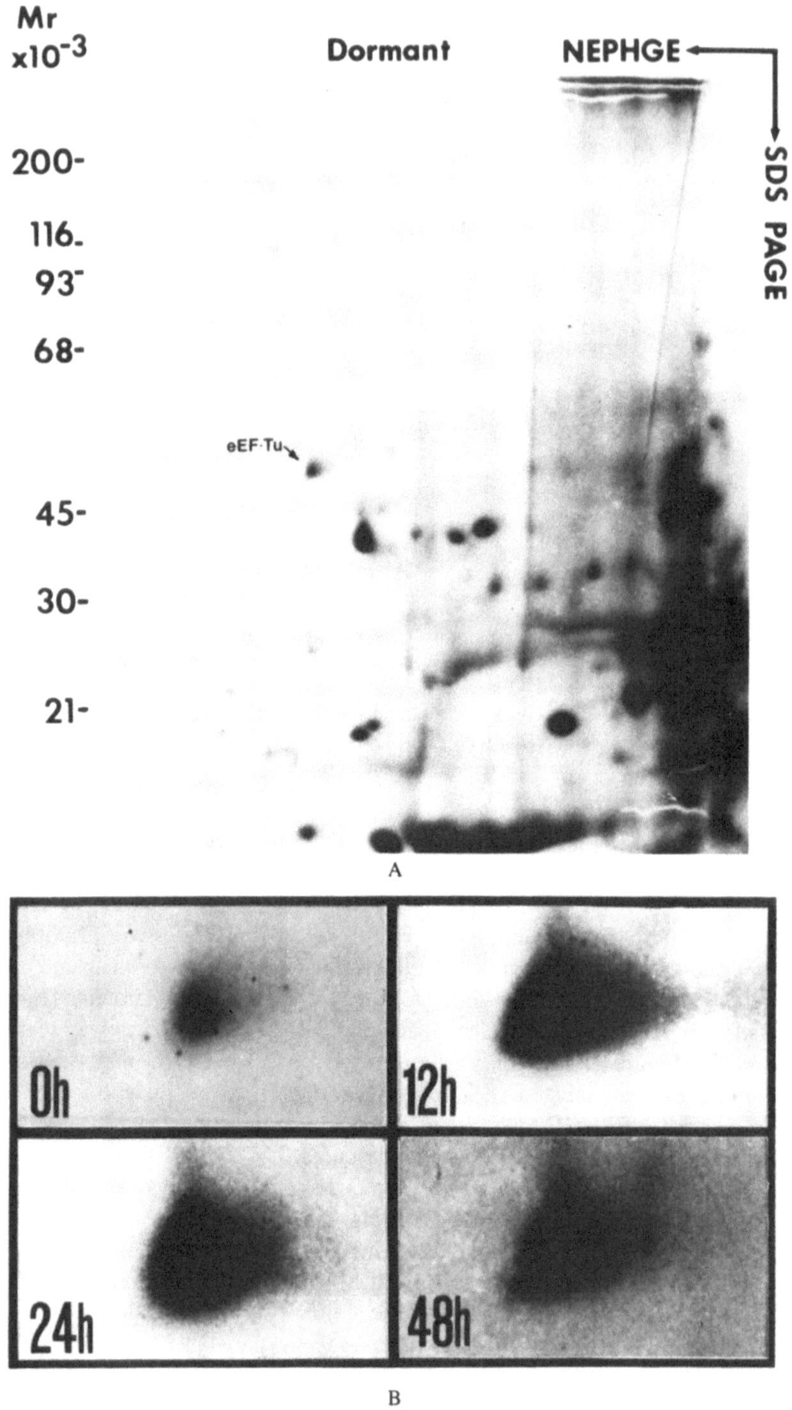

FIGURE 15. Identification of in vitro synthesized eEF-Tu. NEPHGE/dodecyl sulfate-polyacrylamide gel electrophoresis of *Artemia* poly(A)$^+$ RNA-directed in vitro translation products was performed as described in the text. Prior to electrophoresis 5 μg of purified *Artemia* eEF-Tu were added as unlabeled marker to 18,000 cpm of translation products; less than 1% of the applied protein failed to enter the second dimension gel. (A) Fluorograph of a single stage (dormant) is shown for illustration. Arrow indicates eEF-Tu. (B) Fluorographs of in vitro synthesized eEF-Tu with *Artemia* poly(A)$^+$ RNA from different stages of development. Fluorography of all gels was for 10 days at −70°C with Kodak XAR-5 film. (From Daum, III, H. D. et al., *J. Biol. Chem.*, 260, 16347, 1985. With permission.)

TABLE 16
QUANTITATION OF IN VITRO TRANSLATED eEF-Tu AND OF eEF-Tu mRNA LEVELS DURING DEVELOPMENT

Hours after resumption of development	mg total RNA/g hydrated cysts	Fold increase in in vitro translated eEF-Tu[a]	Fold increase in eEF-Tu-specific mRNA[a]
0	1.3—1.4	1.0	1.0
12	1.4—1.6	3.7	1.9
24	1.2—1.7	3.7	2.1
48	1.3—1.6	2.9	2.8

[a] Determined by scanning densitometry of the autoradiogram.

Thus, the primer extension method produced specific priming and the level of non-eEF-Tu-specific cDNA is apparently very low in our preparations.

Potential changes in the appearance of eEF-Tu mRNA at different stages of development were measured by hybridization with this eEF-Tu-specific cDNA probe. To examine the relative changes of eEF-Tu mRNA levels at different stages of development, poly(A)$^+$ RNA from dormant, 12-hr, 24-hr and 48-hr developing embryos was separated on denaturing agarose gels[88] and transferred to nitrocellulose.[89] Hybridization with the labeled cDNA probe revealed a band in the RNA samples from each developmental stage corresponding to an RNA of approximately 1800 bases (Figure 16).[81] The size of this RNA corresponds closely with that of *Artemia* eEF-Tu mRNA.[90] In addition, the level of hybridization to RNAs from different stages (Table 16, column 4) parallels the increase in in vitro synthesis of eEF-Tu seen in Figure 15B and quantitated in Table 16, column 3.

To establish further that the 1800-base RNA which hybridized with the cDNA probe (Figure 16) and with the 14-base oligonucleotide primer (see below) was eEF-Tu message, we separated poly(A)$^+$ RNA from 48-hr developing embryos (nauplii) on a preparative nondenaturing agarose gel and transferred the RNA to poly(U) paper.[91] A portion of the poly(U) paper was hybridized to the eEF-Tu cDNA probe. The autoradiogram of this hybridization revealed a major band migrating faster than 18S RNA (Figure 17A, arrow). A strip containing message co-migrating with the RNA hybridizing with the probe was cut out of the remainder of the poly(U) paper. The mRNA from this strip was eluted and translated in a nuclease-treated reticulocyte lysate. Polyacrylamide gel electrophoresis and autoradiography revealed only three major products (Figure 17B), one of which co-migrated with purified eEF-Tu (M_r = 50,000).[81,82] Finally, precipitation of the in vitro translation reaction with antibodies specific for eEF-Tu revealed the 50,000 M_r product to be the only species precipitated (Figure 17C).

C. Comparison of Interspecies Homologies of eEF-Tu From Various Sources

Because of our interest in the function of eEF-Tu, we investigated interspecies homologies as a potential indicator of functional domains. The amino acid sequence of the region of eEF-Tu from which our oligomer sequence was derived is conserved in *Escherichia coli*[91] and *Artemia*.[90] We, therefore, isolated poly(A)$^+$ RNA from a variety of species to study possible conservation of sequences across other taxonomic boundaries. Following electrophoresis in agarose gels and transfer to nitrocellulose, the RNAs were hybridized with the labeled oligomer. Figure 18 shows that for *Artemia*, mouse, and man the primary hybridizing species is approximately the same size, or about 1800 bases. In addition, further hybridizations have shown that the same is true for *Chironomus*, *Drosophila* and *Bombyx* (data not shown). Thus, for all eukaryotes examined so far, the mRNA for eEF-Tu appears to be virtually identical in size.[82]

Primer extensions with reverse transcriptase and the 14-base oligonucleotide were per-

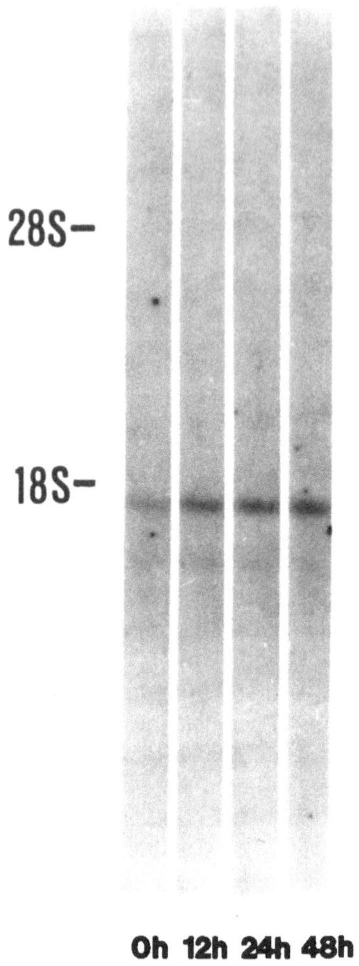

FIGURE 16. Hybridization of eEF-Tu cDNA to *Artemia* poly(A)$^+$RNA. Poly(A)$^+$ RNA (4 μg) from *Artemia* embryos at different stages of development was denatured with 10 m*M* CH_3HgOH and separated on a 10 m*M* CH_3HgOH agarose gel.[88] The RNA was transferred to nitrocellulose and hybridized with 2.0×10^6 cpm of the eEF-Tu-specific cDNA. Exposure for autoradiography was for 72 hr at −70°C. (From Daum, III, H. D. et al., *J. Biol. Chem.*, 260, 16347, 1985. With permission.)

formed using the poly(A)-containing RNAs from different species. The results, shown in Figure 19, demonstrate that the fully extended products from *Artemia*, K562 cells (human), MEL DS-19 cells (mouse), and *Bombyx mori* all had identical sizes on denaturing polyacrylamide gels, co-migrating at approximately 900 bases with the cDNA for *Artemia* eEF-Tu. By analogy with the *Artemia* cDNA preparations, the smaller extension products may be prematurely terminated cDNAs for eEF-Tu. Interestingly, there is heterogeneity in these premature termination products. Bands in some species correspond to those in other species, while others do not. This suggests that although the distance from the site of hybridization of our primer to the 5′-end of the message is conserved, the internal sequences may differ. Different sequences may lead to secondary structures (e.g., hairpin loops)[95] which inhibit the progress of reverse transcriptase, resulting in a premature termination. Sequence data demonstrated that the NH_2 terminal 160 amino acids of *Artemia* eEF-Tu are 96% homologous with those of the mouse sequence.[97]

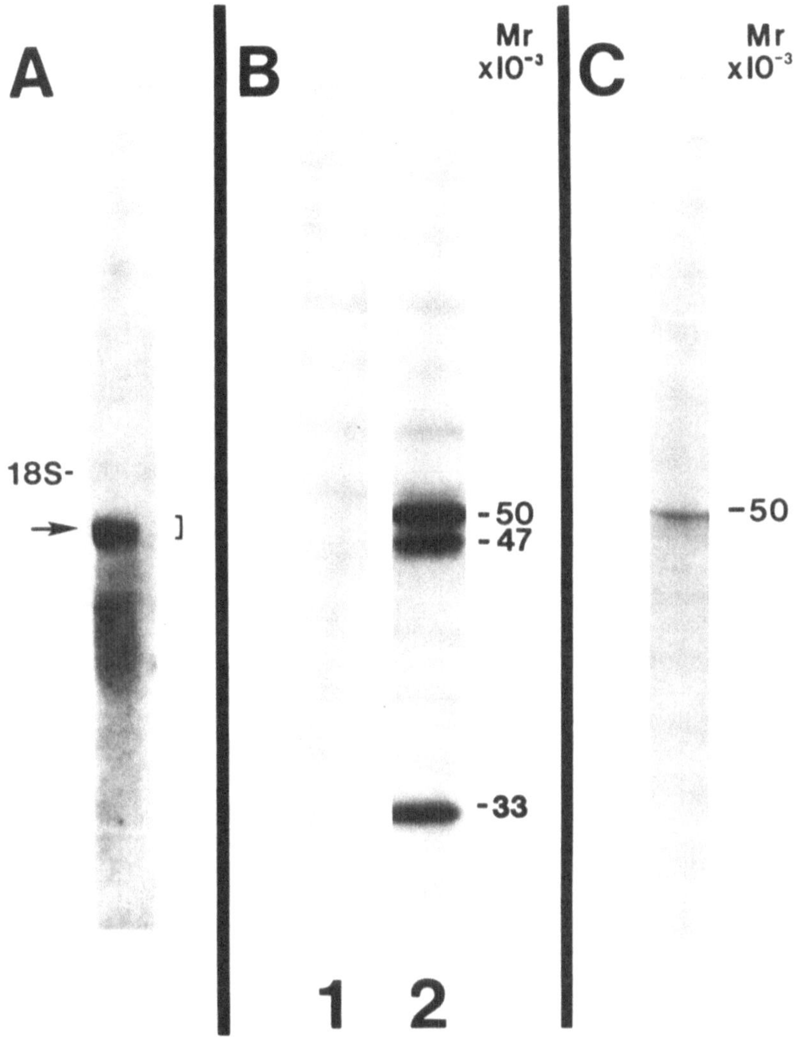

FIGURE 17. Poly(U)-paper selection and translation of eEF-Tu mRNA. Poly(A)⁺ RNA from 48-hr developing embryos was electrophoresed on a preparative 1.5% agarose non-denaturing gel. Following transfer to poly(U)-paper, a strip along the edge was cut off and hybridized with 1.0×10^6 cpm/mℓ of eEF-Tu cDNA. (A) Autoradiography of the hybridized strip, exposed for 16 hr at −70°C. The autoradiograph was aligned with the remainder of the poly(U)-paper, the area (bracket) co-migrating with the hybridizing band (arrow) cut out, and the RNA eluted. (B) Dodecyl sulfate polyacrylamide gel electrophoresis of the in vitro translation products. The 50,000 M_r product co-migrates with *Artemia* eEF-Tu;1, no RNA added; 2, poly(U)-paper selected RNA. (C) Dodecyl sulfate polyacrylamide gel electrophoresis of the eEF-Tu antibody precipitate from in vitro translation products. (From Daum III, H. D. et al., *J. Biol. Chem.*, 260, 16347, 1985. With permission.)

ACKNOWLEDGMENTS

We wish to thank Drs. Jaydev Dholakia and Paul Bragg for helpful discussions. This work was supported in part by United States Public Health Services Grant GM-25451.

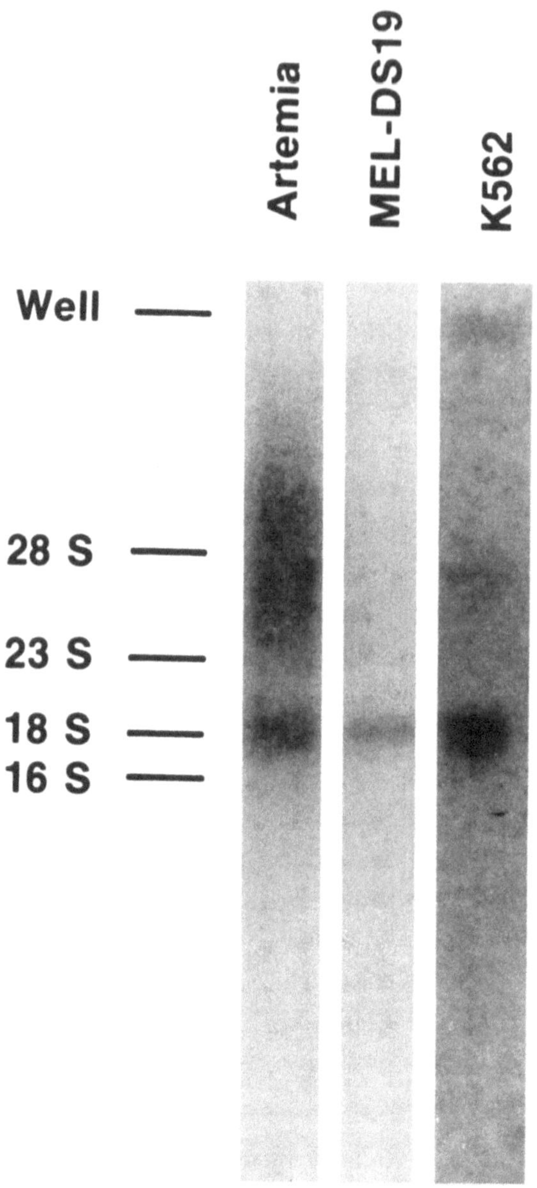

FIGURE 18. Hybridization of the oligonucleotide primer to poly(A)-containing RNA from various species. RNA (4 μg) from *Artemia*, K562 cells (human), and MEL-DS19 cells (mouse) was fractionated on denaturing agarose gels and transferred to nitrocellulose. Exposure for autoradiography was for 24 hr at −70°C.

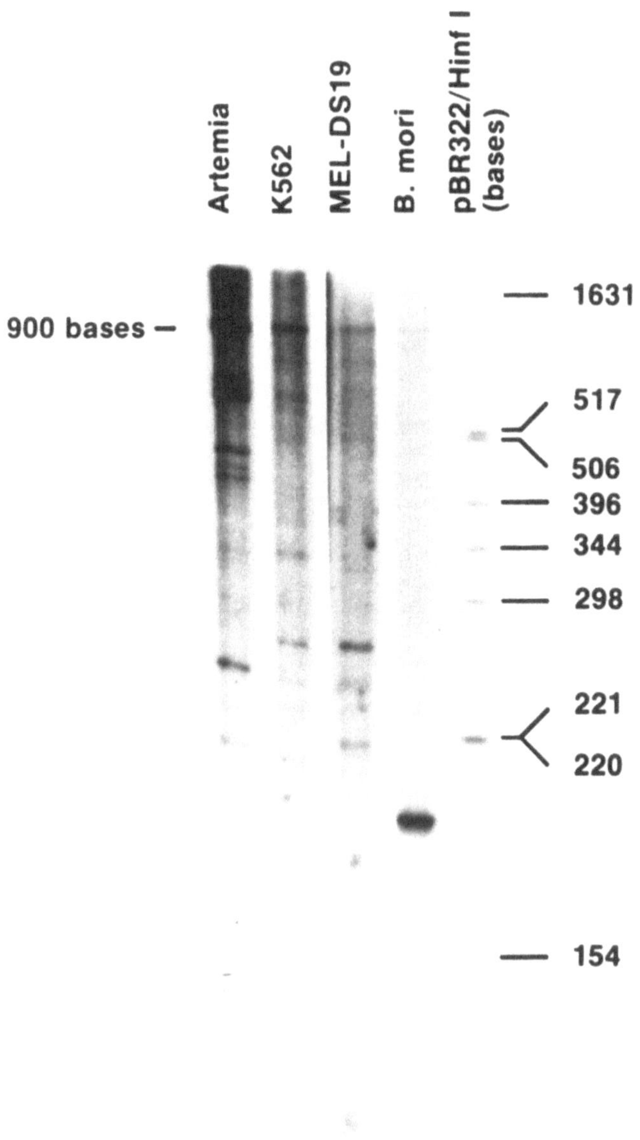

FIGURE 19. Primer extension of mRNA from various species. The mixed-sequence 14-base oligonucleotide was end-labeled and used to prime reverse transcription of 10 μg of poly(A)$^+$ RNA. The cDNA products were fractionated on 8% polyacrylamide 8 *M* urea gels. Marker positions are HinfI-digested and labeled pBR322.

REFERENCES

1. **Nilsson, M. O. and Hultin, T.**, Poly(A)-containing cytoplasmic RNA in dormant cysts of *Artemia salina*, *FEBS Lett.*, 52, 269, 1975.
2. **Sierra, J. M., Filipowicz, W., and Ochoa, S.**, Messenger RNA in undeveloped and developing *Artemia salina* embryos, *Biochem. Biophys. Res. Commun.*, 69, 181, 1976.
3. **Amaldi, P. O., Felicetti, L., and Campioni, N.**, Flow of informational RNA from cytoplasmic poly(A)-containing particles to polyribosomes in *Artemia salina*. Functional and physical changes during development, *Nucleic Acids Res.*, 4, 2109, 1977.
4. **Grosfeld, H., Soreq, H., and Littauer, U. Z.**, Membrane associated cytoplasmic mRNA in *Artemia salina*. Functional and physical changes during development, *Nucleic Acids Res.*, 4, 2109, 1977.
5. **Sierra, J. M., Meier, D., and Ochoa, S.**, Effect of development on the translation of messenger RNA in *Artemia salina* embryos, *Proc. Natl. Acad. Sci. U.S.A.*, 71, 2693, 1974.
6. **Golub, A. and Clegg, J. S.**, Protein synthesis in *Artemia salina* embryos. I. Studies on polyribosomes, *Dev. Biol.*, 17, 644, 1968.
7. **Hultin, T. and Morris, J. E.**, The ribosomes of encysted embryos of *Artemia salina* during cryptobiosis and resumption of development, *Dev. Biol.*, 17, 143, 1968.
8. **Slobin, L. I. and Moller, W.**, Characterization of developmentally regulated forms of elongation factor 1 in *Artemia salina*. I. Purification and structural properties of the enzyme, *Eur. J. Biochem.*, 69, 351, 1976a.
9. **Slobin, L. I. and Moller, W.**, Characterization of developmentally regulated forms of elongation factor 1 in *Artemia salina*. II. Functional properties of the enzymes, *Eur. J. Biochem.*, 69, 367, 1976b.
10. **Reddington, M. A., Fong, A. P., and Tate, W. P.**, The termination of the synthesis of proteins *in vitro* with extracts from the undeveloped cyst of *Artemia salina, Dev. Biol.*, 63, 402, 1978.
11. **Bagshaw, J. C., Finamore, F. J., and Novelli, G.D.**, Changes in transfer RNA in developing brine shrimp, *Dev. Biol.*, 23, 23, 1970.
12. **Filipowicz, W., Sierra, J. M., and Ochoa, S.**, Polypeptide chain initiation in eukaryotes: initiation factor MP in *Artemia salina* embryos, *Proc. Natl. Acad. Sci. U.S.A.*,72, 3947, 1975.
13. **Filipowicz, W., Sierra, J. M., and Ochoa, S.**, Polypeptide chain initiation in eukaryotes: initiation factor requirements for translation of natural messengers, *Proc. Natl. Acad. Sci. U.S.A.*, 73, 44, 1976.
14. **Finamore, F. J. and J. S.**, Biochemical aspects of morphogenesis in brine shrimp, *Artemia salina*, in *The Cell Cycle*, Padilla, G., Whitson, G. L., and Cameron, I. L., Eds., Academic Press, New York, 1969, 399.
15. **Hentschel, C. C. and Tata, J. R.**, The molecular embryology of the brine shrimp, *Trends Biochem. Sci.*, 1, 97, 1976.
16. **Olson, C. S. and Clegg, J. S.**, Cell division during the development of *Artemia salina, Wilhelm Roux's Arch. Dev. Biol.*, 184, 1, 1978.
17. **Clegg, J. S.**, Interrelationships between water and cellular metabolism in *Artemia* cysts. V^{14} CO_2 incorporation, *J. Cell. Physiol.*, 89, 369, 1976.
18. **Chen, Y. C., Woodley, C. L., Bose, K. K., and Gupta, N. K.**, Protein synthesis in rabbit reticulocytes: characteristics of a Met-$tRNA_f$ binding factor, *Biochem. Biophys. Res. Commun.*, 48, 1, 1972.
19. **Dettman, G. L. and Stanley, Jr., W. M.**, Recognition of eukaryotic initiator tRNA by an initiation factor and the transfer of the methionine moiety into peptide linkage, *Biochim. Biophys. Acta*, 287, 124, 1972.
20. **Schreier, M. H. and Staehelin, T.**, Initiation on eukaryotic protein synthesis: Met-$tRNA_f$ 40S ribosomal initiation complex catalysed by purified initiation factors in the absence of mRNA, *Nature (London) New Biol.*, 242, 35, 1973.
21. **Woodley, C. L., Roychowdhury, M., MacRae, T. H., Olsen, K. W., and Wahba, A. J.**, Protein synthesis in brine shrimp embryos: regulation of the formation of the ternary complex (met-$tRNA_f$ eIF-2 GTP) by two purified protein factors and phosphorylation of *Artemia* eIF-2, *Eur. J. Biochem.*, 117, 543, 1981
22. **Maitra, U., Stringer, E. A., and Chaudhuri, A.**, Initiation factors in protein biosynthesis, *Annu. Rev. Biochem.*, 51, 869, 1982.
23. **Gupta, N. K., Grace, M., Banerjee, A. C., and Bagchi, M.**, in *Interaction of Translational and Transcriptional Controls in the Regulation of Gene Expression*, Grunberg-Manago, M. and Safer, B., Eds, Elsevier, Amsterdam, 1982, 339.
24. **Benne, R. and Hershey, J. W. B.**, The mechanism of action of protein synthesis initiation factors from rabbit reticulocytes, *J. Biol. Chem.*, 253, 3078, 1978.
25. **Shatkin, A. J.**, Capping of eukaryotic mRNA, *Cell*, 9, 645, 1976.
26. **Muthukrishman, S., Morgan, M., Banerjee, A. K., and Shatkin, A. J.**, Influence of 5'-terminal m^7G and 2'-o-methylated residues on mRNA binding to ribosomes, *Biochemistry*, 15, 5761, 1976.
27. **Nuss, D. L. Furuichi, Y., Koch, G., and Shatkin, A. J.**, Detection in HeLa cell extracts of a 7-methyl guanosine specific enzyme activity that cleaves m^7GpppNm, *Cell*, 6, 21, 1975.

28. **Sonenberg, N., Guertin, D., and Lee, K. A.,** Capped mRNAs with reduced secondary structure can function in extracts from poliovirus-infected cells, *Mol. Cell. Biol.*, 2, 1633, 1982.
29. **Sonenberg, N., Guertin, D., Cleveland, D., and Trachsel, H.,** Probing the function of the eukrayotic 5′ cap structure by using a monoclonal antibody directed against cap binding proteins, *Cell*, 27, 563, 1981.
30. **Chang, J. C., Temple, G. F., Poon, R., Neumann, K. H., and Kan, Y. W.,** The nucleotide sequences of the untranslated 5′ regions of human α- and β-globin mRNAs, *Proc. Natl. Acad. Sci. U.S.A.*, 74, 5145, 1977.
31. **Baralle, F. E.,** Structure-function relationship of 5′ non-coding sequence of rabbit α- and β-globin mRNA, *Nature (London)*, 267, 279, 1977.
32. **Grosfeld, H. and Littauer, U. Z.,** The translation *in vitro* of mRNA from developing cysts of *Artemia salina*, *Eur. J. Biochem.*, 70, 589, 1976
33. **Tahara, S. M., Morgan, M. A., and Shatkin, A. J.,** Two forms of purified m^7G-cap binding protein with different effects on capped mRNA translation in extracts of uninfected and poliovirus-infected HeLa cells, *J. Biol. Chem.*, 256, 7691, 1981.
34. **Hansen, J. L., Etchison, D. O., Hershey, J. W. B., and Ehrenfeld, E.,** Localization of cap-binding protein in subcellular fractions of HeLa cells, *Mol. Cell. Biol.*, 2, 1639, 1982.
35. **Trachsel, H., Erni, B., Schreier, M. H., and Staehelin, T.,** Initiation of mammalian protein synthesis. II. The assembly of the initiation complex with purified initiation factors, *J. Mol. Biol.*, 116, 755, 1977.
36. **Kozak, M.,** Role of ATP in binding and migration of 40S ribosomal subunits, *Cell*, 22, 459, 1980.
37. **Marcus, A.,** Tobacco mosaic virus ribonucleic acid-dependent amino acid incorporation in a wheat germ system *in vitro*, *J. Cell Biol.*, 245, 962, 1970.
38. **Kramer, G., Konecki, D., Cimadevilla, J. M., and Hardesty, B.,** ATP requirement for binding ^{125}I-labeled globin mRNA to *Artemia salina* ribosomes, *Arch. Biochem. Biophys.*, 174, 355, 1976.
39. **Grifo, J. A., Tahara, S. M., Leis, J. P., Morgan, M. A., Shatkin, A. J., and Merrick, W. C.,** Characterization of eukaryotic initiation factor 4A, a protein involved in ATP-dependent binding of globin mRNA, *J. Biol. Chem.*, 257, 5246, 1982.
40. **Benne, R., Brown-Luedi, M. L., and Hershey, J. W. B.,** Purification and characterization of protein synthesis initiation factors eIF-1, eIF-4C, eIF-4D and eIF-5 from rabbit reticulocytes, *J. Biol. Chem.*, 253, 3070, 1978.
41. **Peterson, D. T., Safer B., and Merrick W. C.,** Role of eukaryotic initiation factor 5 in the formation of 80S initiation complexes, *J. Biol. Chem.*, 254, 7730 1979.
42. **Trachsel, H. and Staehelin, T.,** Binding and release of eukaryotic initiation factor eIF-2 and GTP during protein synthesis initiation, *Proc. Natl. Acad. Sci. U. S. A.*, 75, 204, 1978.
43. **Merrick, W. C.,** Evidence that a single GTP is used in the formation of 80S initiation complexes, *J. Biol. Chem.*, 254, 3708, 1979.
44. **Panniers, R. and Henshaw, E. C.,** A GDP/GTD exchange factor essential for eukaroytic initiation factor 2 cycling in Ehrlich ascite tumor cells and its regulation by eukaryotic initiation factor 2 phosphorylation, *J. Biol. Chem.*, 258, 7928, 1983.
45. **Siekierka, J., Mauser, L., and Ochoa, S.,** Mechanism of polypeptide chain initiation in eukaryotes and its control by phosphorylation of the α subunit of initiation factor 2, *Proc. Natl. Acad. Sci. U.S.A.*, 79, 2537, 1982.
46. **Voorma, H. O. and Amesz, H.,** The control of the rate of protein synthesis initiation, in *Interaction of Translation and Transcriptional Controls in the Regulation of Gene Expression*, Grunberg-Manago, M. and Safer, B., Eds., Elsevier/ North-Holland, Amsterdam, 1982, 297.
47. **Safer, B., Jagus, R., Konieczny, A., and Crouch, D.,** The mechanism of translational inhibition in hemin-deficient lysates, in *Interaction of Translation and Transcriptional Controls in the Regulation of Gene Expression*, Grunberg-Mango, M. and Safer, B., Eds., Elsevier/ North-Holland, Amsterdam 1982, 311.
48. **Bagchi, M. K., Banerjee, A. C., Roy R., Chakrabarty, I., and Gupta, N. K.,** Protein synthesis in rabbit reticulocytes: characteristics of Co-eIF-2 protein complex, *Nucleic Acids Res.*, 10, 6501, 1982.
49. **Salimans, M., Posno, M., Benne, R., and Voorma, H. O.,** Regulation of protein synthesis in eukaryotes, eukaryotic initiation factor eIF-2 and eRF from neuroblastoma cells, *Biochim. Biophys. Acta*, 825, 384, 1985.
50. **Valenzuela, D. M., Chadhuri, A., and Maitra, U.,** Eukaryotic ribosomal subunit anti-association activity of calf liver is contained in a single polypeptide chain protein of M_r = 25,500 (eukaryotic initiation factor 6), *J. Biol. Chem.*, 257, 7712, 1982.
51. **Russell, D. W. and Spremulli, L. L.,** Purification and characterization of a ribosome dissociation factor (eukaryotic initiation factor 6) from wheat germ, *J. Biol. Chem.*, 254, 8796, 1979.
52. **MacRae, T. H., Roychowdhury, M., Houston, K. J., Woodley, C. L., and Wahba, A. J.,** Protein synthesis in brine shrimp embryos; dormant and developing embryos of *Artemia salina* contain equivalent amounts of chain initiation factor 2, *Eur. J. Biochem.*, 100, 67, 1979.

53. **Mehta, H. B., Woodley, C. L., and Wahba, A. J.,** Protein synthesis in brine shrimp embryos and rabbit reticulocytes: the effect of Mg^{2+} on binary and ternary complex formation, *J. Biol. Chem.*, 258, 3438, 1983.
54. **Warner, A. M., MacRae, T. H., and Wahba, A. J.,** The use of *Artemia salina* for developmental studies: preparation of embryos, tRNA, ribosomes, and initiation factor 2, in *Methods in Enzymology*, Vol. 60, Moldave, K. and Grossman, L. Eds., Academic Press, New York, 1979, 832.
55. **Dholakia, J. N., Mueser, T. C., Woodley, C. L., Parkhurst, L. J., and Wahba, A. J.,** The association of NADPH with the guanine nucleotide exchange factor from rabbit reticulocytes: a role of pyridine dinucleotides in eukaryotic polypeptide chain initiation, *Proc. Natl. Acad. Sci. U.S.A.*, 83, 6746, 1986.
56. **Laemmli, U.K.,** Cleavage of structural proteins during the assembly of the head of bacteriophage T4, *Nature (London)*, 227, 680 1979.
57. **Mehta, H. B., Dholakia, J. N., Roth, W. W., Parekh, B. S., Montelaro, R. C., Woodley, C. L., and Wahba, A. J.,** Structural studies on the eukaryotic chain initiation factor 2 from rabbit reticulocytes and brine shrimp *Artemia* embryos, *J. Biol. Chem.*, 261, 6705, 1986.
58. **Kramer, G., Cimadevilla, J. M., and Hardesty, B.,** Specificity of the protein kinase activity associated with the hemin-controlled repressor of rabbit reticulocyte, *Proc. Natl. Acad. Sci. U.S.A.*, 73, 3078, 1976.
59. **de Haro, C., Datta, A., and Ochoa, S.,** Mode of action of the hemin-controlled inhibitor of protein synthesis, *Proc. Natl. Acad. Sci. U.S.A.*, 75, 243, 1978.
60. **de Haro, C. and Ochoa, S.,** Further studies on the mode of action of the heme-controlled translational inhibitor, *Proc. Natl. Acad. Sci. U.S.A.*, 76, 1741, 1979.
61. **Crouch, D. and Safer, B.,** The association of eIF-2 with Met-$tRNA_i$ or eIF-2B alters the specificity of eIF-2 phosphatase, *J. Biol. Chem.*, 259, 10363, 1984.
62. **Zardeneta, G., Kramer, G., and Hardesty, B.,** Structure and function of the peptide initiation factor 2, *Proc. Natl. Acad. Sci. U.S.A.*, 79, 3158, 1982.
63. **Kudlicki, W., Wettenhall, R. E. H., Kemp, B. E., Szyszka, R., Kramer, G., and Hardesty, B.,** Evidence for a second phosphorylation site on eIF-2α from rabbit reticulocytes, *FEBS Lett.*, 215, 16, 1987.
64. **Gross, M., Rynning, J., and Knish, W. M.,** Evidence that the phosphorylation of eukaryotic initiation factor 2α by the hemin-controlled translational repressor occurs at a single site, *J. Biol. Chem.*, 256, 589, 1981.
65. **Ernst, H., Duncan, R. F., and Hershey, J. W. B.,** Cloning and sequencing of complementary DNAs encoding the α-subunit of translational initiation factor eIF-2, *J. Biol. Chem.*, 262, 1206, 1987.
66. **Wahba, A. J. and Woodley, C. L.,** Molecular aspects of development in the brine shrimp *Artemia*, *Prog. Nucleic Acids Res. Mol. Biol.*, 31, 484, 1984.
67. **Brown-Luedi, M. L., Meyer, L. J., Milburn, S. C., Yau, P. M.-P., Corbett, S., and Hershey, J. W. B.,** Protein synthesis initiation factors from human HeLa cells and rabbit reticulocytes are similar, *Biochemistry*, 21, 4202, 1982.
68. **Altman, M., Edery, I., Sonenberg, N., and Trachsel, H.,** Purification and characterization of protein synthesis initiation factor eIF-4E from the yeast *Saccharomyces cerevisiae*, *Biochemistry*, 24, 6085, 1985.
69. **Walton, G. M. and Gill, G. N.,** Nucleotide regulation of a eukaryotic protein synthesis initiation complex, *Biochim. Biophys. Acta*, 390, 231, 1975.
70. **Siekierka, J., Mitsui, K.-I., and Ocha, S.,** Mode of action of the heme-controlled translational inhibitor, *Proc. Natl. Acad. Sci. U.S.A.*, 78, 220, 1981.
71. **Matts, R. L., Levin, D. H., and London, I. M.,** Effect of phosphorylation of the α-subunit of eukaryotic initiation factor 2 on the function of reversing factor in the initiation of protein synthesis, *Proc. Natl. Acad. Sci. U.S.A.*, 80, 2559, 1983.
72. **Goss, D. J., Parkhurst, L. J., Mehta, H. B., Woodley, C. L., and Wahba, A. J.,** Studies on the role of eukaryotic nucleotide exchange factor in polypeptide chain initiation, *J. Biol. Chem.*, 259, 7374, 1984.
73. **Legon, S., Jackson, R. J., and Hunt, T.,** Control of protein synthesis in reticulocyte lysates by haemin, *Nature (London) New Biol.*, 241, 150, 1973.
74. **Farrell, P. J., Balkow, K., Hunt, T., Jackson, R. J., and Trachsel, H.,** Phosphorylation of initiation factor eIF-2 and the control of reticulocyte protein synthesis, *Cell*, 11, 187, 1977.
75. **Levin, D. and London, I. M.,** Regulation of protein synthesis: activation by double-stranded RNA of a protein kinase that phosphorylates eukaryotic initiation factor 2, *Proc. Natl. Acad. Sci. U.S.A.*, 75, 1121, 1978.
76. **Levin, D. H., Petryshyn, R., and London, I. M.,** Characterization of double-stranded-RNA-activated kinase that phosphorylates α-subunit of eukaryotic initiation factor 2, *Proc. Natl. Acad. Sci. U.S.A.* 77, 832, 1980.
77. **Ranu, R. S.,** Regulation of protein synthesis in rabbit reticulocyte lysates, *Biochem. Biophys. Res. Commun.*, 97, 252, 1980.
78. **O'Farrell, P. Z., Goodman, H. M., and O'Farrell, P. H.,** High resolution two-dimensional electrophoresis of basic as well as acidic proteins, *Cell*, 12, 1133, 1977.
79. **Lax, S. R., Osterhout, J. J., and Ravel, J. M.,** Factors from wheat germ that enhance the activity of eukaryotic initiation factor eIF-2, *J. Biol. Chem.*, 257, 8233, 1982.

80. **Osterhout, J. J., Lax, S. R., and Ravel, J. M.,** Isolation and characterization of Co-eIF-2α, *J. Biol. Chem.*, 258, 8285, 1983.
81. **Daum, III, H.D., Bragg, P. W., Sittman, D. B., Dholakia, J. N., Woodley, C. L., and Wahba, A. J.,** The expression of a gene for eukaryotic elongation factor Tu in *Artemia* during development, *J. Biol. Chem.*, 260, 16347, 1985.
82. **Bragg, P. W., Roth, W. W., and Wahba, A. J.,** The expression of a gene for eukaryotic elongation factor Tu in *Artemia*, in *Artemia Research and its Applications*, Vol. 2, Decleir, W., Moens, L., Slegers, H., and Sorgeloos, P., Eds., Universa Press, Wetteren, Belgium, 1987, 369.
83. **Oakley, B. R., Kirsch, D. R., and Morris, N. R.,** A simplified ultrasensitive silver stain for detecting proteins in polyacrylamide gels, *Anal. Biochem.*, 105, 361, 1980.
84. **Berger, S. L. and Berkenmeier, C. S.,** Inhibition of intractable nucleases with ribonucleoside-vanadyl complexes: isolation of messenger ribonucleic acid from resting lymphocytes, *Biochemistry*, 18, 5143, 1979.
85. **Perry, R., LaTorre, J., Kelley, D. E., and Greenberg, J. R.,** On the lability of poly(A) sequences during extraction of messenger RNA from polyribosomes, *Biochim. Biophys. Acta*, 262, 220, 1972.
86. **Bellamy, A. R. and Ralph, R. K.,** Recovery and purification of nucleic acids by means of cetyltrimethylammonium bromide, in *Methods in Enzymology*, Vol. 12B, Grossman, L. and Moldave, K., Eds., Academic Press, New York, 1968, 972.
87. **Aviv, H. and Leder, P.,** Purification of biologically active globin messenger RNA by chromatography on oligothymidylic acid-cellulose, *Proc. Natl. Acad Sci. U.S.A.*, 69, 1408, 1972.
88. **Bailey, J. M. and Davidson, N.,** Methylmercury as a reversible denaturing agent for agarose gel electrophoresis, *Anal. Biochem.*, 70, 75, 1976.
89. **Thomas, P. S.,** Hybridization of denatured RNA and small DNA fragments transferred to nitrocellulose, *Proc. Natl. Acad. Sci. U.S.A.*, 77, 5201, 1980.
90. **van Hemert, F. J., Amons, R., Pluijms, W. J. M., van Ormondt, H., and Moller, W.,** The primary structure of elongation factor EF-1α from the brine shrimp *Artemia*, *EMBO J.*, 2, 1109, 1984.
91. **Wreschner, D. H. and Herzberg, M.,** A new blotting medium for the simple isolation and identification of highly resolved messenger RNA, *Nucl. Acids Res.*, 12, 1349, 1984.
92. **Denhardt, D. T.,** A membrane-filter technique for the detection of complementary DNA, *Biochem. Biophys. Res. Commun.*, 23, 641, 1966.
93. **Arai, K.-I., Clark, B. F. C., Duffy, L., Jones, M. D., Kaziro, Y., Larsen, R. A., L'Italien, J., Miller, D. L., Nagarkatti, S., Nakamura, S., Nielsen, K. M., Peterson, T. C., Takahashi, K., and Wade, M.,** Primary structure of elongation factor Tu from *Escherichia coli*, *Proc. Natl. Acad. Sci. U.S.A.*, 77, 1326, 1980.
94. **Maniatis, T., Fritsch, E. F., and Sambrook, J.,** *Molecular Cloning; a Laboratory Manual*, Cold Spring Harbor Laboratory, Cold Spring Harbor, N.Y., 1982.
95. **Broker, T. R.,** Animal virus RNA processing, in *Processing of RNA*, Apirion, D., Ed., CRC Press, Boca Raton, Fla., 1984, 352.
96. **Wahba, J. A.,** unpublished observations.
97. **Roth, W. W., Bragg, P. W., Corrias, M. V., Reddy, N. S., Dholakia, J. N., and Wahba, A. J.,** Expression of a gene for mouse eukaryotic elongation factor Tu during murine erythroleukemic cell differentiation, *Mol. Cell. Biol.*, 7, 3929, 1987.

Chapter 4

PURINE NUCLEOTIDE METABOLISM IN *ARTEMIA*

Antonio Sillero and María A. Günther Sillero

TABLE OF CONTENTS

I. INTRODUCTION

The expression of different genes in an organism plays a key role in its development. The mechanisms controlling that expression are largely unknown and probably result from the interaction of multiple factors, one of which is the availability of the proper amount of nucleotides. Another example is afforded by the enzyme ribonucleotide-diphosphate reductase (EC 1.17.4.1) which is responsible for the synthesis of deoxyribonucleoside diphosphates. This enzyme is regulated allosterically by nucleoside triphosphates and it can be inhibited or switched toward the synthesis of one or another type of deoxynucleotide depending on the concentration and relative amount of other (deoxy)-nucleosidetriphosphates present in the cell.[1,2] The activity of this enzyme is also correlated with the rate of cell proliferation.[3] Still another example of the influence of nucleotides is provided by experiments showing that *Artemia* may have a different pattern of development depending on the relative amount of nucleotides present in the culture medium.[4] In spite of the apparent importance of nucleotides in embryos, little attention has been paid to the nucleotide content of developing systems. As with other molecules, nucleotides are simply supposed to be there, in the embryo, because they are the primary building blocks upon which DNA and RNA are synthesized.

Artemia presents two very unique characteristics: (1) the first stages of development from encysted gastrula to larva take place without DNA synthesis or cell division[5] and (2) embryos contain a large amount of diguanosine tetraphosphate (Gp_4G).

In this chapter we will deal with the metabolism of purine nucleotides, a topic in which Gp_4G occupies a central position. A review on the interconversion of purine nucleotides in *Artemia* was recently presented in the *Proceedings of the 2nd Int. Symp. on the Brine Shrimp Artemia*.[6]

II. OCCURRENCE OF DIGUANOSINE TETRAPHOSPHATE AND OTHER POLYPHOSPHATES IN *ARTEMIA*

Encysted gastrulae contain around 10 μmol of Gp_4G per 200,000 animals, which corresponds to a concentration of about 12 m*M*. The level of Gp_4G decrease to 25% of the original value after 3 days of development,[7,8] and in adult male animals this nucleotide is undetectable.[9] The situation is different in adult females. Gravid females accumulate Gp_4G exclusively in the developing ovarian eggs over a span of 3 to 4 days.[9] The mature oocytes then migrate to the ovisac where fertilization occurs. After ovoviviparous development to the gastrula stage, the encysted gastrulae, containing Gp_4G at the concentration stated above, emerge from the female which then become depleted of the nucleotide. The absence of Gp_4G in *Artemia* adults reared axenically (including males and immature females) has been confirmed recently in our laboratory using more sensitive techniques of high pressure liquid chromatography, in collaboration with A. Hernandorena (Laboratoire du Museum National d'Histoire Naturelle, Biarritz, France).

Besides Gp_4G, the following nucleotides have been reported in *Artemia* extracts: diguanosine triphosphate (Gp_3G),[10] diguanosine diphosphate (Gp_2G),[11] diadenosine tetraphosphate (Ap_4A),[12,13] diadenosine triphosphate (Ap_3A),[13] and guanosine 5′-triphosphate 5′-adenosine (Gp_3A).[11] Very little is known about the metabolism and function of these nucleotides in *Artemia*,[13,14] but all of the compounds are similar in that they are composed of dinucleoside structures with two or more inner phosphates inaccessible to alkaline phosphatase.[15] Several of these dinucleoside polyphosphates have been reported in relatively high concentrations in *Daphnia magna*,[16] *Branchipus stagnalis*,[17] *Salmonella typhimurium*,[18] human blood platelets,[19,20] Ehrlich ascites tumor cells,[21] *Physarum polycephalum*,[22] and *Drosophila*.[23]

The most studied of the dinucleoside compounds in *Artemia* and other systems is Ap_4A. It had been implicated in the onset of DNA replication,[12,24-28] in blood platelet function,[19,29] as an inhibitor of certain protein kinases,[30] and as a regulator in the interconversion of purine nucleotides.[31-33]

Since embryos of *Artemia* and other crustacea are unique in having a high concentration of dinucleoside polyphosphates,[6,7] they have served as good experimental systems for the study of these compounds.

III. METABOLISM OF DIGUANOSINE TETRAPHOSPHATE

A. Synthesis of Gp_4G

The synthesis of Gp_4G is carried out by the enzyme GTP:GTP guanylyltransferase (EC 2.7.7.45) isolated from a yolk platelet-enriched fraction of encysted embryos.[8,34,35] The enzyme uses 2 mol of GTP to synthesize 1 mol of Gp_4G and 1 mol of pyrophosphate in a freely reversible reaction. The K_m values for GTP, Gp_4G, and PPi are 2.2, 1.1, and 0.8 m*M*, respectively. This enzyme could also act in vivo catalyzing the pyrophosphorolysis of 1 mol of Gp_4G to 2 mol of GTP.[34,35] The enzyme also catalyzes, at a lower rate (8%), the synthesis of Gp_3G and GTP from equivalent amounts of GDP and Gp_4G. It requires Mg^{2+} and dithiothreitol for optimal activity. An estimated molecular weight of 490,000 was obtained for the enzyme by exclusion chromatography on Sepharose 6B.

B. Hydrolysis of Gp_4G

The nucleotide Gp_4G is cleaved to GTP and GMP by the cytosolic enzyme in *Artemia*, *bis*-(5′-guanosyl) tetraphosphatase (EC 3.6.1.17).[36,37] This enzyme has also been reported in a variety of other tissues and organisms including liver, kidney, muscle, brain, and intestinal mucosa of the rat;[38,39] mouse liver;[40] tumor cells;[41-43] and higher plants.[44] The enzyme cleaves Gp_4G, Ap_4A, Up_4U,[37] and Xp_4X[37] to the corresponding nucleosides tri- and monophosphates with similar V_{max} and K_m values.[37] In *Artemia* embryos the tetraphosphatase seems to be present in excess of the amount required to account for the hydrolysis of Gp_4G at the in vivo rate of about 2 nmol/min/g dry cysts.[45] However, the differential compartmentation of the enzyme and substrate in *Artemia*,[6] and the strong inhibition of the enzyme by nucleoside 5′-tetraphosphates must be considered in vivo.[42,45]

Diguanosinetetraphosphatase (another name of bis (5′-guanosyl) tetraphosphatase) does not cleave Gp_3G which is also present in *Artemia* embryos. The search for a dinucleoside triphosphate hydrolyzing activity in *Artemia* and in other sources led us to the identification of a specific dinucleosidetriphosphatase (EC 3.6.1.29) in rat liver.[15] Dinucleosidetriphosphatase and most probably the tetraphosphatase are located both in the cytosol and mitochondrial fraction of rat liver.[46] The occurrence of their substrates (Ap_3A, Ap_4A) has not been described in mitochondria. If present, an interesting question is whether Ap_4A and Ap_3A are involved in the control of some typical mitochondrial processes such as mitochondrial DNA replication or oxidative phosphorylation, and if they participate in the regulation of the adenylic charge.[47] Studies on the subcellular location of these enzymes during *Artemia* development have not been reported.

Dinucleoside tri- and tetraphosphates are hydrolyzed nonspecifically by phosphodiesterase I (EC 3.1.4.1) present in the plasma membrane of rat liver and other tissues.[48,49] The level of this enzyme is very low in *Artemia* cysts, as is usually the case for membrane-associated activities in cysts, that increase when morphogenesis starts.[50]

The decrease in the level of Gp_4G during development from the encysted gastrula to free swimming larva is accompanied by an increase in the total content of adenine.[7,51,52] These observations and others have resulted in arguments for three metabolic pathways for the synthesis of adenine nucleotides from Gp_4G in *Artemia*. These are (1) the conversion of Gp_4G to Ap_4G followed by hydrolysis of this compound to ATP and GMP;[53] (the enzymes responsible for the proposed purine exchange or conversion have not been described, while radiolabeling studies support this pathway[6]); (2) the pyrophosphorolysis of Gp_4G to 2 mol of GTP;[8,54] and (3) the hydrolysis of Gp_4G to GMP and GTP.[36,37] In (1) and (3) GMP is generated from Gp_4G and then AMP is formed by a pathway known to generate AMP. This latter pathway requires the sequential action of GMP reductase (EC 1.6.6.8), adenylosuccinate synthetase (EC 6.3.4.4), and adenylosuccinate lyase (EC 4.3.2.2).

IV. A SURVEY OF *ARTEMIA* ENZYMES INVOLVED IN PURINE NUCLEOTIDE METABOLISM

Information on the pathway for purine·nucleotide metabolism in developing systems is rather scanty, with fragmentary data on the corresponding enzymes, frequently obtained under different experimental conditions and in different biological systems. *Artemia* is one of the few examples in which a systematic study of some of the enzymes implicated in purine nucleotide metabolism has been undertaken. The main kinetic properties including K_m and V_{max} of the Artemia enzymes acting in the interconversion of purine nucleotides are summarized in Table 1. Data on the same enzymes from different sources have also been included in Table 1 for comparison.

A. GMP Reductase

The enzyme GMP reductase catalyzes the conversion of GMP to IMP using NADPH as cofactor. It was first described in microorganisms by Mager and Magasanik.[56] Its presence in human and rabbit erythrocytes was demonstrated by the ability of these cells or hemolysates therefrom to incorporate either guanosine or GMP into IMP.[92] The enzyme was first characterized in eukaryotes in the cytosolic fractions of calf thymus[93] then in *Artemia*.[94] The *Artemia* enzyme (0.05 U/g dry cysts) has apparent K_m values of 5 and 50 μM towards GMP, and xanthosine 5′-phosphate (XMP) is a very strong inhibitor (K_i = 10 nM) of the reaction. In the presence of 1.5 μM XMP hyperbolic kinetics were found with a K_m value of around 50 μM. Gp_4G activated the enzyme (K_a = 30 nM), concomitantly changing the kinetics of the enzyme to a hyperbolic form with a K_m of about 5 μM. Also, Gp_4G counteracted effectively the inhibition of the enzyme by XMP, while the structural analogs Ap_4A and Gp_3G had no measurable effect on the enzyme when tested at micromolar concentrations. GTP in millimolar amounts counteracted the inhibition of GMP reductase by XMP. The effect of Gp_4G on the enzyme is very specific;[31] other dinucleotides with either equal inner phosphates (Ap_4A) or the same guanine residues (Gp_3G) have no appreciable effect on the reductase at concentrations 1000-fold higher than Gp_4G.

GMP reductase is present at concentrations less than 0.001 U/g wet tissue in human erythrocytes and has properties similar to the *Artemia* enzyme.[55] GMP reductase is more active in Enterobacteriaceae, but the enzyme in these organisms has kinetic properties different from GMP reductase in eukaryotes. This enzyme is inhibited by ATP and displays hyperbolic kinetics with a higher K_m value (96 μM) for the substrate. In spite of being the only enzyme known to transform GMP into IMP (and hence AMP), GMP reductase from eukaryotic cells has, to our knowledge, been characterized only in *Artemia*,[31] calf thymus,[93] and erythrocytes.[55,92]

B. Adenylosuccinate Synthetase

The enzyme adenylosuccinate synthetase transforms IMP into adenylosuccinate using GTP and L-aspartate as substrates. Two enzymatically active forms of adenylosuccinate synthetase have been isolated from *Artemia* extracts by chromatography on DEAE-cellulose.[57] The isozymes, designated I and II, have very similar kinetic properties, but differ slightly in pI values (6.5 vs. 6.2) and molecular weight (90,000 vs. 83,000). GMP, GDP, Gp_4G, and fructose 1,6-bisphosphate inhibited both forms of the enzyme with similar K_i values.[57]

The presence of two isozymes of adenylosuccinate synthetase had been reported previously only in rat liver.[95] The liver isozymes (termed M and L) have only slight kinetic differences.

Table 1
KINETIC PARAMETERS OF ENZYMES ACTING IN THE PURINE NUCLEOTIDE METABOLISM OF SEVERAL BIOLOGICAL SYSTEMS

	GMP reductase			Adenylosuccinate synthetase					Adenylosuccinate lyase		
	V^amax			V_{max}		K_m(μ*M*)			V_{max}		
	U^b/g tissue	U/g protein	K_m (μ*M*) GMP	U/g tissue	U/g protein	IMP	GTP	Aspartate	U/g tissue	U/g protein	K_m (μ*M*) Adenylosuccinate
Artemia	0.05	0.7	5^c, 50^d(31)	0.04	0.5	23	30	650 (57)	0.2	3	1 (66)
Erythrocytes, human	0.0009	0.002	5^c, 8^d(55)	—	—	—	—	—	—	0.6	2 (67)
Kidney cortex, rat	—	—	—	0.02	0.3 (58)	—	—	—	0.12	1.5 (58)	—
Liver, chicken	—	—	—	—	—	—	—	—	—	—	—
Liver, rat	—	—	—		0.5	30	15	360 (59,60)	0.5	3.6	10 (68,69)
Placenta, human	—	—	—	—	—	37	31	950 (61)	—	—	—
Skeletal muscle, rat	—	—	—	0.53	12	380	700	250 (62)	0.5	10.4	13 (68,69)
Skeletal muscle, rabbit	—	—	—	0.15	4.5	200	10	300 (63)	—	—	—
Skeletal muscle, chicken	—	—	—	—	—	170	12	30 (60)	—	—	—
Yeast	—	—	—	—	50	200	20	1500 (64)	0.6	60	12 (70)
Ascites tumor cells	—	—	—	—	1.3	12	12	840 (59,60)	—	2.4	8.4 (71)
Hepatoma	—	—	—	—	—	—	—	—	—	—	—
Kidney tumor	—	—	—	0.07	1.0 (58)	—	—	—	0.2	2.6 (58)	—
Aerobacter aerogenes	—	—	—	—	—	—	—	—	—	—	—
Bacillus subtilis	—	—	—	—	—	—	—	—	—	—	—
Enterobacteriaceae	1.5	13	96 (56)	0.1	1.2	30	40	100 (60,65)	—	—	—

Table 1 (continued)
KINETIC PARAMETERS OF ENZYMES ACTING IN THE PURINE NUCLEOTIDE METABOLISM OF SEVERAL BIOLOGICAL SYSTEMS

	IMP dehydrogenase				Cytosolic 5′-nucleotidase			Adenosine deaminase			AMP deaminase		
	V_{max}		$K_m(\mu M)$		V_{max}			V_{max}			V_{max}		
	U/g tissue	U/g protein	IMP	NAD⁺	U/g tissue	U/g protein	$K_m(\mu M)$ IMP	U/g tissue	U/g protein	$K_m(\mu M)$ Adenosine	U/g tissue	U/g protein	$K_m(\mu M)$ AMP
Artemia	0.005	0.089	15	200 (72)	0.07	2.2	2800 (78)	$0.07^e/0.5^f$	$2.5^e/38^f$	40 (83)	<0.001 (6)	—	—
Erythrocytes, human	—	—	—	—	—	—	—	—	0.2	25 (84)	—	9	1,000 (87)
Kidney cortex, rat	0.002	0.03 (58)	—	—	—	—	—	7.7	101 (58)	—	3.3	44 (58)	—
Liver, chicken	—	—	—	—	1.85	20	1400 (79,80)	—	—	—	—	—	—
Liver, rat	0.003	0.035	11	24 (73)	0.76	6	1200 (81,82)	—	14	26 (85)	5	50 (88,89)	—
Placenta, human	—	—	—	—	—	—	—	—	—	—	—	—	—
Skeletal muscle, rat	—	—	—	—	—	—	—	—	—	—	130	3300	450 (90)
Skeletal muscle, rabbit	—	—	—	—	—	—	—	—	—	—	15	380	600 (91)
Skeletal muscle, chicken	—	—	—	—	—	—	—	—	—	—	—	—	—
Yeast	—	—	—	—	—	—	—	—	—	—	—	—	—
Ascites tumor cells	—	—	—	—	—	—	—	—	—	—	—	—	—
Hepatoma	0.04	0.4	11	24 (73)	—	—	—	—	43	28 (85)	—	—	—
Kidney tumor	0.006	0.08 (58)	—	—	—	—	—	—	1.5	21 (58)	5	80 (58)	
Aerobacter aerogenes	1.6	31	21	1100 (74)	—	—	—	—	—	—	—	—	—
Bacillus subtilis	0.1	2 (75)	23	200 (76)	—	—	—	—	—	—	—	—	—
Enterobacteriaceae	10	190	11	300 (77)	—	—	—	19	142	75 (86)	—	—	—

Note: Numbers in parentheses are Reference numbers.

[a] Maximum velocity.
[b] One unit (U) is the amount of enzyme to transform one μmol of substrate per min.
[c] In the presence of diguanosine tetraphosphate.
[d] In the absence of diguanosine tetraphosphate.
[e] Cysts.
[f] Larvae.

Matsuda et al.[95] showed that isozyme L is important in the regulation of purine nucleotide synthesis, whereas isozyme M participates in the regulation of glycolysis. Compared to rat liver isozymes, the physiological significance of *Artemia* adenylosuccinate synthetase isoforms is not clear since their kinetic properties are so similar. Adenylosuccinate synthetase has been studied from human placenta,[61] rat[62,96] and rabbit muscle,[63] and rat liver.[95] The enzyme from *Artemia* seems to be most similar to the enzyme from human placenta as both have relatively low and high K_m values for IMP and aspartate, respectively. Maximum velocities of the enzyme from different sources are similar (Table 1).

C. Adenylosuccinate Lyase

The enzyme adenylosuccinate lyase splits adenylosuccinate into fumarate and AMP and represents the last step in the conversion of Gp_4G to AMP. The enzyme from *Artemia* cysts[66] has marked differences in molecular mass depending on the pH of the medium. Forms from 200 kdaltons to greater than 500 kdaltons were detected after chromatography on Sepharose CL-6B using elution buffers of pH 8.7 and 6.5, respectively. The existence of other forms was observed at intermediate pH values. The 200-kdalton form obtained by chromatography at pH 8.7 did not revert to a larger size upon rechromatography at pH 6.6. However, the higher molecular mass form obtained at pH 7.1 reverted to 200 kdaltons after rechromatography at pH 8.7. These observations could mean that, among several possibilities, the 200-kdalton form is the more stable form of the enzyme or, upon chromatography on Sepharose CL-6B, a factor is lost which is essential for the conversion of the 200-kdalton form to higher molecular forms. Such conversion was observed in the 27,000 $\times$ *g* supernatants of *Artemia* embryos.[66] It is noteworthy that the greatest change in the molecular forms of the enzyme occurs in the pH range 7 to 7.5. pH dependent changes in molecular forms have not been described for adenylosuccinate lyase from other sources. Recently Hand and Carpenter[97] showed that trehalase (EC 3.2.1.28) isolated from *Artemia* embryos exists in two active forms that interconvert when exposed to physiological transitions in pH.

Maximum velocity of adenylosuccinate lyase from *Artemia* cysts is 0.2 U/g (Table 1). The purified enzyme has a molecular mass of 200 kdaltons and a single subunit of 56 kdaltons as revealed by electrophoresis on sodium dodecyl sulfate polyacrylamide gel. The 200-kdalton form may represent a tetramer, while larger molecular forms could be due to association of monomers with or without participation of a factor essential for the association.

The occurrence of 150, 100, or 55-kdaltons forms, in native conditions has not been detected because they are either unstable or inactive.[66] Adenylosuccinate lyase catalyzes the elimination of fumarate from both adenylosuccinate and 5-amino-4-imidazole *N*-succinocarboxamide ribonucleotide,[98] the last compound being an intermediate of the *de novo* pathway of purine ribonucleotides.

D. IMP Dehydrogenase

The enzyme IMP dehydrogenase (EC 1.1.1.205) catalyzes the conversion of IMP to XMP using NAD^+ as cofactor. As for adenylosuccinate lyase,[66] IMP dehydrogenase occurs in different molecular forms in *Artemia*.[72] The enzyme from the high speed supernatant of cell-free extracts has a molecular size of around 450 kdaltons, whereas the fraction obtained between 30 to 40% ammonium sulfate saturation shows sizes ranging from around 450 kdaltons to more than 1000 kdaltons as determined by gel filtration on Sepharose 4B. In the presence of 2 *M* NaCl a single form with a molecular weight around 260 kdaltons was obtained. This anomalous behavior could be due to ionic interactions between enzyme molecules or with other macromolecules, probably enhanced by sequestration of water molecules from the microenvironment of the protein by ammonium sulfate. Depending on the experimental conditions, molecular forms of 130, 220, 260, 375, 450, and more than 1000 kdaltons have been detected in *Artemia* IMP dehydrogenase. The purified enzyme has a molecular weight of 260 kdaltons. Sodium dodecyl sulfate polyacrylamide gel electrophoresis of this preparation showed one major band of 55 kdaltons. The occurrence of different molecular forms of the native enzyme could be due to the tendency of the monomer to form aggregates. The activity of the enzyme is around 0.005 U/g dry cysts (Table 1).

The maximum velocity of IMP dehydrogenase is greater in tumors than in normal tissues.[58,73] As for other enzymes of purine nucleotide interconversion, IMP dehydrogenase is more abundant in prokaryotes than it is in eukaryotes (Table 1).

E. Cytosolic 5′-Nucleotidase

Artemia cysts contain a nucleotidase[78] similar to the cytosolic 5′-nucleotidase (EC 3.1.3.5) previously reported in rat liver,[81,82] chicken liver,[79,80] and human malignant lymphocytes.[99] The enzyme from all sources cleaves preferentially IMP and GMP to their corresponding nucleosides; nucleoside 2′- or 3′-phosphates are not substrates of the enzyme. The rate of hydrolysis of IMP (0.5 m*M*) by cytosolic 5′-nucleotidase from *Artemia* embryos was increased several-fold by 10 μ*M* Ap_4A.[33] Half-maximal activation of the enzyme was achieved with 5 μ*M* Ap_4A. The K_m ($S_{0.5}$) values of the nucleotidase for IMP, GMP, AMP, XMP, and CMP decreased about ten-fold in the presence of 10 μ*M* Ap_4A. The maximum velocity of the enzyme was not affected by Ap_4A. ATP had been previously described as an activator of the enzyme. However, compared with Ap_4A, concentrations of ATP two orders of magnitude higher are needed to elicit similar effects on the enzyme.[33]

F. Adenosine Deaminase

The specific activity of adenosine deaminase in the 150,000 × *g* supernatant fraction increases 15-fold during the first 45 hr of development.[83] Gel filtration of the above supernatant fractions on a Sephacryl S-300 column showed the presence of two different molecular forms of the enzyme at 30 and 135 kdaltons. The smallest form of the enzyme is present in cysts and larvae and consists of 3 isozymes with pI values of 4.9, 5.0, and 5.2. Their total activity decreases threefold during the first 45 hr of incubation. The larger molecular weight form of the enzyme has a pI value of 4.2 and is undetectable in cysts, reaching a value of 0.5 U/g of wet tissue after 45 hr of development. The amount present in cysts is accounted for only by the smallest form of the deaminase (Table 1).

In some biological systems and tissues adenosine deaminase exists exclusively as a small molecular mass form of 35 kdaltons[84,86] showing different degrees of polymorphism.[100] In other tissues a larger molecular form of the enzyme predominates.[101] The small form of adenosine deaminase associates with an adenosine deaminase-binding protein, a glycoprotein with a native molecular mass of 200 kdaltons in a molar ratio of 2:1, to produce the large form of the enzyme.[102] Our attempts to find an adenosine deaminase-binding protein in cysts and larvae of *Artemia* have been unsuccessful.

The occurrence of adenosine deaminase in *Artemia* may explain why the quantity of ammonia liberation in adults is higher when the dietary requirements are met by AMP than by either IMP or GMP.[103] These nucleotides are probably transported inside the cell as the corresponding nucleosides, and the deaminase may liberate ammonia as the first step in the utilization of adenosine.

G. Other Enzymes

Reports of partial purification of adenine phosphoribosyl transferase (EC 2.4.2.7), hypoxanthine phosphoribosyl transferase (EC 2.4.2.8), and adenosine kinase (EC 2.7.1.20) have appeared recently.[104,105] The specific activities of these enzymes in cysts are around 7, 0.5, and 0.1 mU/mg of protein, respectively.[104] The K_m values of adenine phosphoribosyl transferase for adenine and phosphoribosyl pyrophosphate are 4 and 20 μ*M*, respectively, with AMP acting as a competitive inhibitor (K_i = 0.1 m*M*) towards phosphoribosyl pyrophosphate.[105] AMP deaminase (EC 3.5.4.6) activity is very low in extracts of *Artemia* cysts (less than 0.001 U/g tissue) (Table 1),[6] and it is activated by Ap_4A. The kinetic characterization of the activation by Ap_4A has been explored with the enzyme from rat muscle. The enzyme from this source showed kinetics suggestive of positive cooperativity or Michaelian in the presence or absence of Ap_4A, respectively, [32] Ap_4A doubled the activity of AMP deaminase from rat muscle in the presence of 0.05 m*M* AMP with an activation constant of 0.005 m*M*. Other dinucleoside polyphosphates (Gp_4G, Gp_3G, Ap_3A) were not activators of the reaction.[32] The activation of AMP deaminase by Ap_4A appears to favor the transformation of AMP into IMP and hence into guanine nucleotides.

V. RELATIVE CONTENT OF ADENINE AND GUANINE NUCLEOTIDES IN *ARTEMIA*

In various cell types most nucleotides participate in similar reactions as transference coenzymes, regulatory metabolites, and building blocks for the synthesis of DNA and RNA. The purine nucleotides, adenosine and guanosine 5′-mono-, di-, and triphosphates, are mainly distributed in the cytosolic and mitochondrial pools. Except in the case of very specialized cells, the relative content of these nucleotides is similar in the majority of cell types. However in this regard, the nucleotide pool of *Artemia* cysts differs from that of rat liver, a typical and well-characterized tissue. Therefore, it seems of interest to compare both biological systems.

The reported nucleotide concentrations in liver vary slightly depending on the method used,[106-110] but the approximate values measured in μmol/g wet weight are ATP 1.8; ADP 0.6; AMP 0.3; GTP 0.2; GDP 0.06; and GMP 0.05. The adenylic charge calculated as ATP + $^1/_2$ADP/ATP + ADP + AMP is 0.78 and the corresponding value for guanine-containing nucleotides is 0.74. The ratios ATP/ADP and GTP/GDP are 3.0 and 3.3, respectively, and the ratio ATP + ADP + AMP/GTP + GDP + GMP is 9. The percent distributions between the cytosolic and mitochondrial fractions are respectively: ATP (75, 25); ADP (55, 45); AMP (55, 45); GTP (80, 20); GDP (55, 45); and GMP (55, 45). ATP and GTP are present mainly in the cytosol, whereas the other nucleotides are evenly distributed between both compartments. Thus, the ratios ATP/ADP and GTP/GDP are different in the cytosol (4.0 and 4.8) and mitochondrial (1.7 and 1.5) compartments.

The main reactions involved in the interchange and transphosphorylation of purine nucleotides in most tissues and in *Artemia* are presented schematically in Figure 1. The nucleotides are near equilibrium due to the presence of an array of high maximum velocity enzymatic reactions, the most abundant of which are nucleoside monophosphate kinase (EC 2.7.4.4), nucleoside diphosphate kinase (EC 2.7.4.6), and adenylate kinase (EC 2.7.4.3). Guanylate kinase (EC 2.7.4.8) and nucleosidetriphosphate-adenylate kinase (EC 2.7.4.10) also participate in this interchange. The subcellular distribution of the enzymes is not indicated in Figure 1 because some of the enzymes have different locations in different species.[111] Additionally, only fragmentary information is available for *Artemia*. Although not pictured in Figure 1 there are numerous other reactions driven by ATP and GTP consuming high energy phosphate bonds. This energy is restored by oxidative phosphorylation which implies the synthesis of ATP from ADP and the translocation of ATP from the mitochondrial matrix space to the cytosol. As soon as these processes are inhibited, the ratio of ATP/ADP in the cytosol decreases. The transphosphorylation between adenine and guanine nucleotides does not imply changes in the concentration of the adenine and guanine pools. For example, incubation of rat hepatocytes for 30 min with 50 μ*M* carbonyl cyanide *m*-chlorophenyl hydrazone (C1-CCP) resulted in a ratio ATP/ADP of 0.24 (from a value of 5.64) without an appreciable change in the ratio of adenine to guanine nucleotides.[108] This probably occurs because the enzymes involved in the interconversion of adenine and guanine nucleotides are

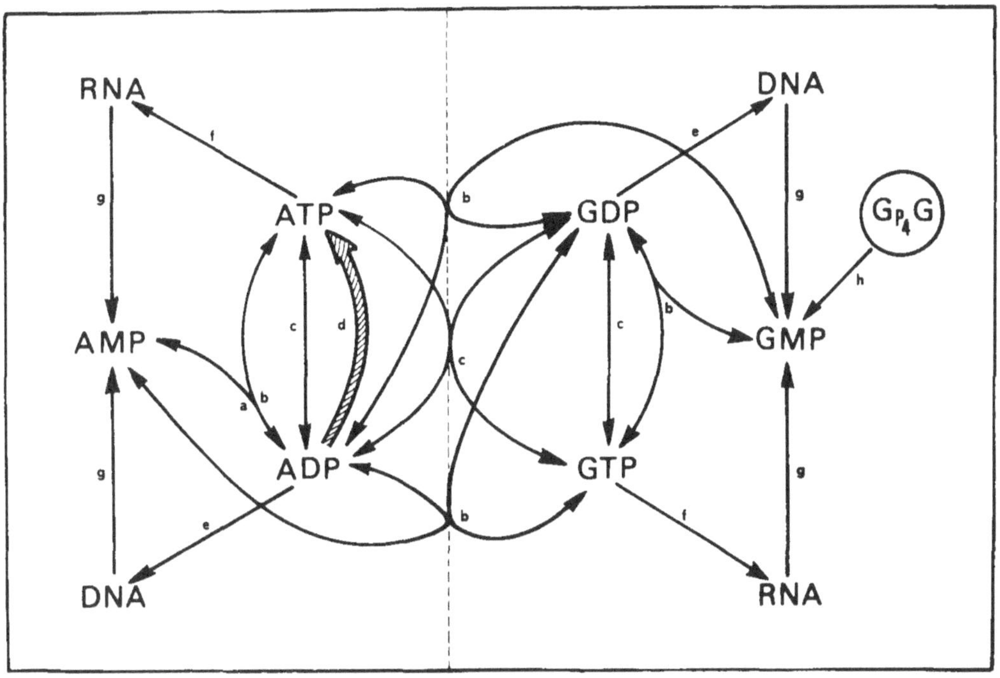

FIGURE 1. Purine nucleotide transphosphorylation. (a) Adenylate kinase (EC 2.7.4.3); (b) nucleoside-phosphate kinase (EC 2.7.4.4); (c) nucleoside diphosphate kinase (EC 2.7.4.6); (d) oxidative phosphorylation; (e) ribonucleoside-diphosphate reductase (EC 1.17.4.1), nucleoside-diphosphate kinase and DNA polymerase (EC 2.7.7.7); (f) RNA polymerase (EC 2.7.7.6); (g) Several hydrolytic enzymes plus purine salvage pathway; (h) *bis*-(5'-guanosyl)-tetraphosphatase (EC 3.6.1.17). The dashed line separates the adenine and guanine nucleotide pools. The oxidative phosphorylation process is indicated by a thick arrow to singularize it as the main mechanism for the generation of phosphoanhydride bonds of ATP.

present at much lower concentration and hence operate at a lower rate than those involved in transphosphorylation reactions between nucleotides.

The level of purine nucleotides in *Artemia* has been determined at different stages of development.[51,52] The ratio of ATP/ADP is 0.4 in cysts. It increases slightly after cysts hydratation and reaches a plateau value of 5 in the first 10 hr of development. This increase is attributable to the increase in mitochondrial activity.[112-115] The ratio ATP + ADP + AMP/GTP + GDP + GMP is 0.19 in the cysts, and increases steadily during development, reaching values of 0.5 and 2.5 after 24 and 80 hr of development.[51,52,116] These higher ratios in larvae are due to an increase in the adenine nucleotides and a decrease in the guanine nucleotides, which implies a shift within the pools. The mechanism of this interchange is more obscure than the mechanism for transphosphorylation among nucleotides. The main pathways involved in the transfer between adenine and guanine nucleotides are shown in Figure 2.

The access to the guanine pool (dashed box at the right of Figure 2) proceeds exclusively through the sequential action of IMP dehydrogenase and GMP synthetase (EC 6.3.5.2). Another important source of guanine nucleotides in *Artemia* embryos is stored in the form of diguanosine tri- and tetraphosphates. Two possibilities of transfer of guanine to the adenine pool can be considered in *Artemia*. One is dependent on the enzyme GMP reductase[31] and the other, postulated by Van Denbos and Finamore[53] (not represented in Figure 2) predicts that ATP is generated from Gp_4G by an unknown mechanism.[6,53] The enzyme guanine deaminase (EC 3:5.4.3) which catalyzes the transformation of guanine into xanthosine in most tissues as part of a degradative pathway may also contribute to maintenance of the

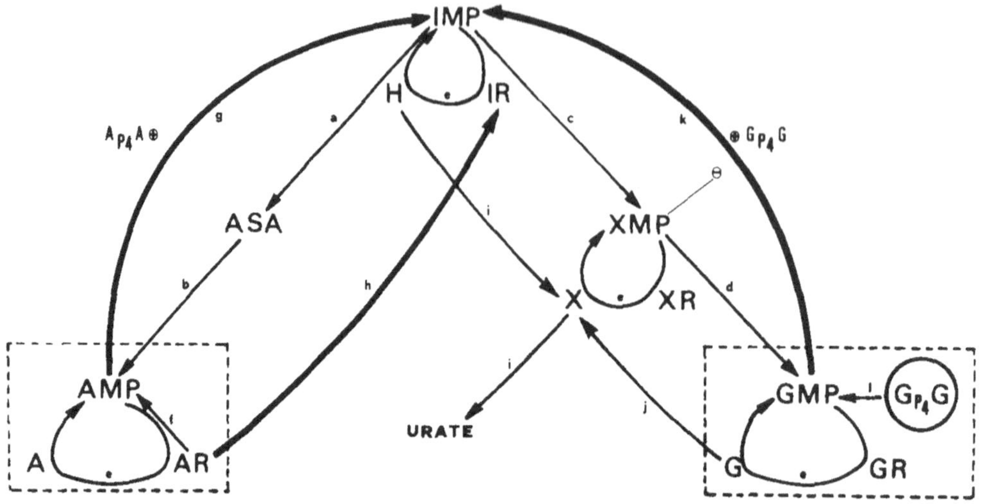

FIGURE 2. Interconversion of purine nucleotides. (a) Adenylosuccinate (ASA) synthetase (EC 6.3.4.4); (b) adenylosuccinate lyase (EC 4.3.2.2); (c) IMP dehydrogenase (EC 1.1.1.205); (d) GMP synthetase (EC 6.3.5.2); (e) purine salvage pathway; (f) adenosine kinase (EC 2.7.1.20); (g) AMP deaminase (EC 3.5.4.6); (h) adenosine deaminase (EC 3.5.4.4); (i) xanthine oxidase (EC 1.1.3.22); (j) guanine deaminase (EC 3.5.4.3); (k) GMP reductase (EC 1.6.6.8); (l) *bis*-(5′-guanosyl)-tetraphosphatase (EC 3.6.1.17). Boxes in dashed lines indicate the pools of adenine and guanine nucleotides. Thick arrows show the main pathways for the exit of the adenine and guanine moieties from their respective pools.

proper level of the guanine nucleotide pool. However, its occurrence in *Artemia* has not been explored.

The access to the adenine pool (dashed box at the left of Figure 2) proceeds through the sequential action of adenylosuccinate synthetase[57] and adenylosuccinate lyase.[66] There are at least two enzymes involved in the transformation of adenine into guanine nucleotides: adenosine deaminase[83] and AMP deaminase. Whereas the level of the first enzyme is in the range of 0.07 to 0.5 U/g of tissue (Table 1), the activity of AMP deaminase is barely detectable in *Artemia* cysts (less than 0.001 U/g). The occurrence of adenine deaminase (EC 3.5.4.2) and ADP deaminase (EC 3.5.4.7) activities (not represented in Figure 2) have also been described in some biological systems. Both enzymes may contribute to the transfer of adenine to guanine nucleotides. It should be noted that no enzyme has been described which is able to catalyze the direct interchange between adenine and guanine nucleotides such as that proposed by Van Denbos and Finamore.[53] As far as it is known, such an interchange takes place through the pool of the hypoxanthine nucleotides.

In summary, *Artemia* embryos may represent an example in which a singular transfer between the guanine and adenine pool occurs during a certain period in a developing organism. The control of these complex pathways, in which Ap_4A and Gp_4G participate, are not well known. Ap_4A has at least a dual role on the exit of the adenine moiety from its pool: it activates AMP deaminase and cytosolic 5′-nucleotidase. Its effect on the last enzyme tends to enhance the level of adenosine and inosine. Gp_4G also facilitates the transfer of guanine to the adenine pool through activation of the GMP reductase (Figure 2). Ap_4A and Gp_4G appear to act as signal molecules to maintain the appropriate levels of the pools of adenine and guanine nucleotides, respectively.

ACKNOWLEDGMENTS

This work was supported by grants from the Comisión Asesora de Investigación Científica y Técnica. We thank Daniel Sillero Günther for his recommendations and Pablo Lozano for typing the manuscript.

REFERENCES

1. **Thelander, L. and Reichard, P.**, Reduction of ribonucleotides, in *Annu. Rev. Biochem.*, 48, 133, 1979.
2. **Thelander, L., Eriksson, S., and Akerman, M.**, Ribonucleotide reductase from calf thymus, *J. Biol. Chem.*, 255, 7426, 1980.
3. **Weber, G., Olah, E., Lui, M. S., Kizaki, H., Tzeng, D.Y., and Takeda, E.**, Biochemical commitment to replication in cancer cells, in *Adv. Enzyme Regul.*, 18, 3, 1980.
4. **Hernandorena, A.**, Relationships between purine and pyrimidine dietary requirements and *Artemia salina* morphogenesis, *Comp. Biochem. Physiol.*, 62B, 7, 1979.
5. **Nakanishi, Y. H., Iwasaki, T., Okigaki, T., and Kato, H.**, Cytological studies of *Artemia salina*. I. Embryonic development without cell multiplication after the blastula stage in encysted dry eggs, *Annot. Zool. Jpn.* 35, 223, 1962.
6. **Sillero, A. and Sillero, M. A. G.**, Interconversion of purine nucleotides in *Artemia:* a review, in *Artemia Research and its Applications,* Vol. 2, Decleir, W., Moens, L., Slegers, H., Sorgeloos, P., and Jaspers, E., Eds., Universa Press, Wetteren, Belgium, 1987, 541.
7. **Finamore, F. J. and Clegg, J. S.**, Biochemical aspects of morphogenesis in the brine shrimp, *Artemia salina*, in *The Cell Cycle. Gene-Enzyme Interactions,* Padilla, G. M., Whitson, G. L., and Cameron, I. L., Eds., Academic Press, New York, 1969, chap. 12.
8. **Warner, A. H.**, The biosynthesis, metabolism and function of dinucleoside polyphosphates in *Artemia* embryos: a compendium, in *The Brine Shrimp Artemia*, Vol. 2, Persoone, G., Sorgeloos, P., Roels, O., and Jaspers, E., Eds., Universa Press, Wetteren, Belgium, 1980, 636.
9. **Warner, A. H. and McClean, D. K.**, Studies on the biosynthesis and role of diguanosine tetraphosphate during growth and development of *Artemia salina, Dev. Biol.*, 18, 278, 1968.
10. **Warner, A. H. and Finamore, F. J.**, Isolation, purification and characterization of P^1, P^3-diguanosine 5′-triphosphate from brine shrimp eggs, *Biochim. Biophys. Acta*, 108, 525, 1965.
11. **Gilmour, S. J. and Warner, A. H.**, The presence of guanosine 5′-diphospho-5′-guanosine and guanosine 5′-triphospho-5′-adenosine in brine shrimp embryos, *J. Biol. Chem.*, 253, 4960, 1978.
12. **McLennan, A. G. and Prescott, M.**, Diadenosine 5′, 5‴-P^1, P^4-tetraphosphate in developing embryos of *Artemia, Nucl. Acids Res.*, 12, 1609, 1984,
13. **Miller, D. and McLennan, A. G.**, Changes in intracellular levels of Ap_3A and Ap_4A in cysts and larvae of *Artemia* do not correlate with changes in protein synthesis after heat-shock, *Nucl. Acids Res.*, 14, 6031, 1986.
14. **Sillero, A., Cameselle. J. C., and Sillero, M. A. G.**, Dinucleosido-polifosfatos, in *Bioquímica y Biología Molecular,* Leloir, L., Ochoa, S., Oró, J., and Sols, A., Eds., Salvat, Barcelona, Spain, 1986, 586.
15. **Sillero, M. A. G., Villalba, R., Moreno, A., Quintanilla, M., Lobatón, C. D., and Sillero, A.**, Dinucleosidetriphosphatase from rat liver, *Eur. J. Biochem.*, 76, 331, 1977.
16. **Oikawa, T. G. and Smith, M.**, Nucleotides in the encysted embryos of *Daphnia magna, Biochemistry*, 5, 1517, 1966.
17. **Zagalsky, P. F. and Gilchrist, B. M.**, Isolation of a blue canthaxanthin lipovitellin from the yolk platelets of *Branchipus stagnalis* (L.) (Crustacea: anostraca), *Comp. Biochem. Physiol.*, 55B, 195, 1976.
18. **Lee, P. C., Bochner, B. R., and Ames, B. N.**, Diadenosine 5′, 5‴-P^1, P^4-tetraphosphate and related adenylylated nucleotides in *Salmonella typhimurium*, *J. Biol. Chem.*, 258, 6827, 1983.
19. **Flodgaard, H. and Klenow, H.**, Abundant amounts of diadenosine 5′, 5‴-P^1, P^4-tetraphosphate are present and releasable, but metabolically inactive, in human platelets, *Biochem. J.*, 208, 737, 1982.
20. **Lüthje, J. and Ogilvie, A.**, The presence of diadenosine 5′, 5‴-P^1,P^3-triphosphate (AP_3A) in human platelets, *Biochem. Biophys. Res. Commun.*, 115, 253, 1983.
21. **Ogilvie, A. and Jakob, P.**, Diadenosine 5′, 5‴-P^1, P^3-triphosphate in eukaryotic cells: identification and quantitation, *Anal. Biochem.*, 134, 382, 1983.
22. **Garrison, P. N. and Barnes, L. D.**, Assay of adenosine 5′-P^1-tetraphospho-P^4-5‴-adenosine and adenosine 5′-P^1-tetraphospho-P^4-5‴-guanosine in *Physarum polycephalum* and other eukaryotes, *Biochem. J.*, 217, 805, 1984.
23. **Brevet, A., Plateau, P., Best-Belpomme, M., and Blanquet, S.**, Variation of Ap_4A and other dinucleoside polyphosphates in stressed *Drosophila* cells, *J. Biol. Chem.*, 260, 15566, 1985.
24. **Grummt, F.**, Diadenosine 5′, 5‴-P^1,P^4-tetraphosphate triggers initiation of *in vitro* DNA replication in baby hamster kidney cells, *Proc. Natl. Acad. Sci. U.S.A.*, 75, 371, 1978.
25. **Zamecnik, P. C.**, Diadenosine 5′, 5‴-P^1,P^4-tetraphosphate (Ap_4A): its role in cellular metabolism, *Anal. Biochem.*, 134, 1, 1983.
26. **Probst, H., Hamprecht, K., and Gekeler, V.**, Replicon initiation frequency and intracellular levels of ATP, ADP, AMP and of diadenosine 5′, 5‴-P^1,P^4-tetraphosphate in Ehrlich ascites cells cultured aerobically and anaerobically, *Biochem. Biophys. Res. Commun.*, 110, 688, 1983.
27. **Weinmann-Dorsch, C., Hedi, A., Grummt, I., Albert, W., Ferdinand, F.-J., Friis, R. R., Pierron, G., Moll, W., and Grummt, F.**, Drastic rise of intracellular adenosine (5′) tetraphospho(5′) adenosine correlates with onset of DNA synthesis in eukaryotic cells, *Eur. J. Biochem.*, 138, 179, 1984.

28. **Zourgui, L., Tharaud, D., Solari, A., Litvak, S., and Tarrago-Litvak, L.**, Stimulation of DNA synthesis by microinjection of diadenosine 5′,5‴-P^1,P^4-tetraphosphate (Ap_4A) into *Xenopus laevis* oocytes, *Dev. Biol.*, 103, 409, 1984.
29. **Lüthje, J. and Ogilvie, A.**, Diadenosine triphosphate (Ap_3A) mediates human platelet aggregation by liberation of ADP, *Biochem. Biophys. Res. Commun.*, 118, 704, 1984.
30. **Maness, P. F., Perry, M. E., and Levy, B. T.**, P^1,P^4-Di(adenosine-5′) tetraphosphate inhibits phosphorylation of immunoglobulin G by Rous sarcoma virus pp60, *J. Biol. Chem.*, 258, 4055, 1983.
31. **Renart, M. F., Renart, J., Sillero, M. A. G., and Sillero, A.**, Guanosine monophosphate reductase from *Artemia salina:* inhibition by xanthosine monophosphate and activation by diguanosine tetraphosphate, *Biochemistry*, 15, 4962, 1976.
32. **Fernández, A., Costas, M. J., Sillero, M. A. G., and Sillero, A.**, Diadenosine tetraphosphate activates AMP deaminase from rat muscle, *Biochem. Biophys. Res. Commun.*, 121, 155, 1984.
33. **Pinto, R. M., Canales, J., Sillero, M. A. G., and Sillero, A.**, Diadenosine tetraphosphate activates cytosol 5′-nucleotidase, *Biochem. Biophys. Res. Commun.*, 138, 261, 1986.
34. **Warner, A. H., Beers, P. C., and Huang, F. L.**, Biosynthesis of the diguanosine nucleotides. I. Purification and properties of an enzyme from yolk platelets of brine shrimp embryos, *Can. J. Biochem.*, 52, 231, 1974.
35. **Warner, A. H. and Huang, F. L.**, Biosynthesis of the diguanosine nucleotides. II. Mechanism of action of GTP: GTP guanylyltransferase on nucleotide metabolism in brine shrimp embryos, *Can. J. Biochem.*, 52, 241, 1974.
36. **Warner, A. H. and Finamore, F. J.**, Isolation, purification, and characterization of P^1,P^4- asymmetrical-pyrophosphohydrolase from brine shrimpeggs, *Biochemistry*, 4, 1568, 1965.
37. **Vallejo, C. G., Lobatón, C. D., Quintanilla, M., Sillero, A., and Sillero, M. A. G.**, Dinucleoside-tetraphosphatase in rat liver and *Artemia salina*, *Biochim. Biophys. Acta*, 438, 304, 1976.
38. **Lobatón, C. D., Vallejo, C. G., Sillero, A., and Sillero, M. A. G.**, Diguanosinetetraphosphatase from rat liver: activity on diadenosine tetraphosphate and inhibition by adenosine tetraphosphate, *Eur. J. Biochem.*, 50, 495, 1975.
39. **Cameselle, J. C., Costas, M. J., Sillero, M. A. G., and Sillero, A.**, Bis-(5′-guanosyl) tetraphosphatase in rat tissues, *Biochem. J.*, 201, 405, 1982.
40. **Höhn, M., Albert, W., and Grummt, F.**, Diadenosine tetraphosphate hydrolase from mouse liver, *J. Biol. Chem.*, 257, 3003, 1982.
41. **Ogilvie, A.**, Determination of diadenosine tetraphosphate (Ap_4A) levels in subpicomole quantities by a phosphodiesterase luciferin-luciferase coupled assay: application as a specific assay for diadenosine tetraphosphatase, *Anal. Biochem.*, 115, 302, 1981.
42. **Moreno, A., Lobatón, C. D., Sillero, M. A. G., and Sillero, A.**, Dinucleosidetetraphosphatase from Ehrlich ascites tumour cells: inhibition by adenosine, guanosine and uridine 5′-tetraphosphates, *Int. J. Biochem.*, 14, 629, 1982.
43. **Ogilvie, A. and Antl, W.**, Diadenosine teraphosphatase from human leukemia cells, *J. Biol. Chem.*, 258, 4105, 1983.
44. **Jakubowski, H. and Guranowski, A.**, Enzymes hydrolyzing ApppA and/or AppppA in higher plants, *J. Biol. Chem.*, 258, 9982, 1983.
45. **Vallejo, C. G., Sillero, M. A. G., and Sillero, A.**, Diguanosinetetraphosphate guanylohydrolase in *Artemia salina*, *Biochim. Biophys. Acta*, 358, 117, 1974.
46. **Costas, M. J., Cameselle, J. C., and Sillero, A.**, Mitochondrial location of rat liver dinucleoside triphosphatase, *J. Biol. Chem.*, 261, 2064, 1986.
47. **Atkinson, D. E.**, The energy charge of the adenylate pool as a regulatory parameter. Interaction with feedback modifiers, *Biochemistry*, 7, 4030, 1968.
48. **Cameselle, J. C., Costas, M. J., Sillero, M. A. G., and Sillero, A.**, Two low *Km* hydrolytic activities on dinucleoside 5′,5‴-P^1,P^4-tetraphosphates in rat liver, *J. Biol. Chem.*, 259, 2879, 1984.
49. **Lüthje, J. and Ogilvie, A.**, Catabolism of Ap_3A and Ap_4A in human plasma, *Eur. J. Biochem.*, 149, 119, 1985.
50. **Sillero, A., Vallejo, C. G., Renart, J., Renart, M. F., Lobatón, M. C. D., Quintanilla, M., Moreno, A., Olalla, A., Villalba, C. R., and Sillero, M. A. G.**, *Artemia* como modelo en estudios de diferenciación y desarrollo, in *Abstracts of the First Congr. Spanish Fed. of the Soc. Exp. Biol.*, FESBE, Madrid, 1976, 118.
51. **Warner, A. H. and Finamore, F. J.**, Nucleotide metabolism during brine shrimp embryogeneses, *J. Biol. Chem.*, 242, 1933, 1967.
52. **Clegg, J. S., Warner, A. H., and Finamore, F. J.**, Evidence for the function of P^1, P^4-diguanosine 5′-tetraphosphate in the development of *Artemia salina*, *J. Biol. Chem.*, 242, 1938, 1967.
53. **Van Denbos, G. and Finamore, F. J.**, An unusual pathway for the synthesis of adenosine triphosphate by the purine-requiring organism *Artemia salina*, *J. Biol. Chem.*, 249, 2816, 1974.
54. **Warner, A. H.**, A simple and inexpensive method for the preparation of [β,γ- ^{32}P] guanosine 5′-triphosphate, *Biochim. Biophys. Acta*, 383, 229, 1975.

55. **Spector, T., Jones, T. E., and Miller, R. L.,** Reaction mechanism and specificity of human GMP reductase, *J. Biol. Chem.*, 254, 2308, 1979.
56. **Mager, J. and Magasanik, B.,** Guanosine 5′-phosphate reductase and its role in the interconversion of purine nucleotides, *J. Biol. Chem.*, 235, 1474, 1960.
57. **Faraldo, A., Pinto, R. M., Canales, J., Fernández, A., Sillero, A., and Sillero, M. A. G.,** Presence of two isozymes of adenylosuccinate synthetase in *Artemia salina* embryos. Purification and properties, *Comp. Biochem. Physiol.*, 75B, 221, 1983.
58. **Jackson, R. C., Goulding. F. J., and Weber, G.,** Enzymes of purine metabolism in human and rat renal cortex and renal cell carcinoma, *J. Natl. Cancer Inst.*, 62, 749, 1979.
59. **Clark, S. W. and Rudolph, F. B.,** Regulation of purine metabolism. Adenylosuccinate synthetase from Novikoff ascites tumor cells, *Biochim. Biophys. Acta*, 437, 87, 1976.
60. **Clark, S. W., Berry, S. A., Ewing, R., and Rudolph, F. B.,** Regulation of purine metabolism: a comparative study of the kinetic properties of adenylosuccinate synthetases from various sources, *Comp. Biochem. Physiol.*, 58B, 63, 1977.
61. **Van der Weyden, M. B. and Kelly, W. N.,** Human adenylosuccinate synthetase. Partial purification, kinetic and regulatory properties of the enzyme from placenta, *J. Biol. Chem.*, 249, 7282, 1974.
62. **Ogawa, H., Shiraki, H., Matsuda, Y., Kakiuchi, K., and Nakagawa, H.,** Purification, crystallization, and properties of adenylosuccinate synthetase from rat skeletal muscle, *J. Biochem.*, 81, 859, 1977.
63. **Muirhead, K. M. and Bishop, S. H.,** Purification of adenylosuccinate synthetase from rabbit skeletal muscle, *J. Biol. Chem.*, 249, 459, 1974.
64. **Nagy, M., Djembo-Taty, M., and Heslot, H.,** Regulation of the biosynthesis of purine nucleotides in *Schizosaccharomyces pombe*. III. Kinetic studies of adenylosuccinate synthetase, *Biochim. Biophys. Acta*, 309, 1, 1973.
65. **Lieberman, I.,** Enzymatic synthesis of adenosine 5′-phosphate from inosine 5′-phosphate, *J. Biol. Chem.*, 233, 327, 1956.
66. **Pinto, R. M., Faraldo, A., Fernández, A., Canales, J., Sillero, A., and Sillero, M. A. G.,** Adenylosuccinate lyase from *Artemia* embryos, *J. Biol. Chem.*, 258, 12513, 1983.
67. **Barnes, L. B. and Bishop, S. H.,** Adenylosuccinate lyase from human erythrocytes, *Int. J. Biochem.*, 6, 497, 1975.
68. **Brand, L. M. and Lowenstein, J. M.,** Inhibition of adenylosuccinase by adenylophosphonopropionate and related compounds, *Biochemistry*, 17, 1365, 1978.
69. **Brand, L. M. and Lowenstein, J. M.,** Effect of diet on adenylosuccinase activity in various organs of rat and chicken, *J. Biol. Chem.*, 253, 6872, 1978.
70. **Carter, C. E. and Cohen, L. H.,** The preparation and properties of adenylosuccinase and adenylosuccinic acid, *J. Biol. Chem.*, 222, 17, 1956.
71. **Brox, L. W.** The cleavage of adenylosuccinate and 5-amino-4-imidazole-N-succino-carboxamide ribonucleotide by an adenylosuccinate lyase from Ehrlich ascites tumor cells, *Can. J. Biochem.*, 51, 1072, 1973.
72. **Canales, J., Fernández, A., Faraldo, A., Pinto, R. M., Sillero, A., and Sillero, M. A. G.,** IMP dehydrogenase from *Artemia* embryos: molecular forms, purification and properties, *Comp. Biochem. Physiol.*, 81B, 837, 1985.
73. **Jackson, R. C. and Weber, G.,** IMP dehydrogenase, an enzyme linked with proliferation and malignancy, *Nature (London)*, 256, 331, 1975.
74. **Hampton, A. and Nomura, A.,** Inosine 5′-phosphate dehydrogenase. Site of inhibition by guanosine 5′-phosphate and of inactivation by 6-chloro and 6-mercaptopurine ribonucleoside 5′-phosphates, *Biochemistry*, 6, 679, 1967.
75. **Wu, T.-W. and Scrimgeour, K. G.,** Properties of inosinic acid dehydrogenase from *Bacillus subtilis*. I. Purification and physical properties, *Can. J. Biochem.*, 51, 1380, 1973.
76. **Wu, T.-W. and Scrimgeour, K. G.,** Properties of inosinic acid dehydrogenase from *Bacillus subtilis*. II. Kinetic properties, *Can. J. Biochem.*, 51, 1391, 1973.
77. **Gilbert, H. J., Lowe, C. R., and Drabble, W. T.,** Inosine 5′-monophosphate dehydrogenase of *Escherichia coli*, *Biochem. J.*, 183, 481, 1979.
78. **Pinto, R. M., Canales, J., Faraldo, A., Sillero, A., and Sillero, M. A. G.,** Cytosol 5′-nucleotidase from *Artemia* embryos. Purification and properties, *Comp. Biochem. Physiol.*, 86B, 49, 1987.
79. **Naito, Y. and Tsushima, K.,** Cytosol 5′-nucleotidase from chicken liver. Purification and some properties, *Biochim. Biophys. Acta*, 438, 159, 1976.
80. **Itoh, R., Usami, C., Nishino, T., and Tsushima, K.,** Kinetic properties of cytosol 5′-nucleotidase from chicken liver, *Biochim. Biophys. Acta*, 526, 154, 1978.
81. **Itoh, R.,** Purification and some properties of cytosol 5′-nucleotidase from rat liver, *Biochim. Biophys. Acta*, 657, 402, 1981.
82. **Van den Berghe, G., Van Pottelsberghe, C., and Hers, H-G.,** A kinetic study of the soluble 5′-nucleotidase of rat liver, *Biochem. J.*, 162, 611, 1977.
83. **Fernández, A., Canales, J., Pinto, R. M., Faraldo, A., Sillero, M. A. G., and Sillero, A.,** Adenosine deaminase isozymes in *Artemia*, *Comp. Biochem. Physiol.*, 78B, 481, 1984.

84. **Agarwal, R. P. and Parks, R. E., Jr.**, Adenosine deaminase from human erythrocytes, *Methods Enzymol.*, 51, 502, 1978.
85. **Hunt, S. W. and Hoffee, P. A.**, Adenosine deaminase from deoxycoformycin-sensitive and -resistant rat hepatoma cells, *J. Biol. Chem.*, 257, 14239, 1982.
86. **Nygaard, P.**, Adenosine deaminase from *Escherichia coli*, *Methods Enzymol.*, 51, 508, 1978.
87. **Nathans, G. R., Chang, D., and Deuel, T. F.**, AMP deaminase from human erythrocytes, *Methods Enzymol.*, 51, 497, 1978.
88. **Smith, L. D. and Kizer, D. E.**, Purification and properties of rat liver AMP deaminase, *Biochim. Biophys. Acta*, 191, 415, 1969.
89. **Ogasawara, N., Goto, H., Yamada, Y., and Yoshino, M.**, Subunit structures of AMP deaminase isozymes in rat, *Biochem. Biophys. Res. Commun.*, 79, 671, 1977.
90. **Coffee, C. J.**, AMP deaminase from rat skeletal muscle, *Methods Enzymol.*, 51, 490, 1978.
91. **Smiley, K. L. Jr., Berry, A. J., and Suelter, C. H.**, An improved purification, crystallization, and some properties of rabbit muscle 5′-adenylic acid deaminase, *J. Biol. Chem.*, 242, 2502, 1967.
92. **Hershko, A., Wind, E., Razin, A., and Mager, J.**, Conversion of guanine to hypoxanthine in mammalian red blood cells, *Biochim. Biophys. Acta*, 71, 609, 1963.
93. **Stephens, R. W. and Whittaker, V. K.**, Calf thymus GMP reductase, control by XMP, *Biochem. Biophys. Res. Commun.*, 53, 975, 1973.
94. **Renart, M. F. and Sillero, A.**, GMP reductase in *Artemia salina*, *Biochim. Biophys. Acta*, 341, 178, 1974.
95. **Matsuda, Y., Ogawa, H., Fukutome, S., Shiraki, H., and Nakagawa, H.**, Adenylosuccinate synthetase in rat liver: the existence of two types and their regulatory roles, *Biochem. Biophys. Res. Commun.*, 78, 766, 1977.
96. **Ogawa, H., Shiraki, H., and Nakagawa, H.**, Study on the regulatory role of fructose-1,6-diphosphate in the formation of AMP in rat skeletal muscle. A mechanism for synchronization of glycolysis and the purine nucleotide cycle, *Biochem. Biophys. Res. Commun.*, 68, 524, 1976.
97. **Hand, S. C. and Carpenter, J. F.**, pH-Induced metabolic transitions in *Artemia* embryos mediated by a novel hysteretic trehalase, *Science*, 232, 1535, 1986.
98. **Woodward, D. O. and Braymer, H. D.**, Purification and properties of *Neurospora* adenylosuccinase, *J. Biol. Chem.*, 241, 580, 1966.
99. **Carson, D. A. and Wasson, D. B.**, Characterization of an adenosine 5′-triphosphate- and deoxyadenosine 5′-triphosphate-activated nucleotidase from human malignant lymphocytes, *Cancer Res.*, 42, 4321, 1982.
100. **Schrader, W. P., Stacy, A. R., and Pollara, B.**, Purification of human erythrocyte adenosine deaminase by affinity column chromatography, *J. Biol. Chem.*, 251, 4026, 1976.
101. **Van der Weyden, M. B. and Kelley, W. N.**, Human adenosine deaminase. Distribution and properties, *J. Biol. Chem.*, 251, 5448, 1976.
102. **Daddona, P. E. and Kelley, W. N.**, Human adenosine deaminase. Stoichiometry of the adenosine deaminase-binding protein complex, *Biochim. Biophys. Acta*, 580, 302, 1979.
103. **Hernandorena, A. and Kaushik, S. J.**, Effect of quality and quantity of dietary purines on ammonia excretion rates of *Artemia* sp., *Comp. Biochem. Physiol.*, 76B, 637, 1983.
104. **Rottlán, P., Liras, A., and Llorente, P.**, Biosynthesis of nucleotides from exogenous purines in *Artemia* nauplii, in *Artemia Research and its Applications*, Vol. 2, Decleir, W., Moens, L., Slegers, H., Sorgeloos, P., and Jaspers, E., Eds. Universa Press, Wetteren, Belgium, 1987, 541.
105. **Llorente, P., Montero, C., Argomániz, L., and Rotllán, P.**, Adenina fosforibosil transferasa de *Artemia* sp: purificación y propiedades, in *Abstracts of the 13th Spanish Natl. Congress of Biochem. Zaragoza*, 1986, 287.
106. **Parrilla, R., Jiménez, M. I., and Ayuso-Parrilla, M. S.**, Cellular redistribution of metabolites during glucagon and insulin control of gluconeogenesis in the isolated perfused rat liver, *Arch. Biochem. Biophys.*, 174, 1, 1976.
107. **Williamson, D. H. and Brosnan, J. T.**, Concentrations of metabolites in animal tissues, in *Methods of Enzymatic Analysis*, Vol. 4, Bergmeyer, H. U., Ed., Academic Press, New York, 1974, 2302.
108. **Kleineke, J., Düls, Ch., and Söling, H. D.**, Subcellular compartmentation of guanine nucleotides and functional relationships between the adenine and guanine nucleotide systems in isolated hepatocytes, *FEBS Lett.*, 107, 198, 1979.
109. **Siess, E. A. and Wieland, O. H.**, Isolated hepatocytes as a model for the study of stable glucagon effects on mitochondrial respiratory functions, *FEBS Lett.*, 101, 277, 1979.
110. **Siess, E. A. and Wieland, O. H.**, Subcellular distribution of guanine and uracil nucleotides and of succinate in normal and glucagon-stimulated hepatocytes, *Hoppe Seyler's Z. Physiol. Chem.*, 360, 1677, 1979.
111. **Isihara, N. and Kikuchi, G.**, Studies on the functional relationship between the phosphopyruvate synthesis and the substrate level phosphorylation in guinea-pig liver mitochondria, *Biochim. Biophys. Acta*, 153, 733, 1968.
112. **Vallejo, C. G., Sillero, M. A. G., and Marco, R.**, Mitochondrial maturation during *Artemia salina* embryogenesis. General description of the process, *Cell. Mol. Biol.*, 25, 113, 1979.

113. **Clegg, J. S.,** The control of emergence and metabolism by external osmotic pressure and the role of free glycerol in developing cysts of *Artemia salina*, *J. Exp. Biol.*, 41, 879, 1964.
114. **Conte, F. P.,** Structure and function of the crustacean larval salt gland, *Int. Rev. Cytol.*, 91, 45, 1984.
115. **Clegg, J. S. and Conte, F. P.,** A review of the cellular and developmental biology of *Artemia*, in *The Brine Shrimp Artemia*, Vol. 2, Persoone, G., Sorgeloos, P., Roels, O., and Jaspers, E., Eds., Universa Press, Wetteren, Belgium, 1980, 636.
116. **Canales, J., Fernández, A., Sillero, M. A. G., and Sillero, A.,** unpublished data, 1986.

Chapter 5

PROTEASES AND PROTEASE INHIBITORS IN *ARTEMIA* AND THEIR ROLE IN THE DEVELOPMENTAL PROCESS

Alden H. Warner

TABLE OF CONTENTS

I. INTRODUCTION

Few developmental systems have yielded significant amounts of data on the spectrum of proteases (also known as proteinases) in developing embryos and on the role of proteases in development. Most studies on proteases in gametes have focused on their involvement and mechanism of action at the time of fertilization and little is known of the role of intracellular proteases in developing embryos. However, proteases have been implicated in a variety of metabolic events during early development including activation of "masked" mRNA,[1] activation of metabolically repressed ribosomes,[2,3] and yolk utilization.[4] Moreover, changes in the relative abundance of free amino acids and proteins in embryos and larvae suggest that various types of proteases must be active in the reorganization of proteins during development.[5]

Studies of cells and tissues of mammalian origin indicate that at least two mechanisms operate in intracellular protein breakdown — a lysosomal and nonlysosomal mechanism.[6] The lysosomal mechanism is thought to function in the breakdown of long-lived proteins, while the nonlysosomal mechanism is believed to be important in the breakdown of short-lived proteins.[7,8] An ATP-dependent, ubiquitin-mediated proteolytic system also appears to operate independently for the removal of abnormal proteins from eukaryotic cells.[7]

Most eukaryotic cells contain two general types of proteases and these have been classified as exopeptidases and endopeptidases. The exopeptidases are hydrolases which act at either the N-terminal or C-terminal position of proteins releasing amino acids or short peptides, while endopeptidases (also referred to as proteases or proteinases) are capable of cleaving peptide bonds in more central regions of the protein.[9] Based on the catalytic mechanism of the active centers of endopeptidases (proteases), four subclasses of these hydrolases have been described. The most intensely studied class of endopeptidases is the serine proteases which include trypsin, chymotrypsin, and the elastases. These proteases are usually secreted from eukaryotic cells as proenzymes which are then activated extracellulary.

In the invertebrates the serine proteases are also the best studied of the proteases, and there is considerable information on the properties of crustacean, trypsin-like proteases.[10,11] In crayfish, multiple forms of trypsin-like proteases are believed to function as digestive enzymes.[10] Carboxypeptidases have also been isolated from the cardia, a stomach like organ in adult crayfish, *Astacus fluviatilis*,[10] and from the hepatopancreas and stomach of penaeid shrimps.[12,13] Several studies have shown that intracellular exo- and endopeptidases play important roles in the dismantling of cells and tissues programmed for elimination during metamorphosis.[14-16] In the blowfly, *Aldrichina grahami*, two enzymes thought to be involved in metamorphosis are cathepsin D-like and cathepsin L-like proteases. An alkaline protease has also been isolated from the gut of *A. grahami*,[15] and trypsin and chymotrypsin-like enzymes have been isolated from the midgut of larvae of the hornets, *V. crabro* and *V. orientalis*.[10] In some of the above cases low molecular weight protease inhibitors have also been identified as important intracellular components. These inhibitors are believed to be important in the control of intracellular protease expression[17-19] about which more will be said later.

Notwithstanding the importance of the above studies, few invertebrate systems have been investigated thoroughly for the presence of various types of proteases and protease regulators and for their role in development. In this respect, *Artemia* is the most studied invertebrate system. This system offers the advantage that gram quantities of relatively synchronized populations of embryos, larvae, and adults can be obtained readily at various stages of development and growth. Studies of *Artemia* have revealed that both active and latent proteases are present in embryos and larvae, and that these enzymes may be involved in such diverse functions as yolk metabolism,[20-22] protein synthesis regulation,[22-24] and RNA synthesis regulation.[25]

The first published study on proteolytic activity in *Artemia* embryos and larvae was carried out by Urbani and co-workers in 1952.[26] Subsequently, several additional reports appeared by these researchers (see Green[5] and Bellini[27]) who demonstrated changes in dipeptidase and protease activity during development.

If proteases play a role in the regulation of developmental events in invertebrates and other eukaryotes, then a thorough understanding of these hydrolases including their subcellular localization, mode of regulation, and substrate specificities is of paramount importance.

Artemia, more than any other eukaryotic system, has yielded considerable information on the role and importance of proteases in the developmental program. The main objectives of this chapter are to review current knowledge about proteases and protease inhibitors in *Artemia* embryos and larvae and their importance in the regulation of developmental events that require proteolytic initiation and/or processing. Several prospects for future studies on the molecular biology of these important proteins are also presented.

II. VARIATIONS IN *ARTEMIA* POPULATIONS

The *Artemia* complex consists of at least five sibling species.[28] They are *A. franciscana* (native to North America), *A. persimilis* (Argentina), *A. tunisiana* (native to the Mediterranean region), *A. urmiana* (Iran), and *A. parthenogenetica* (native to Asia, Australia, and Europe). Most biochemical studies have been conducted using cysts believed to be *A. franciscana*, but some recent studies may have been carried out unknowingly on mixed populations of *Artemia* or commercial lots of cysts from different geographic regions. Studies on the gene structure of different populations of *Artemia*, and on different batches of cysts from the same commercial supplier, have shown considerable heterogeneity with respect to restriction sites for both histone and ribosomal RNA genes.[29-31] The first of these studies revealed that certain batches of cysts purchased from San Franciso Bay Brand, Inc. (i.e., California cysts) were of Chinese origin and thus probably parthenogenetic.[29,32] Thus, differences in properties of many enzymes, including the proteases, may be due to commercial packaging methods resulting in mixed populations of cysts and not to differences in methodology or experimental technique. The reader should keep this problem in mind when evaluating various papers reporting differences in enzyme properties using *Artemia* cysts from different commercial sources or populations. Questions as to the origin of commercially available cysts should be directed to the supplier(s). The researcher may also wish to verify the origin of commercially available cysts using samples of cysts obtained from the Artemia Reference Centre, University of Ghent, Ghent, Belgium.

III. CHACTERIZATION OF PROTEASES IN DORMANT GASTRULAE OF *ARTEMIA*

A. Thiol Proteases

Following the initial observations of Urbani et al.[26] that *Artemia* cysts contain peptidase and protease activity, Bellini[27] measured the protease activity in aqueous extracts of *Artemia* at various stages of development using casein as substrate at pH 4.9. These early studies showed that the level of protease activity assayed in an acidic environment increased gradually during development and was highest in the fully formed prenauplius larva and then declined. Recently the "acid" protease first observed by Urbani and Bellini was purified to homogeneity from two populations of *Artemia* cysts and nauplii and characterized.[20,22,33,34] The "acid" protease appears to be a cathepsin B-like thiol protease which accounts for over 90% of the protease activity in crude homogenates of encysted embryos. The protease has been purified from both the cytosol[20,22,33] and particulate fractions.[34] The soluble protease and the particulate protease (which is thought to be associated with lysosomes) have nearly

Table 1
PROPERTIES OF A CATHEPSIN B-LIKE THIOL PROTEASE FROM TWO POPULATIONS OF *ARTEMIA*

Property/treatment	Source of protease	
	California cysts[a]	Utah cysts[b]
Mol wt	68,000	55,000
Subunits	1	2
Isoelectric forms	nd	2
pH optimum	3.5—4.0	3.5—4.5[c]
Pepstatin	No effect (1 μg/mℓ)	No effect (10 μg/mℓ)
Soybean trypsin inhibitor	No effect (10 μg/mℓ)	No effect (1 mg/mℓ)
Antipain	100% inhibition (1 μg/mℓ)	50% inhibition (0.5 μg/mℓ)
Leupeptin	100% inhibition (1 μg/mℓ)	50% inhibition (0.2 μg/mℓ)
Cu^{2+}	90% inhibition (1 m*M*)	95% inhibition (0.1 m*M*)
Ca^{2+}	No effect (10 m*M*)	No effect (10 m*M*)
Zn^{2+}	85% inhibition (1 m*M*)	31% inhibition (10 m*M*)
Iodoacetate	100% inhibition (10 m*M*)	87% inhibition (1 m*M*)
EDTA	10% stimulation (0.5 m*M*)	14% stimulation (1 m*M*)
PMSF	No effect (0.01 m*M*)	No effect (0.13 m*M*)
KI	100% inhibition (1 m*M*)	No effect (1 m*M*) 25% inhibition (10 m*M*)
pCMB	100% inhibition (2.5μ*M*)	12% inhibition (1 m*M*)

Note: n.d., not measured.

[a] From References 20 and 34.
[b] From References 22, 35.
[c] Range using several substates

identical physical properties. However, certain properties of the protease(s) vary depending on the source of the starting population of *Artemia* cysts. For example, the protease isolated from Utah cysts has a molecular weight of 55 kdaltons and contains subunits of 31.5 and 25.8 kdaltons, while the protease isolated from "California" cysts has a molecular weight of 68 kdaltons and is a single polypeptide. The other striking difference between the two proteases is their sensitivity to the chaotropic agent potassium iodide (KI). The protease from Utah cysts is fully active when assayed in the presence of 1 m*M* KI, while the protease from "California" cysts is completely inhibited by 1 m*M* KI. Various properties of the protease from the "California" and Utah populations are shown in Table 1.

Notwithstanding apparent differences between *Artemia* populations, the cyst protease resembles mammalian cathepsin B except for its low pH optimum (pH 3.5 to 4.5). It catalyzes the hydrolysis of benzoyl-DL-arginine-β-naphthylamide (BANA) and carbobenzoxy-arginyl-arginine-β-naphthylamide (Cbz-arg-arg-NA) which are good substrates for mammalian cathepsin B. A unique property of the *Artemia* cathepsin B-like protease is that it catalyzes limited hydrolysis of *Artemia* protein synthesis elongation factor 2 (EF-2) at pH 6.0 to 6.5 while having little effect on many other soluble proteins from *Artemia*. This latter phenomenon explains the presence of low molecular weight fragments of EF-2 in partially purified preparations of EF-2 from encysted embryos of *Artemia*.[24] Whether this activity occurs in vivo in the control of protein synthesis is not known. In the pH range of 3.5 to 4.5 the cyst protease catalyzes the complete hydrolysis of several proteins including *Artemia* lipovitellin, serum albumin, hemoglobin, casein, EF-2 and other soluble proteins from *Artemia* cysts.[22,35]

The cytosol and particulate cathepsin B-like proteases are sensitive to a variety of thiol-blocking reagents such as leupeptin, antipain, chymostatin, Ep-475, iodoacetate, Cu^{2+}, and Hg^{2+}, but they are relatively insensitive to parachloromecuribenzoate and *N*-ethylmaleimide.

Table 2
KINETIC DATA OF *ARTEMIA* CATHEPSIN B-LIKE PROTEASE ON VARIOUS SUBSTRATES[a]

		Substrates	
Constants	**BANA**	**Cbz-arg-arg-NA**	**Serum albumin[b]**
K_m (mM)	0.13	0.24	0.012
K_{cat}	2.09	15.9	273
K_{cat}/K_m	16.1	66.3	22,750

[a] Cytosol protease from Utah cysts was assayed here.
[b] Bovine origin.

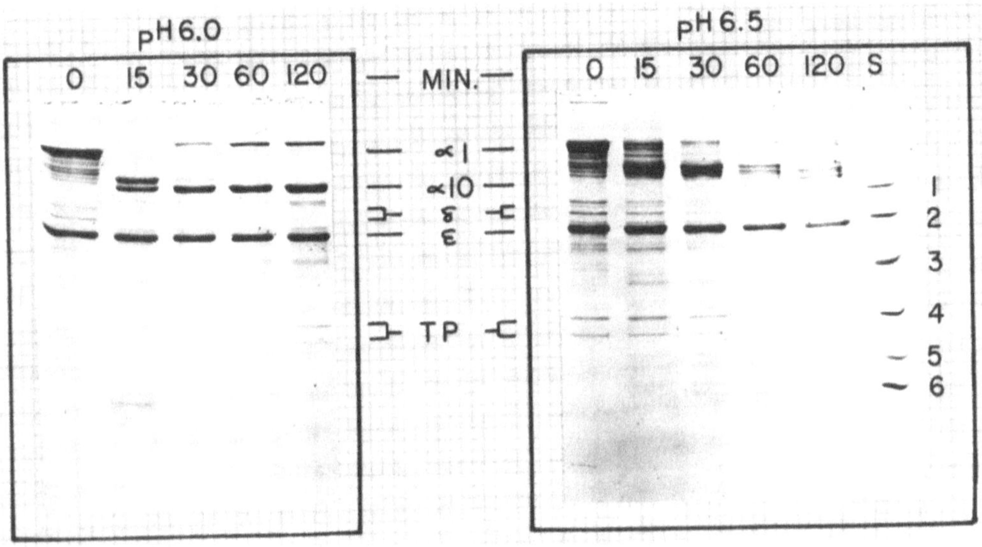

FIGURE 1. Pattern of lipovitellin hydrolysis by the cathepsin B-like protease in the cytosol of *Artemia* embryos. Lipovitellin from encysted embryos of *Artemia* (12μg) was incubated at pH 6.0 and pH 6.5 with *Artemia* cytosol cathepsin B-like protease (1.9μg) at 30°C. Column S represents standard molecular weight protein markers, from top to bottom: (1) phosphorylase (M_r 94,000), (2) BSA (M_r 67,000), (3) ovalbumin (M_r 43,000), (4) carbonic anhydrase (M_r 30,000), (5) soybean trypsin inhibitor (M_r 20,100), and (6) α-lactalbumin (M_r 14,400. TP, cathepsin B-like thiol protease.

Both proteases are also unaffected by pepstatin, soybean trypsin inhibitor, and phenylmethylsulfonylfluoride (PMSF). Kinetic data for leupeptin and Ep-475 show that equimolar amounts of these inhibitors (with enzyme) produce 50% inhibition suggesting that the enzyme is a thiol protease containing two binding sites for these agents which must be filled to completely inactivate the protease.

Table 2 summarizes kinetic data for the cathepsin B-like protease on various substrates. These data indicate that while the *Artemia* thiol protease hydrolyzes substrates normally used in mammalian cathepsin B assays, i.e. BANA and Cbz-arg-arg-NA, it has a higher specificity for proteins as substrate.[22] Similar results were obtained with the cathepsin B-like protease isolated from "California" cysts.[34]

Both the cytosol and particulate cathepsin B-like proteases have the capacity to catalyze the hydrolysis of *Artemia* yolk proteins in vitro, under physiological conditions, with a pattern of hydrolysis closely parallel to that found in yolk protein preparations from gastrulae and nauplii.[20,22,36] The pattern of lipovitellin hydrolysis is shown in Figure 1. The sequential hydrolysis of LV-α1 (M_r 185,000) to smaller lipovitellins and eventually to LV-α10 (M_r 120,000) occurs in vitro at pH 5.0 to 6.5.[20,22] However, the LV-ε (M_r 61,000) is resistant to hydrolysis by the cyst protease even up to 24 hr incubation in vitro.[20]

It is curious that the pattern of lipovitellin hydrolysis (i.e., yolk protein degradation) catalyzed by the cytosol cathepsin B-like protease is very similar to that catalyzed by the trypsin-like protease found tightly associated with the *Artemia* lipovitellin complex (see Section V.A). Perhaps the cathepsin B-like protease functions in the activation of the trypsin-like enzyme sequestered in yolk granules (or inactivation of the lipovitellin associated trypsin inhibitor) which in turn catalyzes lipovitellin breakdown. Slight differences in the pattern of lipovitellin degradation occur in vivo compared with that catalyzed by the cathepsin B-like protease in vitro suggesting that more than one protease may be involved in lipovitellin utilization during development.

Reasons for the differences in various properties of the cathepsin B-like protease isolated from Utah and "California" cysts are not yet clear. Revelations that certain batches of San Francisco Bay Brand cysts purchased between 1976 and 1983 originated from Tientsin, China and not from the San Francisco Bay area may explain the differences observed, since certain populations of cysts from China are parthenogenetic and have different gene structures.[29,32] We have also found that use of NaOCl to sterilize cysts causes artifacts in activity measurements of SH-rich proteins such as the thiol proteases.[22] Thus, while NaOCl-treated cysts are easier to homogenize, thiol proteases prepared from NaOCl-treated cysts have both reduced activity (about 52%) and reduced protein content (76%).

B. Trypsin-Like Proteases

While trypsin-like protease activity has been found in a variety of eggs and embryos including those of insects[37-39] and sea urchins,[40] very little trypsin-like protease activity has been detected, in an active form, in encysted embryos of *Artemia*. The low level of trypsin-like activity found in homogenates of encysted *Artemia* embryos is associated with lipovitellin solubilized from yolk granules.[21] Trypsin-like activity has also been found associated with yolk platelets prepared from ovaries of *Rana pipiens*.[41] In extracts from *Artemia* the trypsin-like protease cannot be dissociated from the high molecular weight lipovitellin by the use of chaotropic salts such as KI, but the enzyme activity is inhibited by antilipovitellin antiserum. The lipovitellin-associated enzyme is also inhibited by leupeptin, antipain, soybean trypsin inhibitor, and ovomucoid and PMSF but not by chymostatin or E-64 a specific thiol protease inhibitor. Thus, encysted *Artemia* embryos contain very little trypsin-like activity in the cytosol, while in yolk platelets, trypsin-like activity is closely associated with lipovitellin and perhaps yolk metabolism.

C. Latent Protease Activity in the Cytosol of *Artemia* Embryos

Various studies in my laboratory during the past 4 to 5 years have demonstrated the presence of at least one additional protease in the cytosol of encysted embryos. This protease shows very little activity in extracts of dormant cysts, but its activity is readily unmasked by incubation of crude extracts at pH 4.2 to 4.5 and 30°C as shown in Figure 2. The activated protease has been partially purified by gel filtration and some properties of both active and inactive forms of the enzyme have been studied. The inactive form(s) of the protease binds to DEAE-cellulose and elutes from the column immediately before the cathepsin B-like protease. In contrast, the fully activated form does not bind to DEAE-cellulose or to CM-cellulose at normal operating pH's for these ion exchanges. The activated form of the enzyme shows optimal activity at pH 7 to 8 and 40 to 45°C. The activity is unaffected by 1 m*M* ATP, GTP, DTT, Ca^{2+}, and iodoacetate or by 2 to 25 μg/mℓ leupeptin and PMSF or 2 μg/mℓ of soybean trypsin inhibitor or ovomucoid. The protease is inhibited slightly (14 to 19%) by 0.2 m*M* *N*-α-*p*-tosyl-lysine chloromethyl ketone (TLCK) and extensively (85%)

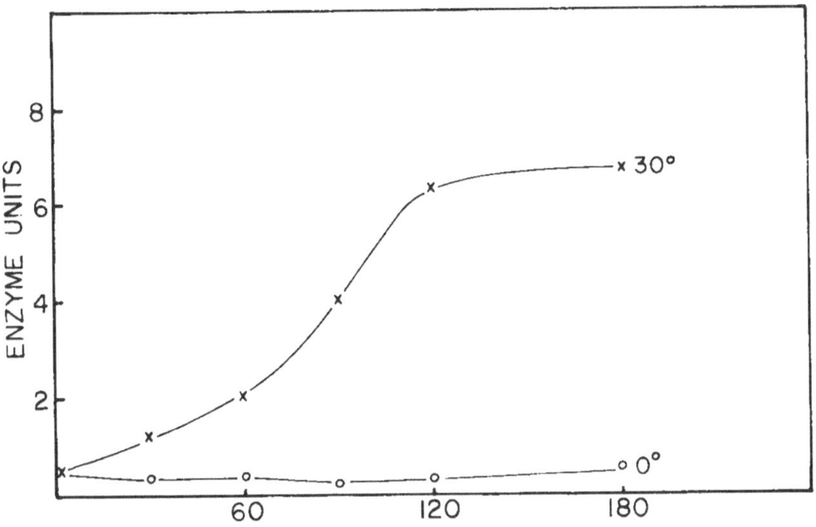

FIGURE 2. Unmasking of latent protease activity in extracts of dormant *Artemia* embryos. A 25 to 75% ammonium sulfate fraction of the postribosomal supernatant fraction of dormant cysts was adjusted to pH 4.2 with acetic acid and incubated for various times at 0 or 30°C. At the times indicated an aliquot of each reaction mixture was neutralized and the soluble material assayed for protease activity at pH 6.8 with dimethyl BSA as substrate. Trinitrobenzene sulfonic acid (TNBS method) was used to measure the liberation of amino groups during the hydrolysis of peptide bonds.[33]

by 0.1% SDS. Aside from these properties the protease is unusual in that it is stable to 0.1 *M* HCl treatment at 90°C (30 min). Moreover, it has an apparent low molecular weight for an enzyme eluting from various gel filtration columns at a position consistent with a molecular weight of 5,000 to 10,000 daltons. The enzyme is most active on soluble acidic proteins from *Artemia* embryos in the 50,000 to 100,000 molecular weight range, but it is slightly active on protamine sulfate and bovine serum albumin (BSA) (see Figure 3). SDS-PAGE analysis of the enzyme reaction mixture suggests that the activated protease is either an exoprotease or aminopeptidase. Considerable additional work needs to be done to more fully characterize this protease and its precursor, and to determine its role during embryonic development, but this enzyme appears to be uniquely stable among proteases which have been isolated and characterized from a variety of other cells and tissues.

IV. LOCALIZATION OF PROTEASE ACTIVITY IN *ARTEMIA* EMBRYOS

The observation that dormant embryos of *Artemia* contain cathepsin B-like protease, an enzyme which is found normally in lysosomes in cells of mammalian origin, has prompted several experiments to identify the site of storage/localization of this enzyme in encysted gastrulae of *Artemia*.

Results of these experiments showed that anywhere from 18 to 86% of the thiol protease activity in homogenates of encysted embryos is sedimentable depending on the population of cysts, nature of homogenization medium, or ratio of homogenization medium to cysts.[20,34,35,41,42] One report also indicated that the subcellular distribution of the protein changed with development ranging from about 10% in the cytosol prior to hatching of nauplii to nearly 70% in the cytosol by 48 hr development.[42] What is clear from these experiments is that homogenization media containing Mg^{2+}(5 m*M*) and Ca^{2+}(0.5 m*M*) and/or Ficoll (15%) enhance the sedimentation properties of the cathepsin B-like protease activity, while media lacking these components, or ones containing EDTA, show a greater proportion of

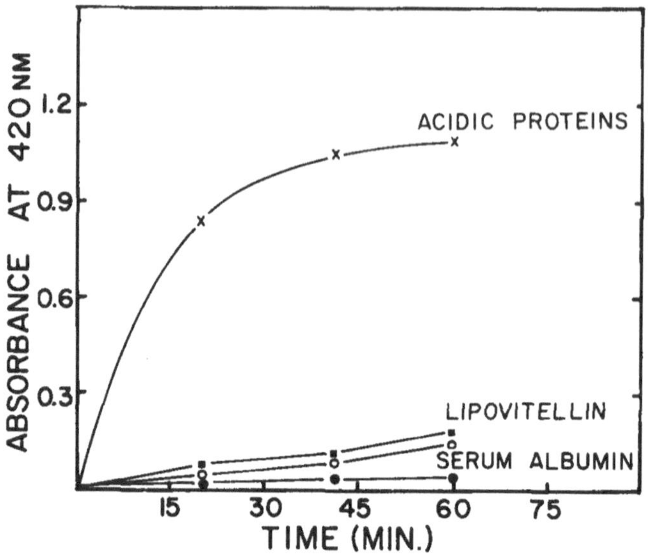

FIGURE 3. Effect of activated latent protease on various potential substrates. Activated latent protease (after 2 hr at pH 4.2) was purified on a column of Sephadex G-50 then added to reaction mixtures containing various substrate proteins (4 mg/mℓ) at pH 6.8 as indicated. (×——×) Acidic proteins of *Artemia* embryos from DEAE-cellulose column, (■——■) lipovitellin purified from *Artemia* embryos, (○——○) BSA, (●——●) control with no enzyme but with acidic proteins. The extent of hydrolysis was determined using the TNBS method.[33]

the protease activity in the cytosol fraction. Data from recent experiments indicate that the cathepsin B-like protease in *Artemia* embryos is associated with particles in the mitochondria/lysosome fraction, but these conclusions need to be confirmed by the use of other techniques, such as immunohistochemical ones, to localize the protease. What is also clear is that the level of protease in the cytosol has been underestimated in many experiments where no attempt was made to separate the protease from the endogenous thiol protease inhibitors in the cytosol.[22,43] The presence of endogenous protease inhibitors in the cytosol may lead to low estimates of protease activity levels in embryos and larva during development.[22] Thus, while it appears that a significant proportion of the cathepsin B-like thiol protease activity is associated with organelles which sediment at 7500 × *g* in Ficoll media, conclusions suggesting that the site of storage of the protease is in lysosomes are still open to question. Resolution of this problem will require the use of immunochemical localization techniques to resolve the problem of protease distribution in the embryo, after which the role of the cathepsin B-like protease(s) during development of *Artemia* may become more clear.

V. ROLE OF EMBRYOS PROTEASES IN EARLY DEVELOPMENT OF *ARTEMIA*

A. Proteases in the Metabolism of Yolk Proteins

In dormant cysts of *Artemia* at least 65% of the embryo protein is present in yolk platelets or granules.[41] Approximately 75,000 yolk platelets are present in each dormant embryo, and these organelles can be obtained easily and in relatively pure form by low speed centrifugation of embryo homogenates prepared using either sucrose of Ficoll containing media depending on the origin and/or source of the *Artemia* cysts.[41,44,45] *Artemia* yolk platelets are highly structured organelles containing a complex outer membrane and numerous internal membraneous vesicles which resemble mitochondria.[46,47] Chemically, *Artemia* yolk platelets are composed mostly of protein (88%), but they also contain carbohydrate (3.3%) and lipids (8.6%) as well. The main component is lipovitellin, a cartenoid-lipoglycoprotein complex of approximately 585,000 daltons. The internal membranous vesicles in *Artemia* yolk platelets appear to be stored mitochondria on the basis of cytological evidence,[46,47] but no clear biochemical or histochemical data exist in support of this conclusion. If the yolk-associated vesicles are indeed mitochondria, they must have arisen as a result of a packaging process during platelet formation in the developing oocyte since mitochondria do not arise *de novo*. Electron microscopic analyses of *Artemia* oocytes during the period of vitellogenesis do not show mitochondria associated with yolk formation. Rather, EM micrographs show that yolk platelet formation in *Artemia* is closely associated with the endoplasmic reticulum.[69]

Protein analyses of the *Artemia* lipovitellin complex from oocytes have shown the presence of two major protein species.[36,48] These proteins have been identified as lipovitellin-αl (LV-αl; 190,000 daltons) and lipovitellin-ε (LV-ε; 68,000 daltons). Two copies of each LV-ε1 and LV-ε are thought to be present in the water-insoluble lipovitellin complex. During development yolk platelets disappear rapidly from the embryo (about 50% of the platelets in the cysts disappear by the time of hatching),[45] and LV-α1 is degraded stepwise, eventually to a 120,000-dalton polypeptide identified as LV-α10.[48,49] In contrast LV-ε, the smallest of the native lipovitellin polypeptides, is resistant to hydrolysis in embryos and young larvae. Since yolk proteins are thought to be the major source of amino acids for protein synthesis during development of *Artemia*, it is of interest to determine the mechanism by which yolk proteins are metabolized in these embryos.

Numerous studies have attempted to link egg and embryo protease activity with yolk utilization, but for the most part, experiments on this subject have not yielded convincing evidence for the role of a specific protease(s) in yolk degradation. The most unequivocal data in this respect have come from work on yolk platelets of *Rana*[50] and *Artemia*.[21,49] In *Artemia*, a trypsin-like protease has been detected in extracts of yolk platelets purified from dormant embryos (cysts).[21] The protease activity in the extract is associated with the lipovitellin and elutes from a gel filtration column with the 585,000-dalton lipovitellin complex. In addition, trypsin-like protease activity has been detected in a 420,000-dalton lipovitellin complex from the cytosol of *Artemia* embryos, and in polypeptide subunits of 70K, 45K, and 30K daltons derived from the 420,000-dalton lipovitellin by mild SDS-DTT treatments.[21] The 70K, 45K, and 30K subunits are thought to be derived from the 420,00-dalton lipovitellin complex in the cytosol by proteolysis since they are not found among the lipovitellin polypeptides in oocyte yolk platelets.[48] The high molecular weight lipovitellin (420,000 daltons) complex in the cytosol is thought to be derived, in some way, from the large lipovitellin complex in yolk platelets. These conclusions are based mainly on immunochemical data using antilipovitellin antibodies prepared from purified embryo yolk platelet lipovitellin.

In newly hatched larvae of *Artemia* (5 hr old), trypsin-like protease activity is associated with polypeptides of 125K and 91K daltons in the cytosol, while in older larvae of *Artemia* (45 hr after hatching), typsin-like activity is associated with 70K and 30K dalton polypeptides.[51] Although somewhat conflicting results have been reported by researchers in Spain on the molecular size of the various trypsin-associated lipovitellins,[21,51] it now appears certain that the 70K and 30K dalton polypeptides in 45-hr *Artemia* larvae are the results of proteolysis of the large lipovitellin subunits found in the cytosol of newly hatched *Artemia* larvae.

The trypsin-like activity in lipovitellin preparations from early and late *Artemia* larvae can be activated by chaotropic salts such as KI, while the protease activity in lipovitellin preparations from *Artemia* embryos is refractory to activation by KI. The KI treatment appears to unmask the protease activity in larval lipovitellin by dissociating the complex.[51]

Recently Ezquieta and Vallejo[49] demonstrated that solubilized *Artemia* embryo lipovitellin has protease inhibitor activity in addition to masked trypsin-like activity. The protease inhibitor activity in the lipovitellin complex is temperature and substrate dependent, and peptide bond cleavage within the lipovitellin is a prerequisite for the lipovitellin to function as a protease inhibitor. Moreover, the lipovitellin-associated inhibitor appears to be specific for the *Artemia* trypsin-like protease. Thus, *Artemia* yolk platelets appear to contain both trypsin-like protease and protease inhibitor, but their interaction, if any, in yolk utilization is not clear. In most of these experiments data interpretation has proven to be difficult because most protease and inhibitor assays were carried out in the presence of high concentrations of NaCl (up to 1 mol) and $MgCl_2$ (5 mmol). These nonphysiological conditions may have affected the enzyme-inhibitor interaction and the type of bonds cleaved by the *Artemia* trypsin-like enzyme compared with the situation in vivo.

Protease inhibitors similar to the one present in *Artemia* yolk platelets have been found in other embryonic systems including *Rana* where trypsin and chymotrypsin inhibitors are covalently linked to the heavy subunit of lipovitellin.[50,52,53] In *Rana* the endogenous protease inhibitors appear to control the hydrolysis of yolk (by an endogenous protease?) and passage of yolk cleavage products into the cytosol in such a way as to produce a slow and even contribution of amino acids to the cytosol metabolic pool during development. Whether similar protease-inhibitor interactions occur in yolk platelets of *Artemia* remains to be demonstrated. The main question raised by these important studies with *Artemia* and *Rana* is whether the lipovitellin-associated trypsin-like protease is responsible for yolk protein catabolism during development. If so, what mechanism controls the activity of the lipovitellin-associated trypsin-like enzyme in the yolk platelets of embryos? This question is especially important in embryos such as *Artemia* which are able to withstand prolonged periods of anaerobiosis, and during which time most metabolic activities, including protein metabolism, are highly regulated. While it has been assumed that yolk platelet-associated protease inhibitors are active in the control of yolk protein utilization, this may not be their only function. These proteins may be stored in yolk platelets and released to the cytosol where they play a role in protein metabolism regulation as development proceeds.

If the trypsin-like protease and trypsin inhibitor are indeed endogenous components of yolk platelets in *Artemia* as the current data suggest, the process of yolk metabolism and/or protease inhibitor storage and release in *Artemia* may be controlled by a mechanism(s) similar to that occurring in *Rana* and other embryos. Data in support of this view could be obtained using antibody probes prepared with purified trypsin-like protease and trypsin inhibitor proteins to ascertain their location in *Artemia* embryos and larvae. These probes could also be used in Western blot analyses to determine the position of the protease and inhibitor domains in various lipovitellin preparations from embryos and larvae. Experiments such as these would show unequivocally whether the trypsin-like protease and trypsin inhibitor in yolk platelets of *Artemia* are associated with the large or small lipovitellin subunit in oocytes and early embryos as they are in *Rana*. Moreover, such analyses would establish whether the site of processing of lipovitellin-associated protease occurs in the yolk granules or after release of the lipovitellin component into the cytosol (by some other mechanism). These results should elucidate whether the increase in trypsin-like activity found in the cytosol of *Artemia* larvae after hatching (see Figure 4 below) is due to activation of precursor molecules (such as lipovitellins) and their subsequent release to the cytosol or to synthesis *de novo*.

Lipovitellin metabolism during development of *Artemia* may also require the participation of other proteases.[20,22] The major protease in dormant cysts and early embryos of *Artemia* is the cathepsin B-like thiol protease described above. Since the pattern of lipovitellin hydrolysis catalyzed by the cathepsin B-like protease is similar in vitro at pH 6 to 6.5 to that observed in vivo, this enzyme may also function in yolk protein utilization. It should be noted that the pattern of lipovitellin hydrolysis catalyzed by the cathepsin B-like protease is nearly identical to that produced by the trypsin-like enzyme. However, incubation of

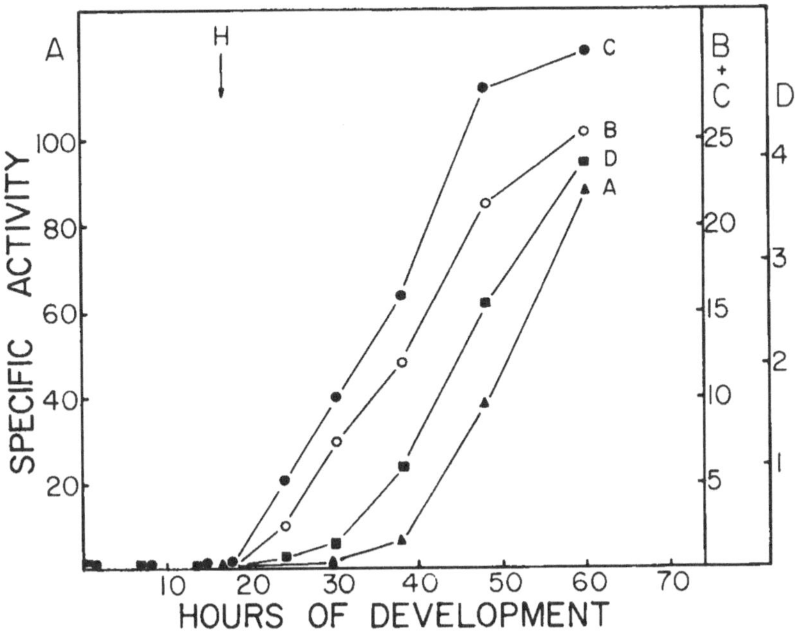

FIGURE 4. Induction of "alkaline" proteases during early development of *Artemia*. The specific activity of protease A (▲——▲) was assayed with CLNE as substrate; protease B (○——○) and protease C (●——●) were assayed with BAPNA as substrate, and protease D (■——■) was measured using casein as substrate in the presence of soybean trypsin inhibitor (100 μg/mℓ). H, onset of hatching.[56]

intact lipovitellin with the cathepsin-B like protease at pH 4 to 5 leads to nearly complete degradation of the lipovitellin fractions.[20,22] LV-ε however, is resistant to hydrolysis by the cathepsin B-like protease at pH values below 6.5.[22] Clearly the embryo cathepsin B-like thiol protease is capable of hydrolyzing yolk proteins under physiological conditions (in vitro) producing products similar to those observed in vivo. Whether this protease functions in vivo in yolk metabolism is still not clear. Perhaps the major function of this enzyme is to act on the large lipovitellin complex (420,000 daltons) in the cytosol. Alternatively, the cathepsin B-like protease may activate the lipovitellin-associated (latent) trypsin-like enzyme permitting the latter to catalyze the hydrolysis of the lipovitellins. Analysis of yolk metabolism in the presence of specific thiol protease and trypsin-like protease inhibitors, under otherwise physiological conditions, may shed more light on the specific role of these enzymes in yolk metabolism. Since lysosomes have been implicated in yolk platelet breakdown in sea urchin embryos,[54,55] the lysosome associated cathepsin B-like protease may be more active in yolk utilization than the cytosol enzyme. Additional studies will be required to delineate the role of the particulate and cytosolic proteases.

VI. DEVELOPMENTAL PROFILE OF CATHEPSIN B-LIKE PROTEASE IN *ARTEMIA*

Attempts to understand developmental changes in the cathepsin B-like thiol protease have yielded conflicting results depending on the method of handling the embryos and cell-free extracts.[20,22,27,35,42] We have observed that cytosol fractions prepared from 0- to 16-hr embryos of NaOCl-treated cysts have highly depressed protease activity at pH 4, while cytosol fractions from swimming nauplii (20 to 36 hr) are unaffected.[22] As well, low molecular weight protease inhibitors in the cytosol of embryos depress protease activity measurements unless the protease is separated from these inhibitors prior to the assay.[43] Avoiding NaOCl treatments and the endogenous protease inhibitors, we have observed that the content of cathepsin B-like enzyme decreases slightly during early development reaching a constant, but lower level in the newly hatched larvae compared with the level in dormant embryos.[22] In contrast to these findings, studies by Spanish researchers have shown that the total cathepsin B-like activity increases dramatically in embryos immediately prior to hatching of nauplius larvae.[20,42] These experiments were carried out on NaOCl-treated cysts as starting material, and the embryo extracts taken for the protease assays contained the endogenous protease inhibitors. Attempts to resolve published discrepancies in protease activity measurements in extracts of *Artemia* embryos and larvae have not proved fruitful, and it appears that newer approaches other than activity measurements will be needed. If the data from the Spanish workers are correct in showing a dramatic increase in cathepsin B-like activity at the time of hatching, then further experiments will be needed to determine whether the increase in activity is due to gene activation, release of specific translational controls, or some other effect.

VII. APPEARANCE OF MULTIPLE PROTEASE ACTIVITIES IN LARVAE OF *ARTEMIA*

When dormant, encysted embryos of *Artemia* are hydrated fully and allowed to develop at 30°C in an aerated environment, the embryos hatch from a transparent membrane attached to the cyst wall beginning around 12 to 14 hr incubation. Shortly after hatching begins, a dramatic increase occurs in "alkaline" protease activity of cytosol fractions.[56] The increase in "alkaline" protease activity is due to the appearance of four proteases which have been designated proteases A, B, C, and D based mainly on their elution positions from a DEAE-cellulose column. The developmental profile of these enzymes is shown in Figure 4. The four proteases have been purified to near homogeneity from 60-hr nauplius larvae and partially characterized.[57,58]

Protease A, the first protease to elute from DEAE-cellulose, hydrolyzes the ester bond in benzyloxycarbonyl-l-leucine-*p*-nitrophenyl ester (CLNE), but not that in *p*-toluenesulfonyl-l-arginine methyl ester (TAME). The enzyme also shows slight lytic activity on *Artemia* yolk platelets, but it is inactive in the hydrolysis of casein and azocasein. Protease A is strongly inhibited by soybean-trypsin inhibitor (STI), ovomucoid, and PMSF, but not by TLCK. The purified protein has a molecular weight of 38,000 daltons, and it is relatively stable to high temperatures requiring heating at 60°C for 5 min to inactivate 50% of the enzyme molecules.[57] Protease A appears to be a chymotrypsin-like serine protease. While this protease appears to be barely detectable prior to hatching (see Figure 4), we have noted the presence of an enzyme with similar properties except for its high molecular weight (around 100,000 daltons) in the cytosol of dormant cysts.[43] Whether this enzyme represents a precursor of the 38,000-dalton enzyme in nauplii remains to be tested, but unmasking of alkaline protease(s) activity seems to occur in *Artemia* during development.[51]

Protease B, the second protease to elute from DEAE-cellulose, acts upon several substrates including α-*N*-benzoyl-DL-arginine *p*-nitroanilide (BAPNA), TAME, casein, azocasein, and *Artemia* lipovitellin. It is inhibited by STI, ovomucoid and TLCK but not by PMSF. It is slightly less stable to heating than protease A requiring only 5 min at 52°C to achieve 50% inactivation of the enzyme molecules.[57] This protease has a molecular weight of 33,000 daltons, and it appears to be a trypsin-like enzyme.

Protease C, the most acidic of the proteases in soluble extracts of 60-hr nauplii, elutes from a DEAE-cellulose column at about 0.85 *M* KCl. The enzyme hydrolyzes BAPNA, TAME, casein, azocasein, and *Artemia* lipovitellin (slowly). It is inhibited by STI, ovomucoid, and TLCK. It has a molecular weight of 34,000 daltons and like protease B, it also appears to be a trypsin-like enzyme.

Protease D elutes from a DEAE-cellulose column at 0.6 *M* KCl. It is active on casein and azocasein and slightly active on *Artemia* lipovitellin, but inactive on a variety of synthetic substrates including CLNE, BAPNA, and TAME. Among a variety of protease inhibitors tested including STI, ovomucoid, PMSF, TLCK, and *p*-hydroxymercuribenzoate, none showed any effect on protease D. In contrast, dithioerythritol (0.7 m*M*) and mercaptoethanol (5 m*M*) showed 50 and 30% inhibition, respectively.

The role of proteases A, B, C, and D in protein metabolism during early development is not known, but their rapid appearance in nauplii has raised several questions. What is (are) the source of each enzyme? Is the appearance of these enzymes the result of gene activation, or are they the products of translational control of stored mRNAs? Are the enzymes present in the embryo as inactive precursors or in the proenzyme states? Are any of these proteases associated with a specific cell type such as cells lining the larval gut? Garesse and co-workers[51] have suggested that proteases A, B, C, and D are present in a masked, high molecular weight complex of about 125,000 daltons in newly hatched nauplii, and that as development proceeds the active forms are released from the complex. While this proposal is attractive and has some experimental support, the active proteases released from the complex by the chaotropic agent, KI, have different molecular weights than the purified proteases found in the cytosol of 60-hr nauplii. If indeed proteases A, B, C, and D are present in an inactive high molecular weight complex, Western blot analyses of protein preparations from embryos/larvae at various stages of development using antibodies to each of the purified enzymes should confirm (or deny) this view.

In summary it seems clear that the role of proteases A, B, C, and D during development will require both immunochemical and immunohistochemical methods to localize these enzymes (or their precursors) in developing larvae. Results of these experiments will permit us to make more accurate statements on the source of these proteases and their probable function during development.

VIII. SIGNIFICANCE OF PROTEASES IN LARVAE OF *ARTEMIA*

Larvae of several invertebrate species have been shown to be rich in a variety of proteases including aspartate, cysteine, and serine proteases.[15,16,37-39,59] In insect larvae, the appearance of proteolytic activity in the alkaline pH range coincides with the beginning of midgut formation and isolated larval guts were found to be rich in alkaline protease activity.[15,19] In larvae of the mosquito, all proteases identified thus far appear to be of the serine-type,[39] while in the gut of blowfly larvae, some acid protease activity has been detected.[15] In insects the greatest increase in acid protease activity appears to be in salivary glands and residual tissues coincident with the onset of metamorphosis during the larva-pupa transition.[15,16] These findings are consistent with the proposal of Lockshin and Williams who stated that enhanced proteolytic activity is an integral part of programmed cell death as occurs during metamorphosis.[14]

In prenauplius larvae of *Artemia* (E-1 stage), the presumptive alimentary tube is clearly apparent 6 to 8 hr before hatching.[60,61] Subsequent to hatching the alimentary tube differentiates into three main regions (foregut, midgut, and hindgut) and acquires the appearance of being fully functional at the time of the first larval moult at about 34 hr development (at 25°C).[62,63] Since the trypsin-like proteases B and C in *Artemia* increase markedly during the period of gut differentiation (see Figure 4), these hydrolases may be digestive enzymes located in cells lining the gut, while the increase in activity of larval proteases and D occurs after gut formation. Whether any of the *Artemia* larval proteases identified thus far are active in the digestive mechanism of *Artemia*, or in the programmed destruction of the larval salt gland (e.g., neck organ[64]) during postlarval development remains to be ascertained using *in situ* immunohistochemical techniques.

IX. PROTEASE ACTIVITY IN *ARTEMIA* ADULTS

Although the literature on protease activity in adults of *Artemia* is limited, available evidence indicates that the larval proteases A, B, C, and D are also present in adults.[45] In addition, latent protease activity has been found in a blue-green glycoprotein called artemocyanin which co-purifies, as a contaminant, with hemoglobin. Two forms of artemocyanin have been described.[65,66] Artemocyanin I is found in the hemolymph of adults and composed of polypeptides of M_r 190,000 (α), 104,000 (β), and 87,000 (δ). Artemocyanin II has been isolated from homogenates of male and female adult *Artemia* and appears to be a modified form of artemocyanin I. It is composed of polypeptides of M_r 104,000, 87,000, and 67,000. Artemocyanin II has latent protease activity which is activated in the presence of 1 to 5% SDS, while artemocyanin I is devoid of protease activity in the presence or absence of protein denaturants such as SDS. Current data support the view that the M_r 67,000 (γ) polypeptide of artemocyanin II is the source of the protease activity.[65,67] The processing of artemocyanin I to artemocyanin II in vivo appears to be at least a two-step process, one leading to the transformation of the M_r 87,000 polypeptide to the M_r 67,000 polypeptide, and a second leading to the modification of the 67,000 polypeptide (or perhaps the 87,000 polypeptide before its cleavage) rendering the product active as a protease. The SDS-activated proteolytic activity in artemocyanin II has been shown to be responsible for the degradation of hemoglobin in vitro, but the physiological significance of latent protease activity in artemocyanin II and its localization in adults is unknown.[66,67]

Two additional SDS-stable proteases of M_r 25,000 and 36,000 have been purified from homogenates of adult *Artemia*, but very little information is available on these enzymes.[67]

X. PROTEASE INHIBITORS IN EMBRYOS, LARVAE, AND ADULTS OF *ARTEMIA*

Regulation of intracellular protease activity is an important event in the maintenance of cell viability and prevention of inappropriate proteolysis. In most cells and tissues, including embryos, control of protease activity is probably exerted at a variety of sites and by several mechanisms including the ionic environment, intracellular pH (pH_i), compartmentation, and presence of inhibitory proteins. Recent studies have shown that endogenous protease inhibitors are present in most cell types and three classes of intracellular protease inhibitors have been identified: thiol (cysteine) protease inhibitors, serine protease inhibitors, and metalloprotease inhibitors.[18,19] In addition several extracellular protease inhibitors have also been identified.[18]

In *Artemia*, protease inhibitors have been found in the cytosol and yolk platelets of embryos and larvae and in adults.[21,33,49] A summary of these studies is presented here.

A. Inhibitors in the Cytosol of Embryos and Larvae

Encysted embryos and prenauplius larvae of *Artemia* contain two low molecular weight protease inhibitors. These inhibitors (Ia and Ib) have been purified to near homogeneity by chromatography on columns of DEAE- and CM-cellulose, Sephadex G-50 and CM-Affi Gel Blue.[43] Both Ia and Ib have molecular weights of about 9,000 by gel filtration, and they show an ID_{50} of 1 to 2 μg/mℓ using the *Artemia* cathepsin B-like thiol protease as test enzyme at pH 5 and 40°C. While Ia and Ib are active in the pH range of 4.0 to 6.5, these inhibitors are most effective in the upper part of this range (pH 5.5 to 6.5) where they are resistant to hydrolysis by the thiol protease.

Protease inhibitor activity has been measured in protease-free preparations from *Artemia* embryos and larvae at various times during development.[43] These measurements showed that the inhibitor activity is maximum shortly before emergence of prenauplius larvae then

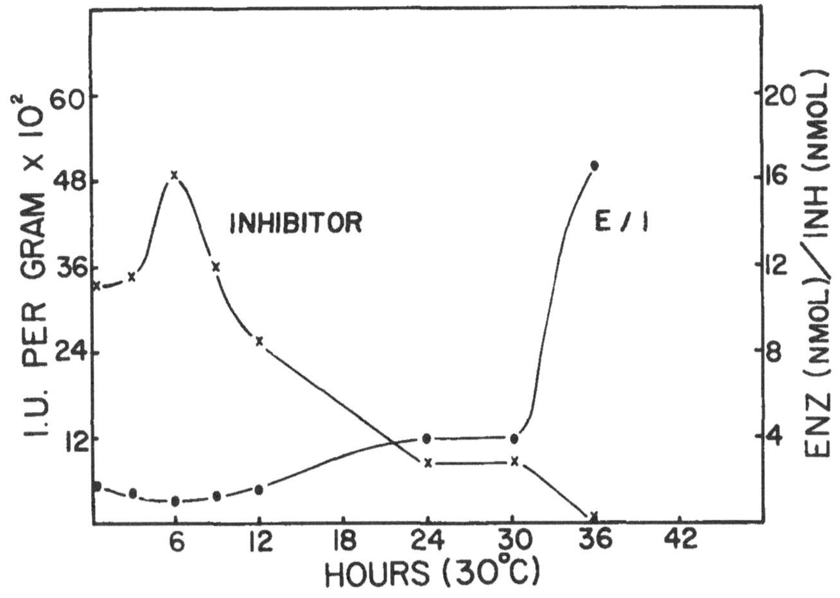

FIGURE 5. Thiol protease inhibitor activity in *Artemia* embryos and larvae. The content of thiol protease inhibitor (×——×) was measured in the protease-free, DEAE-cellulose fraction of the postribosomal fraction prepared from embryos at different times of development. One inhibitor unit (I.U.) is defined as the amount of protein producing 1% inhibition of cathepsin B-like thiol protease of *Artemia* embryos. The I.U. data are expressed per gram wet weight of fully hydrated starting cysts. The curve labeled E/I (●——●) represents the estimated molar ratio of cathepsin B-like thiol protease to inhibitor protein based on protease measurements on the same extract after elution of the enzyme from the DEAE-cellulose column and using specific activity data obtained previously.[22]

declines rapidly over the next 27 to 30 hr of development (30°C). These results are shown in Figure 5. From specific activity data of the target thiol protease[35] and protease inhibitor, the ratio of enzyme-to-inhibitor (E/I) has been found to increase dramatically after hatching of prenauplius larvae, and about the time of the first larval moult, regulation of the thiol protease activity by the protease inhibitor becomes minimal.

Encysted embryos of *Artemia* have acquired the ability, through evolution, to withstand prolonged periods of anoxia (several months) during which time the pH_i falls to near or below pH 6.0.[68] The resistance to anoxia is lost after hatching of nauplii concomitant with loss in protease inhibitor levels in the cytosol and increase in thiol protease to protease inhibitor ratio. Thus, it is tempting to speculate that endogenous protease inhibitors play a protective role during early development by preventing activation of the thiol proteases and hydrolysis of numerous enzymes/proteins which could result in cytolysis and death of young nauplius larvae.

B. Inhibitors in Yolk Granules

Yolk platelets in *Artemia* embryos are storage sites of about two thirds of the embryo's protein.[41] The major component in *Artemia* yolk platelets is lipovitellin a complex consisting of four polypeptides, carbohydrates, and lipids.[48] Solubilized lipovitellin has been found to inhibit the *Artemia* trypsin-like protease by noncompetitive inhibition similar to that found for the α_2-macroglobulins.[49] However, it does not inhibit porcine trypsin or *Artemia* cathepsin B-like protease. The inhibitor effect of lipovitellin is more pronounced when proteins are used as substrates compared with synthetic substrates. During development the inhibitor is believed to autoregulate, in some way, the hydrolysis of lipovitellin by the yolk-associated trypsin. The developmental profile of this inhibitor has not been determined.

C. Inhibitors in Adults

Gunther-Sillero and her co-workers[45] have reported the presence of two protease inhibitors in extracts from adults of two populations of *Artemia*. One inhibitor is active on protease A (present in both larvae and adults), while the second inhibitor is active on proteases B and C. The protease A inhibitor is resistent to heating at 100°C (5 min) and protease B/C inhibitor is resistent to heating at 75°C (5 min), but sensitive to heating at 100°C. The protease A inhibitor is inactivated by pronase and proteinase K suggesting that it is protein in nature. Protease B/C inhibitor also appears to be a macromolecule since it is excluded from a Sephadex G-50 column, but no data are available on whether the inhibitor is a protein. The specificities of these inhibitors for various proteases and their localization in adult tissues are not known.

XI. SUMMARY AND FUTURE PROSPECTS

The occurrence of multiple proteases in embryos and larvae of the brine shrimp, *Artemia* has been described. Five of these hydrolases have been purified to apparent homogeneity including a cathepsin B-like thiol protease in embryos, and two trypsin-like proteases and one chymotrypsin-like protease in larvae. One other protease has been shown to be sequestered in yolk platelets and another is present in an inactive state in the cytosol. Notwithstanding, a relatively large body of data on the biochemistry of these hydrolases, we can still only speculate about the function of these proteases during development. While the trypsin-like protease associated with yolk platelet lipovitellin in *Artemia* appears to be important in yolk protein metabolism, other data also suggest that the cathepsin B-like enzyme in the cytosol and/or lysosomes may also be involved in the hydrolysis of lipovitellin. It is also tempting to speculate that two proteases in *Artemia* larvae (proteases B and C) are digestive enzymes which are synthesized or activated concomitant with gut differentiation.

The presence of low molecular weight thiol protease inhibitors in the cytosol and a lipovitellin-associated trypsin inhibitor in yolk platelets suggest that these proteins may be important in protein turnover regulation in the cytosol and yolk, respectively. Again, direct evidence in support of these views is lacking. Until such time as the various proteases are localized in *Artemia* embryos and larvae using state-of-the-art immunohistochemical techniques and analyzed using specific protease inhibitors, the role of these enzymes in protein metabolism will remain in doubt.

However, biochemical data suggest that at least four proteases and two protease inhibitors in *Artemia* may be developmentally regulated. Work is currently in progress in my laboratory to elucidate the structure and expression of the cathepsin B-like protease and protease inhibitor genes in *Artemia*. Through use of immunochemical and cloning techniques it should be possible to determine whether proteases and protease inhibitors in *Artemia*, particularly those in the cytosol of embryos and larvae, are under transcriptional or post-transcriptional controls. Antibody and nucleic acid probes will also be used to determine which cells in the embryo and larva are active in protease and protease inhibitor gene expression. Results of these studies should provide us with a better understanding of the relationship between protease production, control, and function during development.

ACKNOWLEDGMENTS

The author wishes to thank Dr. D. A. Cotter, Department of Biological Sciences, University of Windsor for a critical reading of the manuscript. Thanks are also due to authors of previous works who allowed their figures to be reproduced here. This work on *Artemia* has been suported by the Natural Sciences and Engineering Research Council of Canada, Grant A-2909.

REFERENCES

1. **Spirin, A. S.,** On "masked" forms of messenger RNA in early embryogenesis and in other differentiating systems, in *Current Topics in Developmental Biology,* Vol.1, Moscona, A. A. and Monroy, A., Eds., Academic Press, New York, 1966, 1.
2. **Metafora, S., Felicetti, L., and Gambino, R.,** The mechanism of protein synthesis activation after fertilization of sea urchin eggs, *Proc. Natl. Acad. Sci. U.S.A.,* 68, 600, 1971.
3. **Monroy, A., Maggio., R., and Rinaldi, M.,** Experimental induced activation of the ribosomes of the unfertilized sea urchin egg, *Proc. Natl. Acad. Sci. U.S.A.,* 54, 107, 1965
4. **Williams, J.,** Yolk utilization, in *The Biochemistry of Animal Development,* Vol. 2., Weber, R., Ed., Academic Press, New York, 1967, 341.
5. **Green J.,** Chemical embryology of the crustacea, *Biol. Rev.,* 40, 580, 1965.
6. **Dean, R. T.,** Regulation and mechanisms of degradation of endogenous proteins by mammalian cells: general considerations, in *Degradative Processes in Heart and Skeletal Muscle,* Wildenthal, K., Ed., Elsevier/North-Holland, Amsterdam, 1980, 3.
7. **Hershko, A. and Ciechanover, A.,** Mechanisms of intracellular protein breakdown, *Annu. Rev. Biochem.,* 51, 335, 1982.
8. **Segal, H. L. and Doyle, D. J., Eds.,** *Protein Turnover and Lysosome Function,* Academic Press, New York, 1978, 1.
9. **Barrett, A. J.,** The many forms and functions of cellular proteinases, *Fed. Proc.,* 39, 9, 1980.
10. **Zwilling, R. and Neurath, H.,** Invertebrate proteases, *Methods Enzymol.,* 80, 633, 1981.
11. **Gates, B. J. and Travis, J.,** Isolation and comparative properties of shrimp trypsin *(Penaeus setiferus), Biochemistry,* 8, 4483, 1969.
12. **Gates, B. J. and Travis, J.,** Purification and characterization of carboxypeptidases A and B from the white shrimp, *Penaeus setiferus, Biochemistry,* 12, 1867, 1973.
13. **Galgani, E. G., Benyamin, Y., and Ceccaldi, H. J.,** Identification of digestive proteinases of *Penaeus kerathurus* (Forskal): a comparison with *Penaeus japonicus* bate., *Comp. Biochem. Physiol.,* 78B, 355, 1984.
14. **Lockshin, R. A. and Williams, C. M.,** Programmed cell death. V. Cytolic enzymes in relation to the breakdown of the intersegmental muscle of silkmoths, *J. Insect Physiol.,* 11, 831, 1965.
15. **Kawamura, M., Wadano, A., and Miura, K.,** Purification and some properties of cathepsin-like thiol protease from pupae of the blowfly *Aldrichina grahami, Comp. Biochem. Physiol.,* 78B, 279, 1984.
16. **Henrikson, P. A. and Clever, U.,** Protease activity and cell death during metamorphosis in the salivary gland of *Chironomus tentans, J. Insect Physiol.,* 18, 1981, 1972.
17. **Katunuma, N., Kominami, E., Banno, Y., Kito K., Aoki, Y., and Urata, G.,** Concept on mechanism and regulation of intracellular enzyme degradation in mammalian tissues, *Adv. Enzyme Regul.,* 14, 325, 1976.
18. **Katunuma, N. and Kominami, E.,** Molecular basis of intracellular regulation of thiol proteinase inhibitors, *Curr. Top. Cell Regul.,* 27, 345, 1985.
19. **Lenny, J. F.,** Inhibitors associated with the proteinases of mammalian cells and tissues, *Curr. Top. Cell Regul.,* 17, 25, 1980.
20. **Perona, R. and Vallejo, C. G.,** Acid hydrolases during *Artemia* development: a role in yolk degradation, *Comp. Biochem. Physiol.,* 81B, 993, 1985.
21. **Ezquieta, B. and Vallejo, C. G.,** The trypsin-like proteinase of *Artemia:* yolk localization and developmental activation, *Comp. Biochem. Physiol.,* 82B, 731, 1985.
22. **Warner, A. H. and Shridhar, V.,** Purification and charterization of a cytosol protease from dormant cysts of the brine shrimp *Artemia, J. Biol. Chem.,* 260, 7008, 1985.
23. **Twardowski, T., Hill J. M., and Weissbach, H.,** Disaggregation of elongation factor 1 by extracts of *Artemia salina, Biochem. Biophys. Res. Commun.,* 71, 826, 1976.
24. **Yablonka-Reuveni, Z. and Warner, A. H.,** Characterization of elongation factor 2 in dormant cysts and developing embryos of *Artemia salina,* in *Biochemistry of Artemia Development,* Bagshaw, J. C. and Warner, A. H., Eds., University Microfilms International, Ann Arbor, Mich., 1979, 22.
25. **Osuna, C., Renart, J. and Sebastian, J.,** Proteolytic origin of a modified form of RNA polymerase I from *Artemia salina* larvae, *Biochem. Biophys. Res. Commun.,* 78, 1390, 1977.
26. **Urbani, E., Rognone, L., and Russo, S.,** Osservazioni sulla vita latente. I. Studio della dipeptidasi di *Artemia salina* L., *Atti Accad. Naz. Lincei Rc.,* 13, 300, 1952.
27. **Bellini, L.,** Studio delle dipeptidasi e proteinasi nello sviluppo di *Artemia salina* Leach, *Atti Accad. Nazion. Lincei Rc.,* 22, 340, 1957.
28. **Bowen, S. T., Fogarino, E. A., Hitchner, K. N., Dana, G. L., Chow, V. H. S., Buoncristiani, M. R., and Carl, J. R.,** Ecological isolation in *Artemia:* population differences in tolerance of anion concentrations, *J. Crust. Biol.,* 5, 106, 1985.

29. **Bagshaw, J. C. and Warner, A. H.,** Restriction fragment length polymorphism in isolated populations of *Artemia, Fed. Proc.*, 45, 1883, 1986.
30. **Bagshaw, J. C.** Population polymorphism of the histone +5S RNA gene repeat unit and other repeated DNA sequenes in *Artemia, Fed. Proc.*, 46, 1947, 1987.
31. **Gallego, M. E., Diaz-Guerra, M., Cruces, J., Sebastian, J., and Renart, J.,** Characterization of two types of rRNA gene repeat units from the crustacean *Artemia, Gene*, 48, 175, 1986.
32. **Warner, A. H. and Bagshaw, J. C.,** Letter to the editor, *Artemia Newsl.* 1, 15, 1986.
33. **Nagainis, P. A. and Warner, A. H.,** Evidence for the presence of an acid protease and protease inhibitors in dormant embryos of *Artemia salina, Dev. Biol.*, 68, 259, 1979.
34. **Perona, R. and Vallejo, C. G.,** The lysosomal proteinase of *Artemia:* purification and characterization, *Eur. J. Biochem.*, 124, 357, 1982.
35. **Warner, A. H. and Shridhar. V.,** Characterization of an acid protease from encysted embryos of *Artemia*, in *The Brine Shrimp Artemia*, Vol. 2, Persoone, G., Sorgeloos, P., Roels, O., and Jaspers, E., Eds., Universa Press, Wetteren, Belgium, 1980, 355.
36. **De Chaffoy de Courcelles, D. and Kondo, M.,** Lipovitellin from the crustacean, *Artemia salina, J. Biol. Chem.*, 255, 6727, 1980.
37. **Kucera, M. and Turner, R. B.,** Changes in activity of proteases during embryogenesis of *Anagasta kuehniella* (Insecta), *Comp. Biochem. Physiol.*, 44B, 577, 1973.
38. **Kuk-Meiri, S., Lichtenstein, N., Shulov, A., and Pener, M. P.,** Cathepsin-type proteolytic activity in the developing eggs of the African migratory locust *(Locusta migratoria migratorioides* R and F), *Comp. Biochem. Physiol.*, 18, 783, 1966.
39. **Kunz, P. A.,** Resolution and properties of the proteinases in the larva of the mosquito, *Aedes aegypti, Insect Biochem.*, 8, 43, 1978.
40. **Carroll, E. J. and Epel, D.,** Isolation and biological activity of the proteases released by sea urchin eggs following fertilization, *Dev. Biol.*, 44, 22, 1975.
41. **Vallejo, C. G., Perona, R., Garesse, R., and Marco, R.,** The stability of the yolk granules of *Artemia*. An improved method for their isolation and study, *Cell Differ.*, 10, 343, 1981.
42. **Marco, R., Vallejo, C. G., Garresse, R., and Perona, R.,** A hypothesis on the activation process of proteolytic activities during *Artemia* early development, in *The Brine Shrimp Artemia*, Vol. 2, Persoone, G., Sorgeloos, P., Roels, O., and Jaspers, E., Eds., Universa Press, Wetteren, Belgium, 1980, 315.
43. **Warner, A. H.,** The role of proteases and their control in *Artemia* development. in *Artemia Research and It's Application*, Vol. 2, Decleir, W., Moens, L., Slegers, H., Sorgeloos, P., and Jaspers, E., Universa Press, Wetteren, Belgium, in press, 1987.
44. **Warner, A. H., Puodziukas, J. G., and Finamore, F. J.,** Yolk platelets in brine shrimp embryos, *Exp. Cell Res.*, 70, 365, 1972.
45. **Sillero, M. A. G., Burillo, S. L., Dominguez, E., Olalla, A., Osuna, C., Renart, J., Sebastian. J., and Sillero, A.,** Multiple proteolytic enzymes in *Artemia*, in *The Brine Shrimp Artemia*, Vol. 2, Persoone, G., Sorgeloos, P., Roels, O., and Jaspers, E., Eds., Universa Press, Wetteren, Belgium, 1980, 345.
46. **Vallejo, C. G., Marco, R., De Chaffoy, D., and Kondo, M.,** Yolk metabolization and mitochondrial release in *Artemia*, in *Endocytobiology*, Vol. 1, Schwemmler, W. and Schenk, H. E. A., Eds., Walter de Gruyter, Berlin, 747, 1980.
47. **Vallejo, C. G., Sillero, M. A. G., and Marco, R.,** Mitochondrial maturation during *Artemia salina* embryogenesis. General description of the process, *Cell. Mol. Biol.*, 25, 113, 1979.
48. **De Choffoy, D., Heip, J., Moens, L., and Kondo, M.,** *Artemia* lipovitellin, in *The Brine Shrimp Artemia*, Vol. 2, Persoone, G., Sorgeloos, P., Roels, O., and Jaspers, E., Eds., Universa Press, Wetteren, Belgium, 1980, 379.
49. **Ezquieta, B. and Vallejo, C. G.,** Lipovitellin inhibition of *Artemia* trypsin-like proteinase: a role for a storage protein in regulating proteinase activity during development, *Arch. Biochem. Biophys.*, 250, 410, 1986.
50. **Slaughter, D. and Triplett, E.,** Amphibian embryo protease inhibitor. II. Biological properties of the inhibitor and its associated protease, *Cell Differ.*, 4, 23, 1975.
51. **Garesse, R., Perona, R., Marco, R., and Vallejo, C.,** The unmasking of proteolytic activity during the early development of *Artemia salina, Eur. J. Biochem.*, 106, 225, 1980.
52. **Slaughter, D., Schoeffel, S., and Triplett, E.,** Amphibian embryo protease inhibitors. IV. Studies on an inhibitor of chymotrypsin and papain, *Cell Differ.*, 6, 273, 1977.
53. **Salisbury, N., Calaprice, N., and Triplett, E. L.,** Amphibian embryo protease inhibiotors. VI. Maternal origin and identity with lipovitellin heavy subunit, *Cell Differ.*, 9, 219, 1980.
54. **Schuel, H., Wilson, W. L., Wilson, J. R., and Bressler, R. S.,** Heterogeneous distribution of lysosomal hydrolases in yolk platelets isolated from unfertilized sea urchin eggs by zonal centrifugation, *Dev. Biol.*, 46, 404, 1975.
55. **Pasteels, J. J.,** Yolk and lysosomes, in *Lysosomes in Biology and Pathology*, Vol. 3, Dingle, J. T., Ed., Joh Wiley & Sons, New York, 1973, 216.

56. **Osuna, C., Olalla, A., Sillero, A., Sillero, M. A. G., and Sebastian, J.**, Induction of multiple proteases during the early larval development of *Artemia salina, Dev. Biol.*, 61, 94, 1977.
57. **Olalla, A., Osuna, C., Sebastian, J., Sillero, A., and Sillero, M. A. G.**, Purification and properties of three proteases from the larvae of the brine shrimp *Artemia salina, Biochem. Biophys. Acta*, 523, 181, 1978.
58. **Burillo, S. L., Sillero, A., and Sillero, M. A. G.**, Partial purification and properties of protease D from larvae of *Artemia salina, Comp. Biochem. Physiol.*, 71B, 89, 1982.
59. **Suzuki, T. and Natori, S.**, Purification and characterization of an inhibitor of the cysteine protease from the hemolymph of *Sarcophaga peregrina* larvae, *J. Biol. Chem.*, 260, 5115, 1985.
60. **Clegg, J. S. and Conte, F. P.**, A review of the cellular and developmental biology of *Artemia*, in *The Brine Shrimp Artemia*, Vol. 2, Persoone, G., Sorgeloos, P., Roels, O., and Jaspers, E., Eds., Universa Press, Wetteren, Belgium, 1980, 11.
61. **Weisz, P. B.**, The histological pattern of metameric development in *Artemia salina, J. Morphol.*, 81, 45, 147.
62. **Hootman, S. R. and Conte, F. P.**, Fine structure and function of the alimentary epithelium in *Artemia salina* nauplii, *Cell Tissue Res.*, 155, 423, 1974.
63. **Anderson, D. T.**, Larval development and segment formation in the branchiopod crustaceans *Limnadia stanleyana* King (Conchostraca) and *Artemia salina* (L) (Anostraca). *Aust. J. Zool.*, 15, 47, 1967.
64. **Conte, F. P., Hootman, S. R., and Harris, P. J.**, Neck organ of *Artemia salina* nauplii, *J. Comp. Physiol.*, 80, 239, 1972.
65. **Krissansen, G. W., Trotman, C. N. A., and Tate, W. P.**, Physical and biochemical characterization of artemocyanin, an abundant glycoprotein complex with latent proteolytic activity from the brine shrimp *Artemia, Biochim. Biophys. Acta*, 747, 151, 1983.
66. **Krissansen, G. W., Trotman, C. N. A., and Tate, W. P.**, Two forms of the glycoprotein artemocyanin from the brine shrimp *Artemia, Biochim. Biophys. Acta*, 747, 159, 1983b.
67. **Krissansen, G. W., Trotman, C. N. A., and Tate, W. P.**, Artemocyanin: a protease zymogen rather than a protease inhibitor, *Comp. Biochem. Physiol.*, 80B, 883, 1985.
68. **Busa, W. B.**, Cellular dormancy and the scope of pH_i-mediated metabolic regulation, in *Intracellular pH: Its Measurement, Regulation and Utilization in Cellular Functions*, Nuccitelli, R. and Deamer, D. W., Eds., Alan R. Liss, New York, 417, 1982.
69. **Warner, A. H.**, unpublished observations.

Chapter 6

THE HEMOGLOBINS OF *ARTEMIA;* OXYGEN CARRIERS WITH MULTIDOMAIN GLOBIN CHAINS

L. Moens, K. De Smet, M.-L. Van Hauwaert, D. Geelen, G. Verpooten, and J. Van Beeumen

TABLE OF CONTENTS

I. INTRODUCTION: THE EXTRACELLULAR HEMOGLOBINS OF THE INVERTEBRATES

In contrast with the hemoglobins of the vertebrates, which are almost invariably intracellular and tetrameric, the intra- and extracellular hemoglobins of the invertebrates show a wide variety in their molecular size ($M_r \sim 16,000$ up to $\sim 1.7 \times 10^6$ and architecture.[1-5] With the exception of the hemoglobins of the chironomide larvae[6] practically all of the extracellular hemoglobins have a high M_r which is probably necessary to avoid elimination from the hemolymph by excretory processes and a too high osmotic pressure.[4]

Despite their heterogeneity, Svedberg and Hedenius[7] suggested that all these pigments are composed of myoglobin-like polypeptide chains of $M_r \sim 16,000$ containing one heme group and able to bind oxygen reversibly. Polypeptide chains or fragments of much longer chains having these characteristics ($M_r \sim 16,000$; one heme) were defined by Vinogradov[5] as "heme-binding domains". Based on the number of domains and subunits in the native molecule, the invertebrate extracellular hemoglobins can be classified into four groups.

- Single-domain, single-subunit hemoglobins consisting of a single polypeptide chain, containing one heme group and having an $M_r \sim 16,000$ (*Chironomus*).
- Single-domain, multisubunit hemoglobins consisting of aggregates of monomeric subunits, some of which are disulfide bonded (*Annelida*).
- Two-domain, multisubunit hemoglobins consisting of aggregates of dimeric polypeptide chains (M_r 30,000 to 40,000), each containing two heme-binding domains (*Arthropoda, Nematoda*).
- Multidomain, multisubunit hemoglobins consisting of 2 or more polypeptide chains each comprising from 8 to 20 heme-binding domains (*Arthropoda, Mollusca*).

Although the majority of Crustacaea have hemocyanin as oxygen-carrier, extracellular hemoglobins are mainly found in the hemolymph of the Branchiopoda. The carapaced branchiopods, such as *Lepidurus, Daphnia, Cyzicus, Caenestheria*, all have hemoglobins with molecular masses varying between 250,000 to 800,000 and consisting of 10 to 24 two-domain (M_r 30,000 to 40,000) subunits.[8-11]

The carapaceless branchiopods, including *Artemia, Parartemia*, and *Streptocephalus* have M_r of 240,000 to 300,000 and consist of two multidomain (6 to 8) subunits. Similar hemoglobins, with M_r of 1.5 to 1.75×10^6 and containing several multidomain (10 to 20) subunits, are also found in gastropod snails and clams.[12-15] The *Artemia* hemoglobins thus represent a vast group of invertebrate extracellular hemoglobins on which very little information is available. In this review we will discuss their structure, ontogenetically programmed synthesis, and function.

II. THE HEMOGLOBIN STRUCTURE

The hemolymph of the brine shrimp, *Artemia*, contains three hemoglobins which can be identified electrophoretically by their order of migration toward the cathode at pH 8 to 9 on cellulose acetate or on polyacrylamide gel supports as Hb I, Hb II, and Hb III. The first detailed physiological studies on this hemoglobin were described by Gilchrist in 1954.[88]

However, the biochemical characterization of the *Artemia* hemoglobins was reported only after 1970, mainly by two laboratories, one in San Francisco (S. Bowen and co-workers, see References 16, 17, 37, 39, 48, 49, 50, 90) and our own in Antwerp. Both groups used mainly the Great Salt Lake variety for the primary characterization of these hemoglobins, whereas for the determination of the globin substructure, the variety Aigle Morte (Le Grau du Roi, France) was used.

A. Isolation

A crude hemoglobin fraction can be prepared by ammoniumsulfate (50%) precipitation of a homogenate of *Artemia* and DEAE chromatography with step elution (125 and 225 m*M* NaCl, pH 7.5). A second chromatography on DEAE Sephadex A50 with gradient elution allows the separation of the three individual hemoglobins[16,17,18] (Figure 1).

However, in *Artemia* extracts a great variety of proteases are present.[20-24] They readily attack the hemoglobins during purification[25-26] resulting in a set of more or less nicked molecules that partially copurify with the native hemoglobins. This was the basis of a different interpretation of the hemoglobin subunit structure, as will be discussed later.[17,19]

As the proteolytic activity in the extract is extremely difficult to inhibit completely, an alternative purification scheme was proposed.[27,28] Using polyethyleneglycol (PEG) instead of ammoniumsulfate precipitation, much more membraneous material is eliminated in the first steps, resulting in a higher binding efficiency, and by avoiding a desalting step a much quicker separation of the general proteases from the hemoglobins by stepwise DEAE Sephadex A50 chromatography could be obtained.

At this stage, the hemoglobins are still contaminated by several other proteins from the hemolymph, including the artemocyanins,[29-33] which are blue-green pigments carrying a biliverdin-related chromophore. Two different molecular forms, artemocyanin I ($S_{20,W}$ = 28.1) and artemocyanin II ($S_{20,w}$ = 27.3) were identified. Artemocyanin I, which is a zymogen, (M_r = 1.2×10^6, polypeptide M_r 190,000) may be converted by limited proteolysis to artemocyanin II (M_r = 1.2×10^6, polypeptide M_r = 104,000/87,000/67,000). The smallest polypeptide of artemocyanin II exhibits latent protease activity, which is specifically activated by denaturing conditions. It is essential to separate it from the hemoglobins. This can be obtained by preparative polyacrylamide gel electrophoresis under nondenaturing conditions which separates the trimer and dimer of artemocyanin II from the hemoglobins. Based on the fact that the artemocyanins are glycoproteins, the monomer of artemocyanin II can be separated from the hemoglobins by affinity chromatography on a concanavalin A Sepharose 4B column.[27,31]

Hemoglobin prepared in this way contains a high percentage of intact globin chains. However, this method has two disadvantages. First, it needs fresh or carefully purified and stored *Artemia* as starting material and it is much less succesful with commercially available frozen *Artemia*. Second, the preparative electrophoresis step is a limiting factor for large scale preparations which are necessary for structural work.

B. Physicochemical Characterization of the Native Hemoglobins

Several physicochemical parameters of the purified native hemoglobins are summarized in Table 1. The absorption spectrum of the isolated pigments, liganded or not, is virtually identical to that of human hemoglobin and those of many other invertebrates.[1] This suggests the presence of protoporphyrin IX as the chromophore.

Based on the iron and heme content, a theoretical minimal molecular weight of one heme group for each $M_r \sim 18,000$ can be calculated.[26] This is in fairly good agreement with many other invertebrate hemoglobins supporting Svedberg's postulate.[1,4,5,7] The apparent M_r and sedimentation coefficient of the native hemoglobins have been estimated to be 230,000 to 270,000 and 11.3 to 11.75 respectively. With the exception of gradient polyacrylamide gel electrophoresis, all results are in good agreement with each other.[35] The M_r obtained for the three hemoglobins by the latter method are significantly higher (Table 1).

Our estimations are obtained by interpolation on a calibration curve constructed by means of standards with known molecular weight.[38] Similar values are obtained by applying the method of Manwell,[41] using an empirical formula based on the hypothetical limiting polyacrylamide gel pore size. Also Bucks and Bowen[39] obtained similar values (M_r = 310,000) by using a retardation coefficient estimated by the negative slope of a Ferguson plot.

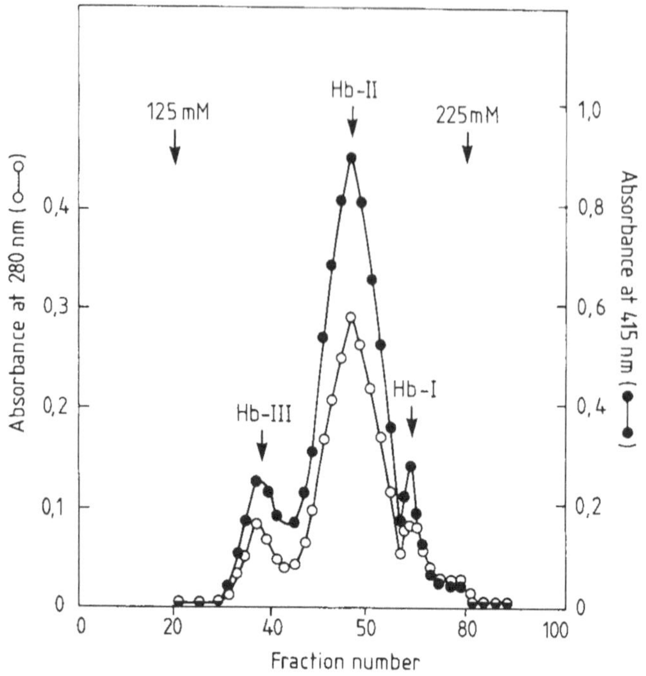

FIGURE 1. Separation of the *Artemia* hemoglobins by ion exchange chromatography. A total hemoglobin fraction isolated from adult *Artemia* was loaded on a DEAE-Sephadex A 50 column equilibrated in 125 m*M* NaCl, 50 m*M* Tris-HCl (pH 7.5). The column was developed with a linear gradient of 125 to 225 m*M* NaCl in 50 m*M* Tris-HCl (pH 7.5). (From Moens, L. and Kondo, M., *Eur. J. Biochem.*, 82, 65, 1978. With permission.)

The observed differences in the M_r determined by conventional methods and by electrophoretic techniques are most likely caused by the hydrodynamic behavior of the hemoglobins in the polyacrylamide network and the electrical field. Although the absolute value determined with these techniques may be a matter of controversy, a small relative difference in molecular weight between Hb I, Hb II, and Hb III was detected (Table 1).

The three *Artemia* hemoglobins have extremely similar amino acid compositions which are comparable to those of other invertebrate hemoglobins (Table 2).[1,4] Like other molluskan and crustacean extracellular hemoglobins, they have a very low cystein content which is in contrast with the giant hemoglobins of the annelids. In the latter globins, the cystein residues allow the formation of interchain bridges which contribute to the stabilization of many low M_r chains into a high M_r aggregate. On the other hand, the domains of the molluskan and crustacean globins are already linked in the native globin chains, so that they don't need extra disulfide bridges. Based on the amino acid composition, a mimimum subunit M_r of 17,500 can be calculated. This is in good agreement with the theoretical minimum M_r based on iron and heme content suggesting the existence of a structural unit (domain) of that size.[17,19,26]

The morphology of the *Artemia* hemoglobin, as observed in the electron microscope, is not so clear-cut as other extracellular respiratory pigments.[35] However, two morphologically distinct features could be identified: a circular lobed structure with a diameter of 120 ± 10 Å and a rectangular one with dimensions 120 ± 10 Å × 70 ± 10 Å composed of two equal parts divided by a dark line. A possible interpretation is that the former is a top view of the hemoglobin molecule and the latter a side view of two stacked, lobed disks. The

Table 1
PHYSICOCHEMICAL CHARACTERISTICS OF THE *ARTEMIA* HEMOGLOBINS

	Hb I	Hb II	Hb III	Ref.
$S_{20,w}$11.3—11.7	11.3—11.7	11.3—11.7	16, 17, 34, 35	
$D_{20,w}$(cm^2/sec)	—	(3.7 ± 0.1) 10^{-7}	—	34
V (cm^3/g)	—	0.737	—	34
Form factor (f/f_0)	—	1.03	—	34
Axial ratio	—	1.7	—	34
E.M. dimensions (Å)	—	120 × 70	—	35
$M_r \times 10^3$		240—270		
Gel filtration	—	240.4 ± 2.6	—	17, 19, 36
Equilibrium sedimentation	—	230—270	—	34, 35
From Svedberg relation	230—270	250	230—270	17, 34, 35
Light scattering	—	300—310	—	34
Gradient gel electrophoresis	270	5.7	330	37, 38, 39
pI	5.6	3.6	5.9	18, 26
Heme content (%)	—	0.29	—	17, 26, 40
Iron content (%)	—	17.09	—	17, 26, 40
E 1% (0.1*M*NaCl, 50 m*M* Tris,		47.71		17, 19
HCl pH 7.5)		4.87		
280 nm		4.30		
412 nm				
540 nm				
575 nm				

Table 2
AMINO ACID COMPOSITION OF THE *ARTEMIA* HEMOGLOBINS AND SOME OF THEIR DOMAINS (EXPRESSED AS NUMBER OF RESIDUES PER M_r 17,000)

	Intact molecules			Domains		
				E_1		E_7
Amino acid	Hb I	Hb II	Hb III	Analyzed	Sequenced	Analyzed
Asp	15.20	14.59	14.56	14.43	14	17.50
Thr	7.34	6.80	7.28	6.87	7	9.19
Ser	8.42	7.82	8.35	7.73	8	8.13
Glu	17.80	17.22	17.08	21.09	21	13.78
Pro	5.61	5.23	5.62	4.83	4	4.11
Gly	10.67	10.09	10.94	10.82	12	10.90
Ala	12.21	11.36	11.50	9.92	10	14.98
Cys	1.14	1.14	1.18	—	0	—
Val	12.86	11.78	11.04	14.12	13	9.07
Met	2.91	2.72	2.84	1.12	2	2.26
Ile	8.06	7.84	8.14	5.32	5	7.96
Leu	16.26	16.12	16.05	17.72	17	17.40
Tyr	2.35	3.31	3.34	3.40	3	1.34
Phe	9.19	9.48	9.49	11.10	12	12.76
Trp	1.14	1.25	1.14	1.03	1	trace
Lys	9.37	9.75	8.95	6.82	7	7.30
His	5.16	5.85	5.31	3.55	3	4.93
Arg	7.86	9.13	7.30	8.21	8	9.50
No. of residues	154	151	150	148	147	153

From Moens, L. and Kondo, M., *Eur. J. Biochem.*, 82, 65, 1978. With permission.

dimensions of the hemoglobin molecules ($120 \times 70 \pm 10$ Å) are in agreement with the calculated axial ratio of 1.7.[34]

C. The Hemoglobin Subunit Structure

All three hemoglobins can be dissociated into subunits by treatment with hot 6 *M* guanidium hydrochloride.[42] The M_r of the globin chains of Hb II obtained by this method were determined by gel filtration and equilibrium sedimentation to be M_r 126,000[19] and M_r 118,000[35] respectively. Similar results were obtained by treatment of the native molecule at low (≤ 4) or high (≥ 11) pH, resulting in polypeptides with a sedimentation coefficient ($S_{20,w}$) of 6.17 to 6.30 and an apparent M_r of 115,000.[17,35]

In contrast SDS-PAGE of partially purified hemoglobins shows a population of polypeptides with M_r ranging from 14,000 to 15,000 up to approximately 130,000. As mentioned in Section II.A, several aspecific and specific proteases, for which the hemoglobins are a ready substrate, are present in *Artemia* extracts. SDS-treatment dissociates the partially nicked hemoglobin molecules and activates, if present, the latent protease activity of artemocyanin II, resulting in a further degradation of the hemoglobins. The series of polypeptides observed after SDS-polyacrylamide gel electrophoresis (SDS-PAGE), is clearly formed by the proteolytic cleavage of the M_r 125,000 to 130,000 chains as shown by their formation in vitro and their similarity in amino acid composition.[19,25,26]

Although the degree of degradation is quite variable, the fragments found tend to occur as multiples of M_r 15,000 to 17,000. Two major bands of M_r 80,000 and 50,000 are mainly present together with a set of rather stable "limit degradation" products of M_r 15,000 to 17,000. The occurrence of these "limit degradation" products is in accordence with the minimum M_r calculated from amino acid composition, heme and iron content (Table 1).

In contrast, carefully prepared hemoglobin containing no proteolytic activity and degradation products,[28,57] or fresh hemolymph denatured within seconds[8] shows, after SDS-PAGE, a single globin band of $M_r \sim 130{,}000$, in agreement with the guanidium-hydrochloride denaturing experiments (Figure 2). This in vitro degradation was the basis of a conflicting interpretation of the hemoglobin subunit structure.[17,19]

The high M_r of the globin chains was recently confirmed by the translation of nauplius poly A^+ RNA in an in vivo (*Xenopus* oocytes) and an in vitro (rabbit reticulocyte lysate) system followed by immunoprecipitation, showing authentic globin chains of $M_r \sim$ 130,000.[43,44]

As the M_r of the native hemoglobins is 240,000 to 260,000 (Table 1) and that of their constituent polypeptides M_r 130,000, the former must be dimers of two globin chains. These chains can be separated by Triton-acid-urea gel electrophoresis[37,45] or by high resolution SDS-PAGE.[46,47,57] Application of these methods to Hb I, Hb II, and Hb III reveals the presence of two globin chains, α and β, which are slightly different in M_r and charge. Hb I and Hb III are respectively α_2 and β_2 homodimers whereas Hb II is an $\alpha\beta$ heterodimer (Figure 3A). This subunit structure is confirmed by reconstruction experiments, immunological cross reactivity, and genetic and physiological evidence:

- Incubation of purified Hb I (α_2) and Hb III (β_2) in a 1:1 ratio at 50°C for 15 min, followed by slow cooling, results in the formation of Hb II ($\alpha\beta$) as revealed by cellulose acetate electrophoresis (Figure 3B).
- As expected, Hb I and Hb III, both react with an affinity purified anti-Hb II antibody (Figure 3C).
- Bowen and co-workers[48-50] demonstrated the presence of two independent genetical loci coding for the three hemoglobins which is in accordance with the proposed model.
- Finally, the oxygen binding characteristics and the inducibility of Hb III at low oxygen partial pressure are also in agreement with the existence of two different globin chains, one (α) with a low oxygen affinity, the other (β) with a high oxygen affinity.

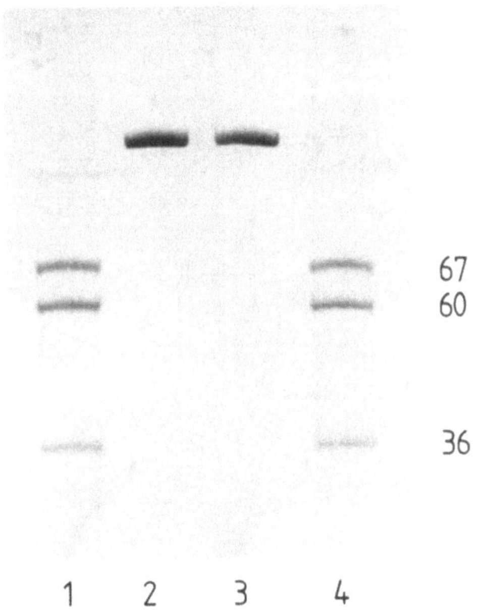

FIGURE 2. SDS-PAGE of Hb II purified in the presence of protease inhibitors. A 12.5% gel, according to Laemmli[46] was used. Lanes 1 and 4: molecular weight markers, lanes 2, 3: Hb II.

The proposed subunit structure was established using the Great Salt Lake variety of *Artemia*.

To extend our observations we screened, with the high resolution SDS-PAGE technique, the hemoglobins of *Artemia* populations from different geographical and ecological origin including Tientsin, Chaplin Lake, Macau, and Galera Zamba. With the exception of hemoglobin Tientsin, all the hemoglobins examined show two globin chains, comparable to the Great Salt Lake α and β chain. Although hemoglobin Tientsin shows on cellulose acetate three hemoglobins indistinguishable from the Great Salt Lake variety, it shows a single globin chain when separated on the basis of M_r.[51] This suggests the possible existence of globin chain variants in *Artemia*.

D. The Globin Substructure

The *Artemia* hemoglobins are dimers of two high M_r subunits or globin chains (see Sections II.B and C). Each globin chain binds eight heme groups, or one heme per domain M_r 17,000, which is in accordance with the minimum molecular weight calculated from amino acid composition (Table 2). This suggests that these globin chains are built up by the covalent linkage of "myoglobin-like" functional units or domains (Section I). Globin chains with similar characteristics are found in mollusks and arthropod hemoglobins. The substructure and the structural and evolutional relationship of these "domains" to the classical globins of the vertebrates are unknown.[52]

1. Cleavage of the Hemoglobins into Domains

If the *Artemia* globin chains contain eight covalently linked myoglobin-like domains, then it must be possible to liberate these domains by chemical or enzymatic methods. Indeed, the domains will form compact structures joined together by stretches of more unfolded "linkers", which are more accessible for proteolytic enzymes. Practically all globins se-

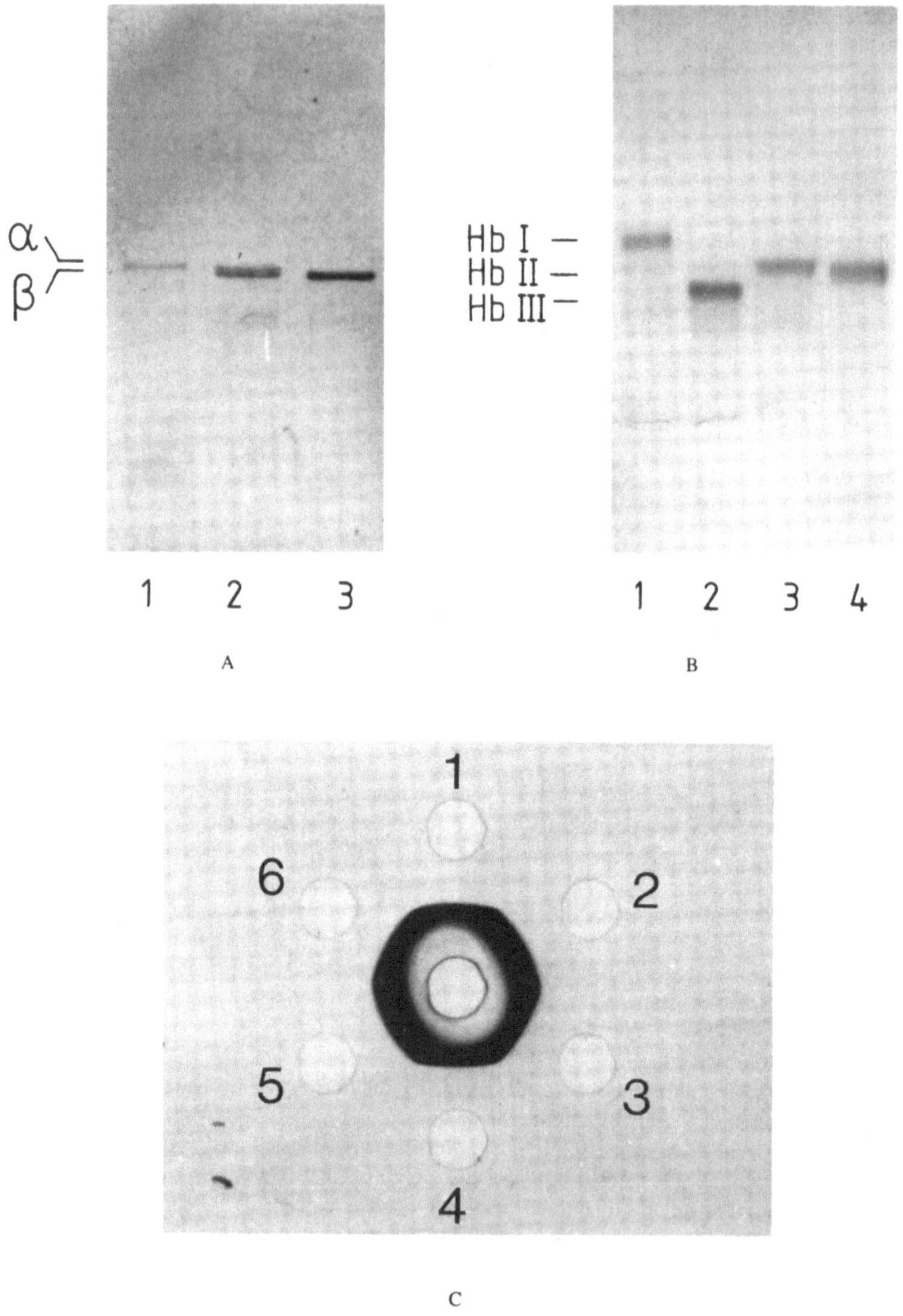

FIGURE 3. Subunit structure of the *Artemia* hemoglobins. (A) Separation of the globin chains by SDS-PAGE (detail). 7.5% gel containing 0.05% *N-N'-bis*-methylene-acrylamide according to Laemmli[46] was used. Gels were run at 12.5 mÅ. Lane 1, Hb I; lane 2, Hb II; lane 3, Hb III. (B) Reconstitution of Hb II. Hb I and Hb II were mixed in a one-to-one ratio and incubated at 50°C for 15 min. After cooling the mixture was analyzed by cellulose acetate electrophoresis according to Bowen et al.[90] Lane 1, Hb I; lane 2, Hb III; lane 3, reconstituted Hb II; lane 4, authentic Hb II. (C) Immunological cross reactivity. Double immunodiffusion was performed on 1% agar in 50 m*M* Tris- HCl pH 7.5, 100 m*M* NaCl. Center well: rabbit anti-*Artemia* hemoglobin serum. Wells 1 and 4: Hb I, wells 2 and 5: Hb II; wells 3 and 6: Hb III. After developing, gels were stained with Coomassie Blue.

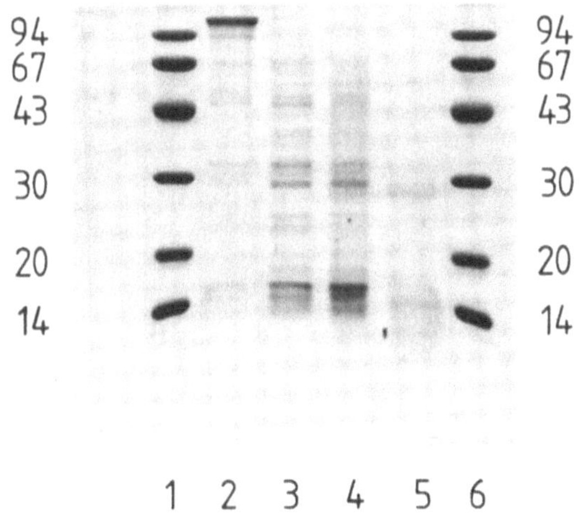

A

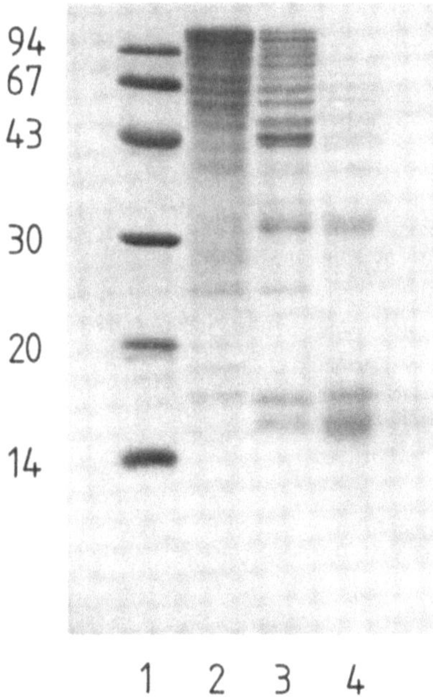

B

FIGURE 4. Chemical and enzymatic cleavage of *Artemia* hemoglobins. (A) Cleavage of Hb II with *N*-bromosuccinimide in 8 *M* urea, 2 *M* acetic acid at 20°C for 20 min at a 5-, 10-, 20-, and 30-fold molar excess (lane 2, 3, 4, 5, respectively) of the reagent to the tryptophan content. Lanes 1 and 6: M_r markers. Peptide fragments were separated on a 15% polyacrylamide gel. (B) Cleavage of the hemoglobin with subtilisin in 40 m*M* glycine/NaOH pH 10.5, 10m*M* EDTA at 25°C at an enzyme/substrate ratio of 1/50 during 0, 30, and 60 min (lane 2, 3, 4, respectively). Lane 1: M_r markers. (From Moens, L. et al., *Biochem. J.*, 223, 861, 1984. With permission.)

quenced up to now contain a very conservative tryptophan residue at position 12 of the A helix.[52-54] The *Artemia* globins contain one tryptophan residue for every 17,000 (Table 2), and cleavage at this unique position must result in fragments of M_r ~17,000.

Chemical[55,56] (*N*-bromosuccinimide) or enzymatic[57-59] cleavage of the globins or hemoglobins results in a population of fragments with decreasing M_r (130,000 to 12,000) as shown by SDS-PAGE (Figures 4A and B). Although their ratio is dependent on the proteolytic or oxidation strength used, bands with M_r 15,000 to 17,000, 30,000 to 32,000, 45,000 to 48,000, 64,000 to 68,000 and 80,000 to 84,000 are clearly predominant.

These bands form a series of integral multiples of M_r ~16,000. Increasing the proteolytic activity (time, enzyme/substrate ratio, temperature) or the oxidation strength results in a decrease of the M_r of the fragments generated and in a partial accumulation of a temporally stable set of fragments (limit degradation products) with M_r 15,000 to 17,000. This cleavage pattern is compatible with the random scission of a "polymer" (M_r 130,000) containing eight "monomers" (domains) of M_r ~16,000.[60] A similar cleavage pattern was also observed in several other invertebrate extracellular hemoglobins containing high M_r globin chains.[1,4,5]

The limited proteolysis of the *Artemia* hemoglobins was optimized in terms of a maximal generation of single domains. This was obtained with the nonspecific enzyme subtilisin at an enzyme/substrate ratio of 1/50 at 25°C for 60 min. The resulting fragment mixture was separated by gel filtration into five heme-containing fractions with apparent M_r of 17,000 to 19,000 (E), 32,000 (D), 50,000 (C), 68,000 (B) and ≥ 100,000 (A) (Figure 5). They clearly correspond to a single (E), two (D), three (C), and four (B) domains, whereas fraction A represents higher multiples and undegraded hemoglobin. Redigestion of these fractions

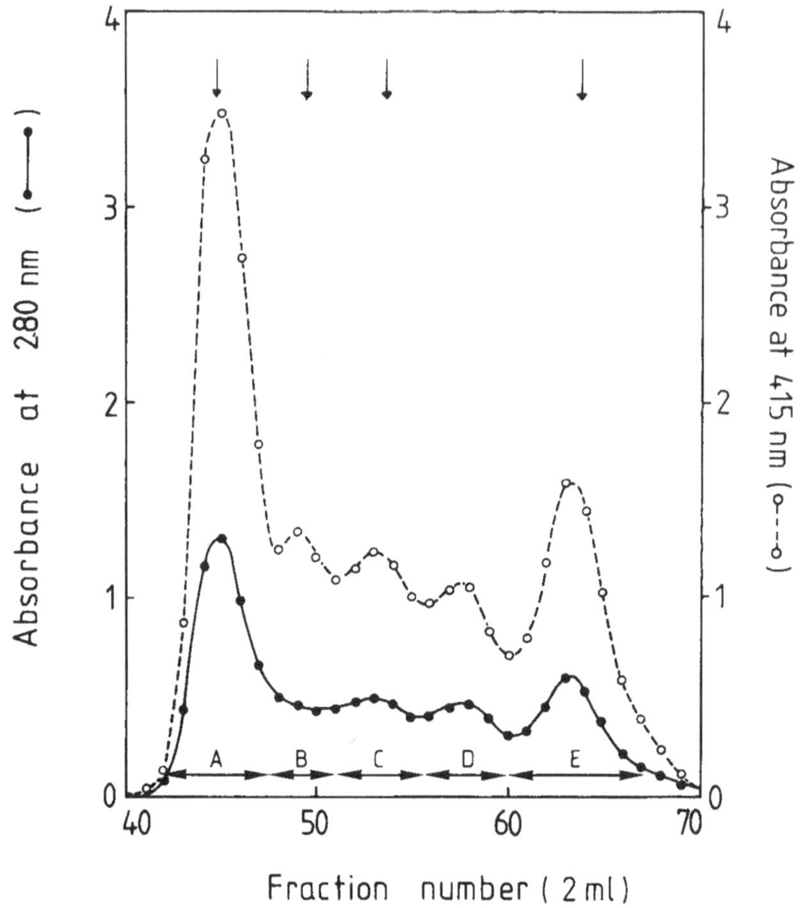

FIGURE 5. Separation of subtilisin-digested hemoglobin by gel filtration. Digestion was performed at an enzyme-to-substrate ratio of 1/50 for 60 min at 20°C. The fragments were separated by chromatography on a Sephacryl S-200 column equilibrated in 40 m*M* glycine/NaOH (pH 10.5) 2.6 m*M* PMSF at a flow rate of 66 mℓ/hr and the fractions were pooled as indicated. (From Moens, L. et al., *Biochem. J.*, 223, 861, 1984. With permission.).

under the same conditions again produces fragments which are multiples of M_r 15,000 to 17,000 proving that the *Artemia* globin chains are constructed by the covalent linking of domains of that size.[59]

The amino acid composition of the fractions A to E are very similar to each other as well as to those of intact hemoglobin,[59] and their calculated minimum M_r all approximate 16,000 to 17,000, indicating that each fraction contains fragments that are multiples of a structural unit of M_r ~17,000. Their iron content, determined by atomic absorption shows a constant iron/protein ratio of one iron atom for every M_r ~16,000 suggesting that the proteolytic cleavage mainly occurred between the domains, leaving the heme-binding environment intact.[40,59] This is confirmed by the comparison of the absorption spectra (oxy, deoxy, carbonmonoxy, cyanmet) and c.d. spectra (200 to 600 nm) of fraction E with those of the intact pigments. These spectra are virtually identical, indicating that the fragments are able to react reversibly with different ligands.[59]

The summarized experiments demomstrate that the *Artemia* globins are constructed by the covalent linking of "myoglobin-like" domains. Accepting this model[25,59] it becomes possible to interpret the banding pattern obtained after SDS-PAGE of hemoglobin preparations partially degraded during isolation by contaminating proteases. This pattern is the

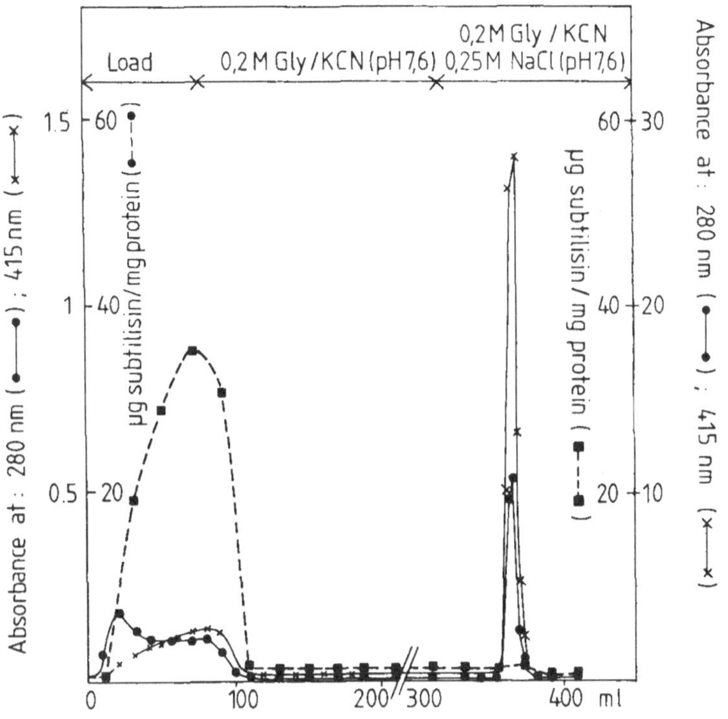

FIGURE 6. Elimination of subtilisin by ion-exchange chromatography. Fraction E (Figure 5A) was adjusted to pH 7.6 and loaded onto a Whatman DE 52 column (1.5 cm × 10 cm) equilibrated in 0.2 *M* glycine-KCN. After loading, the column was washed with the same buffer and the hemoglobin fragments were eluted with 0.25 *M* NaCl in equilibration buffer. Subtilisin activity was measured using Hide power Azur as substrate. (From Moens, L. et al., *Biochem. J.*, 227, 917, 1985. With permission.)

result of a conformation dependent cleavage between the domains within the globin chains.[2,3] Hemoglobin, extensively cleaved during preparation, only shows polypeptides of M_r 15,000 to 17,000 representing single domains. This was the basis of Bowen's interpretation of the globin size being 19,000.[17]

The isolated fraction E (as well as the other fragment fractions) is not homogeneous as shown by electrophoresis under denaturing and nondenaturing conditions. This heterogeneity might have been artificially generated by the subtilisin digestion or may reflect a heterogeneity within the domains and/or linker regions within the authentic hemoglobin molecules. A further characterization of the domains will require their purification to homogeneity.

2. Purification of the Domains

Before the domains could be purified, contaminating subtilisin, used in the cleavage of the globin chains into domains, had to be eliminated. This was accomplished by chromatography of fraction E on a DE 52 column at pH 7.6[61] (Figure 6). Under these conditions the fragments were quickly washed free of protease activity and could be eluted as a very sharp band whereas oxidized material was irreversibly bound. The proteinase-free material was still heterogeneous as shown by two dimensional electrophoresis[62] (Figure 7). At least four major fractions and several minor ones with M_r 15,000 to 17,000 were revealed.

A preparative separation of the domain fraction was obtained by isoelectric focusing on granulated gel[63-65] or by chromatofocusing on a mono-P column[66,67] (Figure 8, Table 3). Five major fractions were revealed by isoelectric focusing. Fraction E_1^{IEF} was fairly ho-

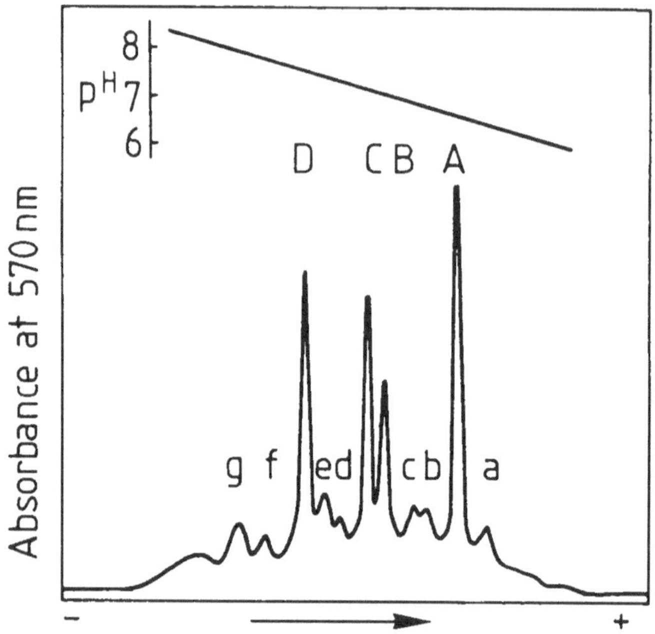

A

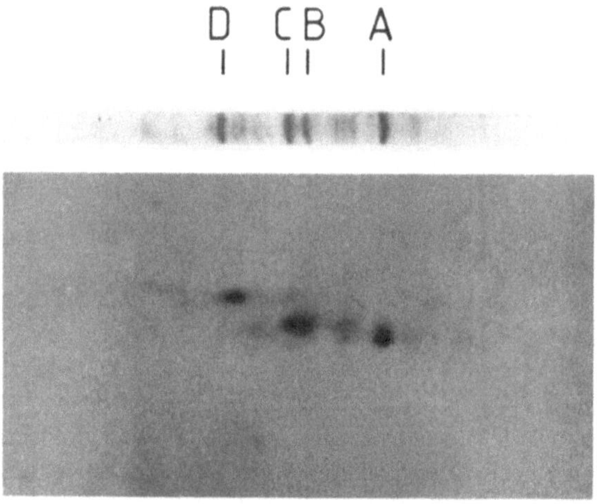

B

FIGURE 7. Analysis of fraction E by the O'Farrell technique.[62] Fraction E (Figures 5 and 6) was analyzed in the first dimension by isoelectric focusing on a 5% (w/v) polyacrylamide gel in a pH 3 to 10 gradient. A 15% polyacrylamide gel was used[46] in the second dimension. Gels were stained with Coomassie Blue R250. (A) Densitometric scanning of a stained one-dimensional gel; (B) Alignment of a one and two dimensional gel. (From Moens, L. et al., *Biochem. J.*, 223, 861, 1984. With permission.)

mogeneous (>90%) whereas the other fractions represent, as revealed by two dimensional electrophoresis, collections of polypeptides with similar charges. Chromatofocusing results in a similar separation but with a better resolution. Fractions E_1^{CF} and E_7^{CF} are homogeneous

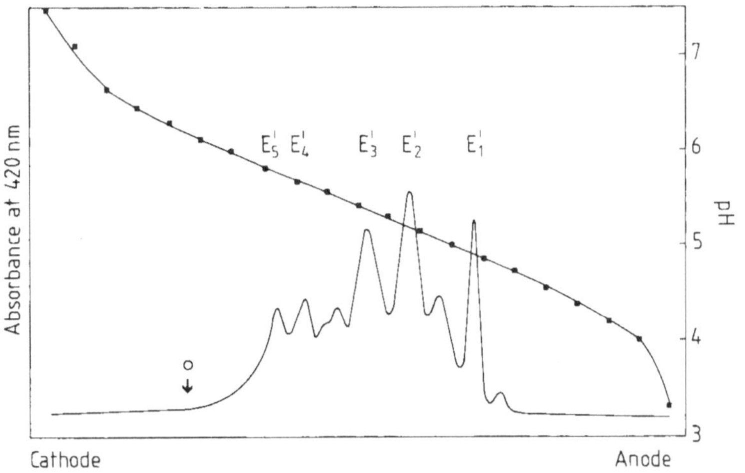

A

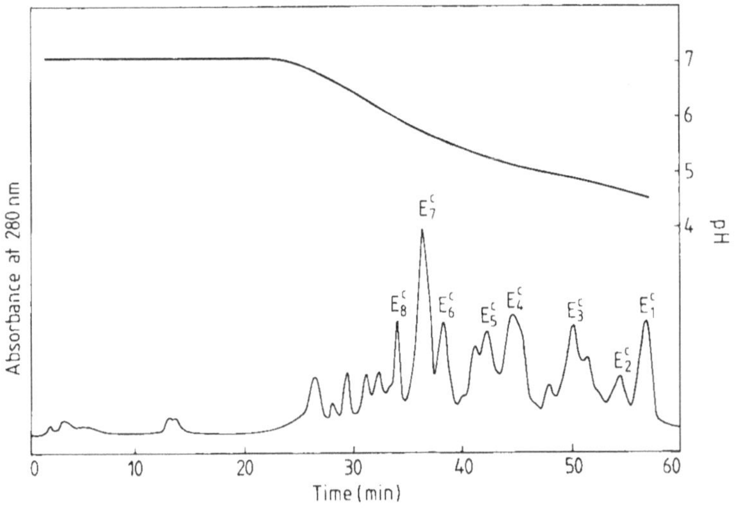

B

FIGURE 8. Separation of fraction E (Figure 5) by isoelectric focusing and chromatofocusing. (A) Isoelectric focusing according to Frey and Radola.[65] The matrix was a Sephadex G 75 (superfine) layer (21 cm × 25 cm × 0.15 cm) containing 4 *M* urea and 2% ampholines (pH range 4.0 to 6.5). The sample was loaded at 10 mg/cm and focused for 4 hr at 10 W. The layer was scanned with a Zeine soft-laser densitometer at 420 nm. (From Moens, L. et al., *Biochem. J.*, 227, 917, 1985. With permission.) (B) Chromatofocusing was performed on a Pharmacia Mono P HR 5/20 column using 0.025 *M* *bis-tris*-iminodiacetic acid at pH 7.1 and 9% poly-buffer 74-iminodiacetic acid at pH 4.0 respectively as starting and elution buffer.

and amenable to sequence analysis. Fraction E_1 isolated by isoelectric focusing (E_1^{IEF}) and chromatofocusing (E_1^{CF}) are identical as shown by their similarity in M_r, pI, amino acid composition, tryptic map, and partial sequence analysis.[75] As E_1 was the first domain that became available for structural work, we determined its amino acid sequence first.

Table 3
ISOELECTRIC POINT AND MOLECULAR WEIGHT OF THE ISOLATED DOMAINS OBTAINED BY ISOELECTRIC FOCUSING OR CHROMATOFOCUSING

Isoelectric focusing (IEF)			Chromatofocusing (CF)		
Fraction	pI	$M_r \times 10^{-3}$	Fraction	pH of elution	$M_r \times 10^{-3}$
E_1	4.80	15.8	E_1	4.60	16
E_2	5.16	15.8/17.2	E_2	4.70	16
E_3	5.30	15.8/17.2	E_3	4.92	16
E_4	5.50	15.8	E_4	5.11	17
E_5	5.62	17.2	E_5	5.38	16
			E_6	5.60	17
			E_7	5.79	16
			E_8	5.98	16/17

3. The Amino Acid Sequence of the Domains E_1 and E_7

The total amino acid sequence of the most acidic structural unit E_1 and a partial sequence of E_7 were determined by automated and manual sequencing.[68-71] In Figure 9 the proposed sequences are aligned with the sequence of the human β chain. Chain E_1 contains 147 amino acid residues (M_r 16,958) and is comparable to the classical globin chains of the vertebrates.[53,54] Compared to the β chain, E_1 shown an extension with six residues at the amino terminus, which is absent in E_7, and a shortening of two residues on the carboxy terminus. Two insertions (residues 25 to 26; 90 to 93) and four deletions (residues 58; 110 to 114; 132; 137 to 138) are necessary to obtain maximal homology (38 or 23.2%). This homology is higher than with any other invertebrate globin sequenced so far.[72] The majority of residues, identical in all vertebrate myoglobin and hemoglobin chains, are also identical or functionally conserved in E_1 and E_7.[73] As these residues are involved in the molecular folding, this strongly suggests that the *Artemia* globin domain may have the myoglobin fold.[74]

The eight types of domains ($E_1 \rightarrow E_8$) composing the globin chains strongly resemble each other as shown by their amino acid composition and tryptic mapping (data not shown). However no more than 40% of the 75 identified residues of E_7 are homologous with E_1 (Figure 9). Assuming a similar sequence difference between the other domains we may conclude that the *Artemia* globin chains are covalent "heteropolymers" of eight myoglobin-like units. The high M_r globin chains (α,β) do not arise by the post-translational linkage of low M_r globin gene products as demonstrated by the translation of naupliar poly A^+-RNA to authentic globin chains (~M_r 130,000) as well as in the rabbit reticulocyte lysate and the *Xenopus* oocyte system.[43,44] The question thus arises about the structure of the *Artemia* globin genes.

4. A Speculative Model for the Artemia Hemoglobin Molecule

Assuming that the globin domains have dimensions comparable to those of the classical globins (an ellipsoid of 4.5 × 4.5 × 2.5 nm) we can now propose a speculative model for the *Artemia* hemoglobin molecule, based on model building. Each globin chain forms a disk with a diameter of about 12 nm and a height of about 3.5 nm (Table 1). Two such disks stacked on each other form a native hemoglobin molecule. Within a chain, the eight domains are organized along an eightfold axis of symmetry (Figure 10A). Relative to each other these domains can be oriented in several ways, but the organization with the shortest linker regions is assumed to be the most probable. We thus propose that each chain contains four pairs of domains, and that each pair is formed by an upright and an upside down domain with the heme groups facing each other (Figure 10B).

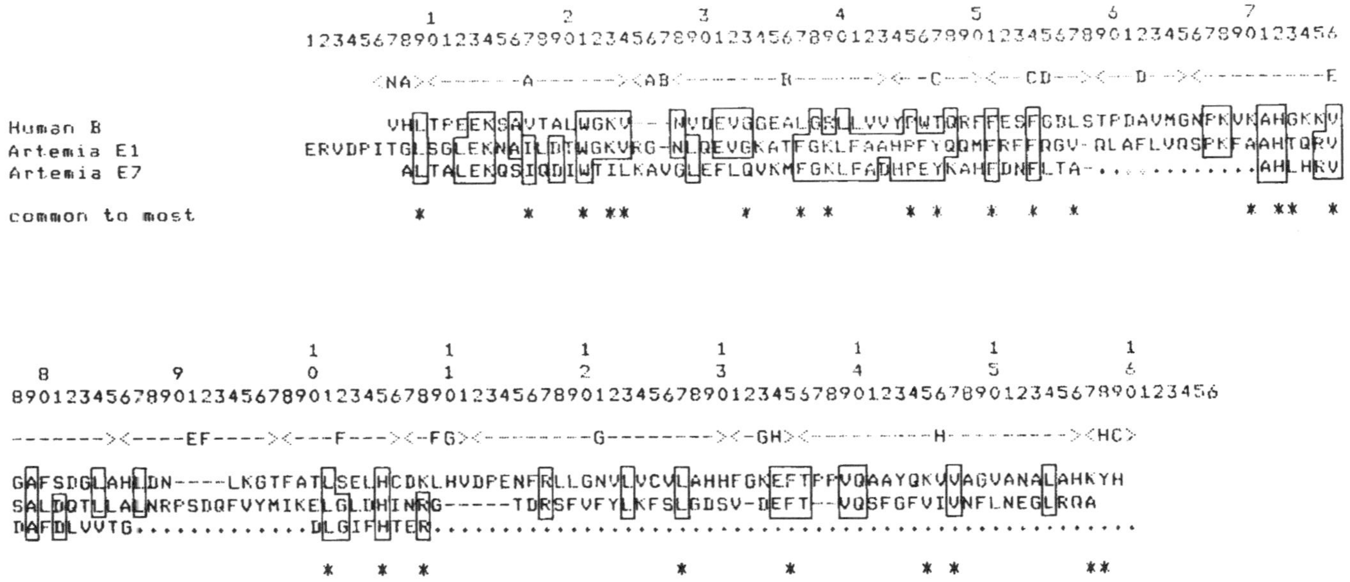

FIGURE 9. Alignment of the amino acid sequences of the *Artemia* domains E_1 and E_7 with the human β chain. The helix notifications are those of the human chain.

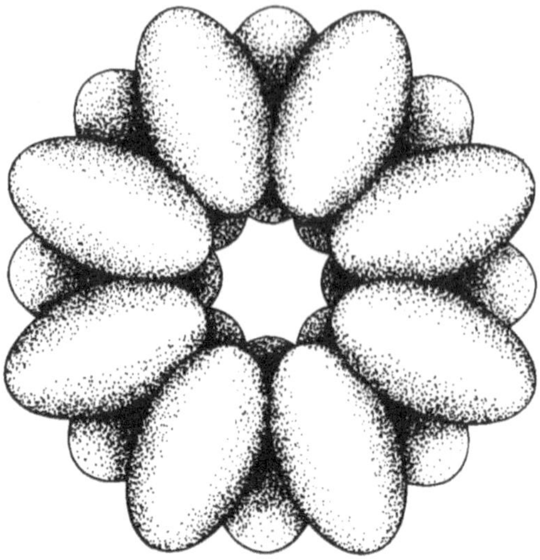

A

B

FIGURE 10. A speculative model for the *Artemia* hemoglobin molecule. (A) Top view; (B) side view.

This model implies that there are two types of linker regions; a long one of about ten amino acids between the domains in a pair, and a short one of about zero to five amino acids between the pairs. The stacking of both globin chains in the native molecule can be performed in at least two ways. The domains of one chain are on top of those of the other chain or one domain of a chain falls between two domains of the other chain (Figure 10B). The latter way will result in a more compact packing, making it more probable.

III. THE OXYGEN BINDING CHARACTERISTICS

The oxygen binding characteristics of the purified hemoglobins (Great Salt Lake variety) were determined using the diffusion chamber technique of Sick and Gersonde[76] (Table 4).

Table 4
IN VITRO OXYGEN-BINDING CHARACTERISTICS OF THE *ARTEMIA* HEMOGLOBINS, GLOBINS, AND THEIR PROTEOLYTIC FRAGMENTS[40,77]

Phenotype	Genotype	$M_r \times 10^3$	Oxygen affinity P_{50} (mmHg)[a]	Hill coefficient n[b]	Bohr factor ϕ[c]	Temperature effect ΔH (kJ/mol)[d]
Native Molecules (Great Salt Lake)						
	α_2	260	5.34	1.62	−0.09	−44.8 to −45.6 (15—35°C; pH 8—9)
	$\alpha\beta$	260	3.70	1.92	−0.21	−44.0 to −54.8 (15—35°C; pH 8—9)
	β_2	260	1.80	1.58	−0.03	−22.6 to −23.0 (10—25°C; pH 8—9)
						−53.6 to −54.8 (15—30°C; pH 7.5—8)
Globin Chains and Globin Fragments (Aigle Morte)[e]						
Globin chains (α β)		130	3.45	1.00	—	—
Fraction B		$68 = [16]_4$	1.47	0.98	—	—
Fraction C		$48 = [16]_3$	1.21	1.03	—	—
Fraction D		$32 = [16]_2$	0.71	0.97	—	—
Fraction E		$16 = [16]$	0.36	0.94	—	—

[a] Oxygen affinity determined in 0.1 *M* Na-borate, pH 8.5 at 25°C; 0.6 *M* NaCl; 0.02 *M* KCl.

[b] Hill coefficient determined from Hill plots, at high oxygen partial pressure, calculated from oxygen dissociation data as in Footnote a.

[c] Bohr factor ($\phi = \frac{\Delta \log P_{50}}{\Delta pH}$); determined in buffer above pH 8 containing 0.06 *M* NaCl, 0.02 *M* KCl at 20°C.

[d] Temperature effect [$\Delta H = 2.303\ R\ (\frac{\Delta \log P_{50}}{\Delta T^{-1}})$]

[e] As determined in Section II.D.1 and Figure 5.

Because this involves in vitro measurements and because nothing is known about the in vivo conditions such as pH, temperature, and ionic strength in the hemolymph of *Artemia*, interpretations have to be made with care.

The three hemoglobins show a typical sigmoidal dissociation curve with significantly different oxygen affinities as illustrated by their P_{50} values.[77,78] The homodimers Hb I (α_2) and Hb III (β_2) show, respectively, a low and a high oxygen affinity, whereas the heterodimer, Hb II ($\alpha\beta$), shows an intermediate oxygen affinity. This suggests that low and high oxygen affinity may be an intrinsic characteristic of the α and β globin chain respectively. The oxygen affinity of the proteolytic fragments (fractions B → E, Figure 5), which contain multiple oxygen binding sites, increases linearly with decreasing M_r.[40] Similar results were obtained when native single domain, multisubunit hemoglobins or hemocyanins are dissociated into their subunits with a concomitant decrease of the M_r.[2,5] The Hill plots of the hemoglobins at pH 8.5 and 25°C were nonlinear. Only at high oxygen partial pressures is the Hill coefficient (n) different from unity, indicating a low to moderate cooperativity. In accordance with other oxygen carriers (hemoglobins and hemocyanins) containing covalently linked multidomain chains, the globin chains of *Artemia* and its subfragments (fractions B → E, Figure 5) do not show any cooperativity.[40] Consequently, cooperativity is only expressed between oxygen binding sites from different globin chains and not within a single chain.

The influence of parameters such as ionic strength, temperature, and pH on the oxygen binding of the hemoglobins in vitro was studied.[77,78] Variation of the ionic strength between 0.2 and 1.0 *M* sodium chloride at pH 7.5 had no influence on the P_{50} values. The effect of pH on P_{50} values (Bohr effect) is, in absolute terms, rather small as expressed by the Bohr factors (Table 4). However, the Bohr effect seems to be temperature dependent. All hemoglobins exhibit a positive Bohr effect above pH 8.5 and at temperature between 10 to 30°C. However Hb III shows a reverse Bohr effect below this pH value and between 10 and 20°C.

Figure 11 illustrates the temperature effect on the oxygen affinities of the three hemoglobins, from which the ΔH values were calculated (Table 4). Hb I and Hb II show a general decrease in oxygen affinity with an increase in temperature. Hb I is rather temperature insensitive at pH 8.5 to 9 but not below this pH value.

Recently Wolf et al.[79] compared the physiological properties, measured by the diffusion chamber technique, of total hemoglobin samples from *Artemia* populations of different geographical and ecological origin including Tientsin, Great Salt Lake, Chaplin Lake, Macau, and Galera Zamba. Measurements at different pH values and temperatures revealed significant differences in oxygen affinity (P_{50} values between 1.25 and 13.56 mmHg), cooperativity (Hill coefficients between 1.07 and 2.18), and free oxygen binding energy (ΔH between −34.7 and +56.2 kJ). All the strains showed a decrease in oxygen affinity at increasing temperature and a positive Bohr effect. *Artemia* from Tientsin, containing a considerable amount of Hb I (36.3%), showed the lowest oxygen affinity, while *Artemia* from Macau and Galera Zamba, containing mostly Hb III (8.6 and 26.6%, respectively), showed the highest affinity. The highest cooperativity was found for *Artemia* from Tientsin and the lowest for *Artemia* from Galera Zamba.

In general, the oxygen binding characteristics of the hemoglobins as measured in vitro can be summarized as follows: Hb I, low oxygen negligible Bohr effect, and no temperature adaptation; Hb II, intermediated oxygen affinity, strong positive Bohr effect, and no temperature adaptation; Hb III, high oxygen affinity, negligible Bohr effect and good temperature adaptation. By changing the ratio of these three hemoglobins, *Artemia* is able to adapt to a variety of environmental conditions.

IV. THE EXPRESSION OF THE HEMOGLOBINS

A. The Expression as a Function of Development

During preemergence development, *Artemia* mainly shows an anaerobic metabolism. Due

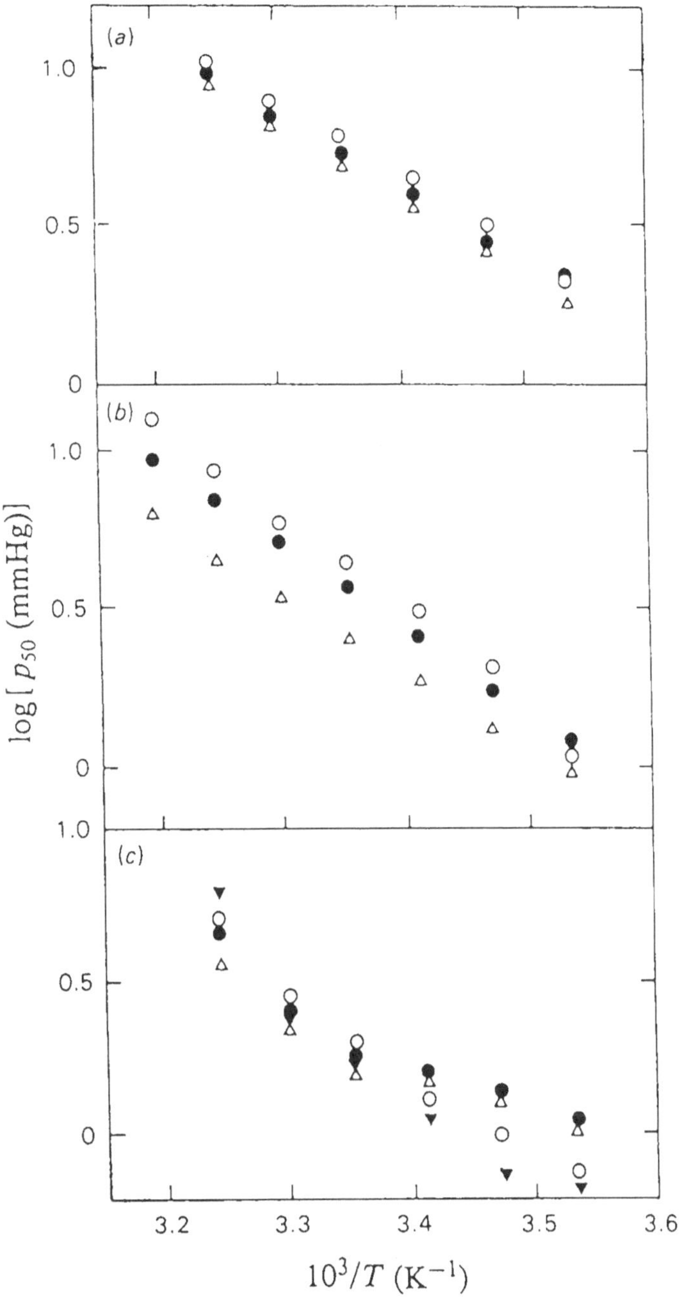

FIGURE 11. Effect of temperature on P_{50} of *Artemia* hemoglobin. Temperature effects on P_{50} of Hb I (a), Hb II (b), and Hb III (c) were measured at pH 7.5 (▼), 8 (●), 8.5 (○), and 9 (△). (From D'Hondt, J. et al., *Biochem. J.*, 171 705, 1978. With permission.)

to a higher energy demand, metabolism changes, around hatching, to an aerobic type, requiring molecular oxygen and consequently an oxygen carrier.[80] The identification of hemoglobin as a major protein, (~20% of soluble protein) induced during postgastrula development, was confirmed by many authors.[25,81-83] Heip et al.[81] studied, by spectroscopic, immunological, and electrophoretic methods, the phenotypic appearance and disappearance of the different hemoglobin types during the life cycle of *Artemia* (Great Salt Lake) cultured

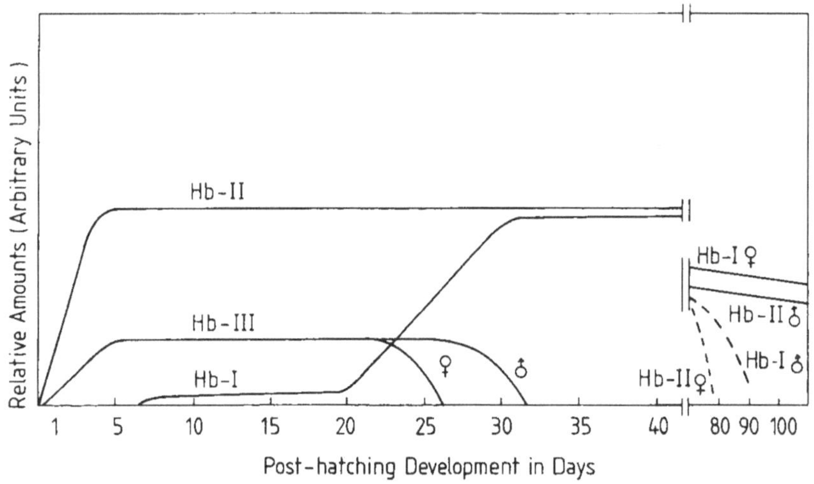

FIGURE 12. A schematic representation of ontogenetically controlled synthesis of hemoglobins of *Artemia*. The presence or absence and the ratio of the individual hemoglobins was determined by cellulose acetate electrophoresis of hemolymph samples from animals cultured under standard conditions. (From Heip, J. et al., *Dev. Biol.*, 64, 73, 1978. With permission.)

under controlled conditions (0.51 *M* salinity: 0.17 to 0.21 m*M* O_2; 28 ± 1°C). Hb II was the first hemoglobin to appear in the swimming nauplius at about 2 hr posthatching, followed by Hb III at about 8 hr posthatching. Hb I was detectable only after 7 to 9 days posthatching when the functional gills were formed. Hb III was always lower in quantity than Hb II and disappeared completely from the hemolymph between 30 to 40 days posthatching. The disappearance of Hb III was earlier in the female than in the male. Hb I and II remain for most of the life period in the hemolymph of both the male and the female, but Hb II decreased gradually or sometimes disappeared in the old female, whereas this hemoglobin continued to be the major species in the old male (Figure 12). It can be concluded that the phenotypic appearance of the three hemoglobins in the hemolymph is carefully ontogenetically controlled as it is in many other invertebrate and vertebrate species.[73,84-86]

The control of the hemoglobin phenotype can be interpreted as a function of the globin chain synthesis. Accepting two independent loci,[48-50] a balanced synthesis of the α and β chain will result in the formation of the stable heterodimer αβ (Hb II), whereas the unbalanced synthesis will result in the formation of the heterodimer (αβ) together with one of the less stable homodimers (α_2 or β_2). Not only biosynthesis but also breakdown will determine the ratio of the hemoglobins in the hemolymph, but no information is available about the turnover of the individual hemoglobin types.

The early hemoglobin synthesis, just after hatching, is not due to the assembly of preexisting globin chains or to the translation of stored mRNA. Screening S_{105} extracts, prepared at different times of development, with a rabbit antihemoglobin antibody ruled out the first possibility.[81] The absence of stored globin mRNA during pre-emergence development was demonstrated by the inability of isolated poly A^+ RNA to synthesize, in the rabbit reticulocyte system, immunoprecipitable globin chains.[87] The latter results have to be interpreted with care as Manning et al.[43,44] have shown that the translation of *Artemia* Poly A^+ RNA to authentic globin chains in the rabbit reticulocyte system is strongly hampered by the tertiary structure of the mRNA.

They were however successful in the faithful translation of poly A^+ RNA to authentic globin chains in the *Xenopus laevis* system, confirming the presence of globin mRNA at 20 hr development.

B. The Expression as a Function of Environmental Factors

Very early,[88-90] it was observed that environmental factors such as salinity, oxygen, and ferric ions, have a significant effect on the hemoglobin content of the hemolymph. Heip et al.[91] studied the effect of salinity and oxygen on the hemoglobin phenotype of *Artemia* (Great Salt Lake), cultured under standard conditions (0.51 *M* salinity; 0.17 to 0.21 in m*M* O_2; 28 ± 1°C). A 20-fold variation of salinity (0.08 to 1.53 *M*) had no significant effect on the ontogenetic pattern of hemoglobin induction. This confirms the oxygen binding properties of the isolated hemoglobin as measured in vitro, where no effect of salinity on the P_{50} value was observed. The independence of the respiratory pigments from the environmental ionicity is absolutely essential for an animal that can adapt to media varying from one tenth the salinity of seawater, which is a hypertonic regulator, to crystallizing brine, which is a very effective hyporegulator.

In contrast to salinity, the variation of the oxygen concentration in the environment has a profound effect on the hemoglobin concentration and phenotype. At 100% O_2, no hemoglobin at all is present in the hemolymph. In contrast, lowering the oxygen concentration of the medium below 0.09 m*M* results in a two- to threefold overall increase in the hemoglobin concentration in the hemolypmph.[37,79,88] Under all conditions Hb II remains the main pigment.

Heip et al.[91] took advantage of the fact that in their culture conditions, adult *Artemia* (30 to 40 days) lack Hb III (β_2). Subjecting such cultures to low oxygen concentration (0.04 to 0.06 m*M* O_2; salinity 0.51 *M*) results in a specific and reversible induction of Hb III. Using two dimensional electrophoresis with one dimension under native and the other under denaturing conditions, Ferry et al.[88] not only confirmed this observation, but clearly demonstrated that an hypoxic stress results in the increase of the *de novo* synthesis of the β polypeptide chain relative to the α polypeptide chain. This is not unexpected as Hb III (β_2) has the highest oxygen affinity in vitro (Table 4) and is thus the best adapted for oxygen binding under hypoxic conditions. It also confirms that the rate of synthesis of both chains has a profound effect on the hemoglobin phenotype.

V. CONCLUSION

The expression of the individual hemoglobins of *Artemia* and their oxygen binding characteristics are regulated in such a way that the animal is able to cope with extreme differences in environmental conditions. The structure of these hemoglobins is different from other invertebrate hemoglobins. The molecules are dimers of two high M_r globin chains, each of which can be considered a "covalent polymer" of eight myoglobin-like structural and functional units or domains. The structure of the *Artemia* globins is not an exception but rather an example of a vast group of invertebrate (mollusks and arthropods) extracellular hemoglobins containing high M_r globin chains. It also raises the question of the structure of their coding genes. This will be the main line for future research.

ACKNOWLEDGMENT

The support of the Fund for Joint Basic Research (F.J.B.R.) contract no. 2.0012.82, is gratefully acknowledged.

REFERENCES

1. **Chung, M. C. M. and Ellerton, M. D.,** The physico-chemical and functional properties of extracellular respiratory hemoglobins and chlorocruorins, *Prog. Biophys. Mol. Biol.*, 35, 53, 1979.
2. **Vinogradov, S. N., Shlom, J. M., Kapp, O. N., and Frossard, P.,** The dissociation of annelid extracellular hemoglobins and their quaternary structure, *Comp. Biochem. Physiol.*, 67B, 1, 1980.
3. **Terwilliger, R. C.,** Structure of invertebrate hemoglobins, *Ann. Zool.*, 20, 53, 1980.
4. **Wood, E. J.,** The oxygen transport and storage proteins of invertebrates, *Essays Biochem.*, 16, 1, 1980.
5. **Vinogradov, S. N.,** The structure of invertebrate extracellular hemoglobins (erythrocruorins and chlorocruorins), *Comp. Biochem. Physiol.*, 82B, 1, 1985.
6. **Goodman, M., Braunitzer, G., Kleinschmidt, T., and Aschauer, H.,** Evolution of monomeric and homodimeric hemoglobins of *Chironomus thummi thummi, Hoppe-Seyler's Z. Physiol. Chem.*, 364, 205, 1983.
7. **Svedberg, T. and Hedenius, A.,** The sedimentation constants of the respiratory pigments, *Biol. Bull.*, 66, 191, 1934.
8. **Ilan, E. and Daniel, E.,** Structural diversity of arthropod extracellular hemoglobins, *Comp. Biochem. Physiol.*, 63B, 303, 1979.
9. **Ilan, E. and Daniel, E.,** Erythrocruorin from the water flea *Daphnia magna*. Quaternary structure and arrangements of subunits, *Biochem. J.*, 207, 197, 1982.
10. **Dangott, L. J. and Terwilliger, R. C.,** Structural studies of a branchiopod crustacean *(Lepidurus belobatus)* extracellular hemoglobin, *Biochim. Biophys. Acta*, 579, 452, 1979.
11. **Daniel, E.,** Subunit structure of arthropod erythrocruorins, *Life Chem. Rep.*, Suppl. 1, 157, 1983.
12. **Wood, E. J. and Mosby, L. J.,** Physicochemical properties of *Planorbis corneus* erythrocruorin, *Biochem. J.*, 149, 437, 1975.
13. **Terwilliger, N. B., Terwilliger, R. C., and Schabtach, E.,** The quaternary structure of a mollusc *(Helisoma trivolvis) Biochim. Biophys. Acta*, 453, 101, 1976.
14. **Terwilliger, N. B. and Terwilliger, R. C.,** Oxygen binding domains of a clam *(Cardita borealis)* extracellular hemoglobin, *Biochim. Biophys. Acta*, 537, 77, 1978.
15. **Terwilliger, R. C. and Terwilliger, N. B.,** Oxygen binding domains in invertebrate hemoglobins, *Life Chem. Rep.*, Suppl. 1, 227, 1983.
16. **Waring, G., Poon, M. C., and Bowen, S. T.,** The hemoglobins of *Artemia* salina II. Isolation of three hemoglobins, *Int. J. Biochem.*, 1, 537, 1970.
17. **Bowen, S. T., Moise, H., Waring, G., and Poon, M. C.,** The hemoglobins of *Artemia salina*. III. Characterization, *Comp. Biochem. Physiol.*, 55B, 99, 1976.
18. **Moens, L. and Kondo, M.,** The structure of *Artemia salina* hemoglobins, *Eur. J. Biochem.*, 67, 397, 1976.
19. **Moens, L. and Kondo, M.,** Evidence for a dimeric form of *Artemia salina* extracellular hemoglobins with high molecular weight subunits, *Eur. J. Biochem.*, 82, 65, 1978.
20. **Osuna, C., Olalla, A., Sillero, A., Sillero, M. G., and Sebastian, J.,** Induction of multiple proteases during the early larval development of *Artemia salina, Dev. Biol.*, 61, 94, 1977.
21. **Olalla, A., Carmen, O., Sebastian, J., Sillero, A., and Sillero, M. G.,** Purification and properties of three proteases from the larvae of the brine shrimp *Artemia salina, Biochim. Biophys. Acta*, 523, 181, 1978.
22. **Nagainis, P. A. and Warner, A. M.,** Evidence for the presence of an acid protease and protease inhibitors in dormant embryo's of *Artemia salina, Dev. Biol.*, 68, 259, 1979.
23. **Garesse, R., Perona, R., Marco, R. and Vallejo, C. G.,** The unmasking of proteolytic activity during the early development of *Artemia salina, Eur. J. Biochem.*, 106, 225, 1980.
24. **Warner, A. H. and Shridda, V.,** Purification and characterization of a cytosol protease from dormant cysts of the brine shrimp *Artemia, J. Biol. Chem.*, 260, 7008, 1985.
25. **Mansfield, B. C., Krissansen, G. W., Smith, M. G., and Tate, W. P.,** Isolation of intact larval hemoglobin from the brine shrimp *Artemia salina, Biochim. Biophys. Acta*, 623, 163, 1980.
26. **Moens, L. and Kondo, M.,** Characterization of the extracellular hemoglobins of *Artemia salina, Biochem. J.*, 165, 111, 1977.
27. **Krissansen, G. W., Trotman, C. N. A., and Tate, W. P.,** A novel protease may explain widely differing models for the structure of *Artemia salina* hemoglobin, *Biochim. Biophys. Acta*, 671, 104, 1981.
28. **Krissansen, G. W., Trotman, C. N. A., and Tate, W. P.,** Purification of stable hemoglobin from the adult brine shrimp *Artemia, Biochem. Int.*, 7, 519, 1983.
29. **Krissansen, G. W., Trotman, C. N. A., and Tate, W. P.,** Two forms of the glycoprotein Artemocyanin from the brine shrimp *Artemia*. A relationship between latent proteolytic activity and structural modifications to the protein, *Biochim. Biophys. Acta*, 747, 159, 1983.
30. **Krissansen, G. W., Trotman, C. N. A., and Tate, W. P.,** Purification of artemocyanin, a hemolymph protein from the brine shrimp *Artemia, Biochem. Int.*, 7, 369, 1983.

31. **Krissansen, G. W., Trotman, C. N. A., and Tate, W. P.,** Physical and biochemical characterization of artemocyanin, an abundant glycoprotein complex with latent proteolytic activity from the brine shrimp *Artemia, Biochim. Biophys. Acta,* 747, 151, 1983.
32. **Krissansen, G. W., Trotman, C. N. A., and Tate W. P.,** Identification of the blue-green chromophore of an abundant biliprotein from the haemolymph of *Artemia, Comp. Biochem. Physiol.,* 77B, 249, 1984.
33. **Krissansen, G. W., Trotman, G. N. A., and Tate, W. P.,** Artemocyanin: a proteinase zymogen rather than a proteinase Inhibitor, *Comp. Biochem. Physiol.,* 80B, 883, 1985.
34. **De Voeght, F. and Clauwaert, J.,** Structural characterization by biophysical methods of extracellular hemoglobin of *Artemia salina,* in *Structure, Active Site and Function of Invertebrate Oxygen Carriers,* Lamy, J., Ed., Marcel Dekker, New York, 1980.
35. **Wood, E. J., Barker, C., Moens, L., Jacob, W., Heip, J., and Kondo, M.,** Biophysical characterization of *Artemia salina* extracellular hemoglobin, *Biochem. J.,* 193, 353, 1981.
36. **Dangott, L. J., and Terwilliger, R. C.,** The subunit structure of *Daphnia pulex* hemoglobin, *Comp. Biochem. Physiol.,* 67B, 301, 1980.
37. **Ferry, J. A., Nichols, R. C., Condon, S. J., Stubbs, J. D., and Bowen, S. T.,** Artemia hemoglobins. Increase in net synthesis of the β polypeptide (relative to the α polypeptide) in hypoxia, *Biochim. Biophys. Acta,* 739, 249, 1983.
38. **Heip, J., Moens, L., Hertsens, R., Wood, E. J., Heyligen, H., Van Broeckhoven, A., Vrints, R., De Chaffoy, D., and Kondo, M.,** *Artemia* extracellular hemoglobins: ontogeny, structure and *in vivo* radiolabeling, in *The Brine Shrimp Artemia,* Vol. 2, Persoone, G., Sorgeloos, P., Roels, O., and Jaspers, E., Eds., Universa Press, Wetteren, Belgium, 1980.
39. **Bucks, D. A. W. and Bowen, S. T.,** The naupliar hemoglobins of three populations of *Artemia,* in *Biochemistry of Artemia Development,* Bagshaw, J. C. and Warner A. H., Eds., University Microfilms Inc., Ann Arbor, Mich., 1979.
40. **Wolf, G., Van Pachtenbeke, M., Moens, L., and Van Hauwaert, M. L.,** Oxygen binding characteristics of *Artemia* hemoglobin domains, *Comp. Biochem. Physiol.,* 76B, 731, 1983.
41. **Manwell, C. A.,** A simplified electrophoretic system for determining molecular weights of proteins, *Biochem. J.,* 165, 487, 1977.
42. **Weber, K., Pringle, J. R., and Osborn, M.,** Measurement of molecular weights by electrophoresis on SDS-acrylamide gel, in *Methods in Enzymology,* Vol. XXVIC, Hirs, C. M. W. and Timasheff, S. N., Eds., Academic Press, New York, 1972.
43. **Manning, A. M., Ting G. S., Trotman, C. N. A., and Tate, W. P.,** Faithful translation of the mRNA for the haemoglobin polypeptide of *Artemia* nauplii in *Xenopus* oocytes, *Proc. Univ. Otago Med. School,* 63, 24, 1985.
44. **Manning, A. M., Ting, G. S., Mansfield, B. C., Trotman, C. N. A., and Tate, W. P.,** The isolation and faithful translation of *Artemia* naupliar hemoglobin mRNA, *Biochem. Int.,* 12, 715, 1986.
45. **Rovera, G., Magarian, C., and Borun, T. W.,** Resolution of hemoglobin subunits by electrophoresis in acid urea polyacrylamide gels containing triton X-100, *Anal. Biochem.,* 85, 506, 1978.
46. **Laemmli, U. K.,** Cleavage of structural proteins during the assembly of the head of bacteriophage T_4, *Nature (London),* 227, 680, 1970.
47. **Heyligen, H., Moens, L., Heip, J., Kondo, M.,** Subunit composition of Artemia hemoglobins, *Arch. Int. Physiol. Biochim.,* 89, B19, 1981.
48. **Bowen, S. T., Sterling, G., and Barkan, J.,** The hemoglobins of *Artemia salina* IV. A model for genetic control of hemoglobin 1, hemoglobin 2, and hemoglobin X, *Biochem. Genet.,* 15, 409, 1977.
49. **Bowen, S. T., Durkin, J. P., Sterling, G., and Clark, L.,** *Artemia* hemoglobins: genetic variation in parthenogenetic and zygogenetic populations, *Biol. Bull.,* 155, 273, 1978.
50. **Sterling, G. and Bowen, S. T.,** The hemoglobins of *Artemia salina.* V. Genetic variation in hemoglobin 1, hemoglobin 2 and hemoglobin 3, *Biochem. Genet.* 15, 423, 1977.
51. **Van Hauwaert, M. L. and Moens, L.,** unpublished results, 1985.
52. **Goodman, M.,** Decoding the pattern of protein evolution, *Prog. Biophys, Mol. Biol.,* 37, 105, 1981.
53. **Dayhoff, M. O.,** *Atlas of Protein Sequences and Structure,* Natl. Biomed. Res. Found., Silverspring, Md., 1972.
54. **Fasman, G. P., Ed.,** *Handbook of Biochemistry and Molecular Biology,* 3rd. ed., Vol. 3, CRC Press, Boca Raton, Fla., 1976.
55. **Feldhoff, R. C. and Peters, T.,** Determination of the number and relative position of tryptophan residues in various albumins, *Biochem. J.,* 159, 529, 1976.
56. **Heyligen, H., Moens, L., Heip, J., and Kondo, M.,** Identification of heme-containing domains in *Artemia* hemoglobins, *Arch. Int. Physiol. Biochim.,* 88, B137, 1980.
57. **Moens, L.,** The extracellular hemoglobins of *Artemia* sp. A biochemical and ontogenetical study, *Acad. Anal.,* 44, 3, 1982.
58. **Geelen, D., Moens, L., Heip, J., Hertsens, R., Donceel, K., and Clauwaert, J.,** The structure of *Artemia* sp. hemoglobins. I. Isolation and characterization of oxygen binding domains obtained by limited tryptic digestion, *Int. J. Biochem.,* 14, 991, 1982.

59. **Moens, L., Geelen, D., Van Hauwaert, M. L., Wolf, G., Blust, R., Witters, R., and Lontie, R.,** The structure of *Artemia* sp. hemoglobin. Cleavage of the native molecules into functional units by limited subtilisin digestion, *Biochem. J.*, 223, 861, 1984.
60. **Tanford, C.,** *Physical Chemistry of Macromolecules*, John Wiley & Sons, New York, 1961.
61. **Schoeder, W. A. and Huisman, H. J.,** The chromatography of hemoglobins, *Clin. Biochem. Anal.*, 9, 1, 1980.
62. **O'Farell, P. H.,** High resolution two-dimensional electrophoresis of proteins, *J. Biol. Chem.*, 250, 4007, 1975.
63. **Radola, B.,** Isoelectric focusing in layers of granulated gels. I. Thin-layer isoelectric focusing of proteins, *Biochim. Biophys. Acta*, 295, 412, 1973.
64. **Radola, B. J.,** Isoelectric focusing in layers of granulated gels. II. Preparative isoelectric focusing, *Biochim. Biophys. Acta*, 386, 181, 1974.
65. **Frey, M. D. and Radola, B. J.,** High resolution preparative isoelectric focusing in layers of granulated gels, *Electrophoresis*, 3, 216, 1982.
66. **De Smet, K., Van Hauwaert, M. L., Moens, L., and Van Beeumen, J.,** The structure of the *Artemia* hemoglobins. A comparison of the structural units composing the *Artemia* sp. globin chains, in *Artemia, Research and its Applications*, Vol. 2, Decleir, W. et al., Eds., Universa Press, Wetteren, Belgium, 1987, 41.
67. **Moens, L., Van Hauwaert, M.-L., and Wolf, G.,** The structure of *Artemia* sp. hemoglobins. III. Purification of a structural unit to homogeneity, *Biochem. J.*, 227, 917, 1985.
68. **Hewick, R. M., Hunkapiller, M. W., Hood, L. E., and Dreyer, W. J.,** A gas-liquid solid phase peptide and protein sequenator, *J. Biol. Chem.*, 256, 7990, 1981.
69. **Chang, J. Y., Creaser, E. H., and Bentley, K. W.,** 4-N,N'-dimethylaminoazobenzene-4'-isothiocyanate. A new chromophoric reagent for protein sequence analysis, *Biochem. J.*, 153, 607, 1976.
70. **Moens, L., Van Hauwaert, M.-L., Geelen, D., De Smet, K., Verpooten, G. and Van Beeumen, J., Wodak, S., Alarol, P., and Trotman, C.,** A structural domain of the covalent polymer globin chains of *Artemia* interpretation of amino acid sequence data, *J. Biol. Chem.*, 263, 4679, 1988.
71. **De Smet, K., Van Hauwaert, M.-L., Moens, L., and Van Beeumen, J.,** The structure of *Artemia* sp. hemoglobin, *Biol. Chem. Hoppe-Seyler*, 367, Abstr. No., 05.01.46, 1986.
72. **Runnegar, B.,** Derivation of the globins from type b cytochrome, *J. Mol. Evol.*, 21, 33, 1984.
73. **Dickerson, R. E. and Geis, I.,** *Hemoglobin: Structure, Function, Evolution and Pathology*, Benjamin Cummings, Menlo Park, Calif., 1983.
74. **Lesk, A. M. and Chothia, C.,** How different amino acid sequences determine similar protein structures: the structure and evolutionary dynamics of the globins, *J. Mol. Biol.*, 136, 225, 1980.
75. **De Smet, K.,** Vergelijkende studie van de strukturele eenheden van het hemoglobine van *Artemia* sp., *Licentiaatsverhandeling*, Universiteit Antwerpen (U.I.A.), Antwerp, Belgium, 1985.
76. **Sick, H. and Gersonde, K.,** Method for continuous registration of oxygen-binding curves of hemoproteins by means of a diffusion chamber, *Anal. Biochem.*, 32, 362, 1969.
77. **D'Hondt, J., Moens, L., Heip, J., D'Hondt, A., and Kondo, M.,** Oxygen-binding characteristics of the extracellular hemoglobins of *Artemia* sp., *Biochem. J.*, 171, 705, 1978.
78. **Van Den Brande, C., D'Hondt, J., Moens, L., and Decleir, W.,** Functional properties of the hemoglobins of *Artemia* sp., *Comp. Biochem. Physiol.*, 60A, 185, 1978.
79. **Wolf, G., Van Pachtebeke, M., Hens, F., Decleir, W., Moens, L., and Van Hauwaert, M.-L.,** Study of the oxygen binding properties of *Artemia* hemoglobins, in *Artemia, Research and its Application*, Decleir, W., et al., Eds., Universa Press, Wetteren, Belguim, 1987, 221.
80. **Finamore, F. J. and Clegg, A. G.,** Biochemical aspects of morphogenesis in the brine shrimp *Artemia salina*, in *The Cell Cycle. Gene-Enzyme Interactions*, Padilla, G. M., Whitson, G. L., and Cameron, L. L., Eds., Academic Press, New York, 1969.
81. **Heip, J., Moens, L., Joniau, M., and Kondo, M.,** Ontogenetical studies on extracellular hemoglobins of *Artemia salina*, *Dev. Biol.*, 64, 73, 1978.
82. **Bagshaw, J. C. and Acey, R.,** Stage-specific gene expression in *Artemia*, in *Biochemistry of Artemia Development*, Bagshaw, J. C. and Warner, A. M., Eds., University Microfilm Int., Ann Arbor, Mich.,1979.
83. **Moens, L. and Kondo, M.,** Polysome dependent synthesis of embryonic proteins of *Artemia salina* during cell differentiation and analysis of heme-containing protein, *Dev. Biol.*, 49, 457, 1976.
84. **Ingram, V. M.,** *The Hemoglobins in Genetics and Evolution*, Columbia University Press, New York, 1963.
85. **Moss, B. and Ingram, V. M.,** Hemoglobin synthesis during amphibian metamorphosis, *J. Mol. Biol.*, 32, 481, 1968.
86. **Masters, C. J. and Holmes, R. S.,** *Hemoglobins and Their Ligands*, North-Holland, Amsterdam, 1975.
87. **Moens, L., Heip, J., Vrints, R., Pollet, D., Piot, E., Jacobs, G., and Kondo, M.,** Absence of hemoglobin mRNA in the developing gastrulae of *Artemia salina*, *Arch. Int. Physiol. Biochem.*, 86, 920, 1978.
88. **Gilchrist, B. M.,** Haemoglobin in *Artemia*, *Proc. R. Soc. London Ser. B*, 143, 136, 1954.
89. **Fox, H. M.,** Hemoglobin in crustacea, *Nature (London)*, 164, 58, 1949.

59. **Moens, L., Geelen, D., Van Hauwaert, M. L., Wolf, G., Blust, R., Witters, R., and Lontie, R.**, The structure of *Artemia* sp. hemoglobin. Cleavage of the native molecules into functional units by limited subtilisin digestion, *Biochem. J.*, 223, 861, 1984.
60. **Tanford, C.**, *Physical Chemistry of Macromolecules*, John Wiley & Sons, New York, 1961.
61. **Schoeder, W. A. and Huisman, H. J.**, The chromatography of hemoglobins, *Clin. Biochem. Anal.*, 9, 1, 1980.
62. **O'Farell, P. H.**, High resolution two-dimensional electrophoresis of proteins, *J. Biol. Chem.*, 250, 4007, 1975.
63. **Radola, B.**, Isoelectric focusing in layers of granulated gels. I. Thin-layer isoelectric focusing of proteins, *Biochim. Biophys. Acta*, 295, 412, 1973.
64. **Radola, B. J.**, Isoelectric focusing in layers of granulated gels. II. Preparative isoelectric focusing, *Biochim. Biophys. Acta*, 386, 181, 1974.
65. **Frey, M. D. and Radola, B. J.**, High resolution preparative isoelectric focusing in layers of granulated gels, *Electrophoresis*, 3, 216, 1982.
66. **De Smet, K., Van Hauwaert, M. L., Moens, L., and Van Beeumen, J.**, The structure of the *Artemia* hemoglobins. A comparison of the structural units composing the *Artemia* sp. globin chains, in *Artemia, Research and its Applications*, Vol. 2, Decleir, W. et al., Eds., Universa Press, Wetteren, Belgium, 1987, 41.
67. **Moens, L., Van Hauwaert, M.-L., and Wolf, G.**, The structure of *Artemia* sp. hemoglobins. III. Purification of a structural unit to homogeneity, *Biochem. J.*, 227, 917, 1985.
68. **Hewick, R. M., Hunkapiller, M. W., Hood, L. E., and Dreyer, W. J.**, A gas-liquid solid phase peptide and protein sequenator, *J. Biol. Chem.*, 256, 7990, 1981.
69. **Chang, J. Y., Creaser, E. H., and Bentley, K. W.**, 4-N,N'-dimethylaminoazobenzene-4'-isothiocyanate. A new chromophoric reagent for protein sequence analysis, *Biochem. J.*, 153, 607, 1976.
70. **Moens, L., Van Hauwaert, M.-L., Geelen, D., De Smet, K., Verpooten, G. and Van Beeumen, J., Wodak, S., Alarol, P., and Trotman, C.**, A structural domain of the covalent polymer globin chains of *Artemia* interpretation of amino acid sequence data, *J. Biol. Chem.*, 263, 4679, 1988.
71. **De Smet, K., Van Hauwaert, M.-L., Moens, L., and Van Beeumen, J.**, The structure of *Artemia* sp. hemoglobin, *Biol. Chem. Hoppe-Seyler*, 367, Abstr. No., 05.01.46, 1986.
72. **Runnegar, B.**, Derivation of the globins from type b cytochrome, *J. Mol. Evol.*, 21, 33, 1984.
73. **Dickerson, R. E. and Geis, I.**, *Hemoglobin: Structure, Function, Evolution and Pathology*, Benjamin Cummings, Menlo Park, Calif., 1983.
74. **Lesk, A. M. and Chothia, C.**, How different amino acid sequences determine similar protein structures: the structure and evolutionary dynamics of the globins, *J. Mol. Biol.*, 136, 225, 1980.
75. **De Smet, K.**, Vergelijkende studie van de strukturele eenheden van het hemoglobine van *Artemia* sp., *Licentiaatsverhandeling*, Universiteit Antwerpen (U.I.A.), Antwerp, Belgium, 1985.
76. **Sick, H. and Gersonde, K.**, Method for continuous registration of oxygen-binding curves of hemoproteins by means of a diffusion chamber, *Anal. Biochem.*, 32, 362, 1969.
77. **D'Hondt, J., Moens, L., Heip, J., D'Hondt, A., and Kondo, M.**, Oxygen-binding characteristics of the extracellular hemoglobins of *Artemia* sp., *Biochem. J.*, 171, 705, 1978.
78. **Van Den Brande, C., D'Hondt, J., Moens, L., and Decleir, W.**, Functional properties of the hemoglobins of *Artemia* sp., *Comp. Biochem. Physiol.*, 60A, 185, 1978.
79. **Wolf, G., Van Pachtebeke, M., Hens, F., Decleir, W., Moens, L., and Van Hauwaert, M.-L.**, Study of the oxygen binding properties of *Artemia* hemoglobins, in *Artemia, Research and its Application*, Decleir, W., et al., Eds., Universa Press, Wetteren, Belguim, 1987, 221.
80. **Finamore, F. J. and Clegg, A. G.**, Biochemical aspects of morphogenesis in the brine shrimp *Artemia salina*, in *The Cell Cycle. Gene-Enzyme Interactions*, Padilla, G. M., Whitson, G. L., and Cameron, L. L., Eds., Academic Press, New York, 1969.
81. **Heip, J., Moens, L., Joniau, M., and Kondo, M.**, Ontogenetical studies on extracellular hemoglobins of *Artemia salina, Dev. Biol.*, 64, 73, 1978.
82. **Bagshaw, J. C. and Acey, R.**, Stage-specific gene expression in *Artemia*, in *Biochemistry of Artemia Development*, Bagshaw, J. C. and Warner, A. M., Eds., University Microfilm Int., Ann Arbor, Mich., 1979.
83. **Moens, L. and Kondo, M.**, Polysome dependent synthesis of embryonic proteins of *Artemia salina* during cell differentiation and analysis of heme-containing protein, *Dev. Biol.*, 49, 457, 1976.
84. **Ingram, V. M.**, *The Hemoglobins in Genetics and Evolution*, Columbia University Press, New York, 1963.
85. **Moss, B. and Ingram, V. M.**, Hemoglobin synthesis during amphibian metamorphosis, *J. Mol. Biol.*, 32, 481, 1968.
86. **Masters, C. J. and Holmes, R. S.**, *Hemoglobins and Their Ligands*, North-Holland, Amsterdam, 1975.
87. **Moens, L., Heip, J., Vrints, R., Pollet, D., Piot, E., Jacobs, G., and Kondo, M.**, Absence of hemoglobin mRNA in the developing gastrulae of *Artemia salina, Arch. Int. Physiol. Biochem.*, 86, 920, 1978.
88. **Gilchrist, B. M.**, Haemoglobin in *Artemia, Proc. R. Soc. London Ser. B*, 143, 136, 1954.
89. **Fox, H. M.**, Hemoglobin in crustacea, *Nature (London)*, 164, 58, 1949.

90. **Bowen, S. T., Lebherz, H. G., Poon, M. C., Chow, V. M. S., and Grigliatti, T. A.,** The hemoglobins of *Artemia salina*. I. Determination of phenotype by genotype and environment, *Comp. Biochem. Physiol.*, 31B, 733, 1969.
91. **Heip, J., Moens, L., and Kondo, M.,** Effect of concentrations of salt and oxygen on the synthesis of extracellular hemoglobins during development of *Artemia salina*, *Dev. Biol.*, 63, 247, 1978.

Chapter 7

CAROTENOIDS IN RELATION TO *ARTEMIA* DEVELOPMENT

H. J. Nelis, P. Lavens, L. Moens, P. Sorgeloos, and A. P. De Leenheer

TABLE OF CONTENTS

I. CAROTENOIDS: INTRODUCTION

A. Physicochemical Aspects

Carotenoids are among the most abundant pigments in nature. The chemical structure of these polyenes consists of a number of isoprenoid units, arranged in such a way that the two central methyl groups are in the 1,6 position, while the remaining methyl groups are positioned 1,5 relative to each other. A series of conjugated double bonds constitutes a chromophore of variable length, resulting in characteristic yellow to red colors. Most natural carotenoids possess the all-*trans* configuration, although in recent years the genuine presence of *cis* isomers has been documented in certain organisms. Unsubstituted hydrocarbons, e.g., α - and β-carotene are commonly designated as "carotenes". The collective term "xanthophylls" covers oxygenated derivatives including alcohols (e.g., lutein, zeaxanthin), epoxides (e.g., violaxanthin, antheraxanthin) and keto derivatives such as echinenone, canthaxanthin, and astaxanthin. Hydroxyl groups may further be esterified with fatty acids. The chemical structures of the compounds which are of concern in the context of this chapter are presented in Figure 1. More comprehensive information on the physicochemical properties of carotenoids can be found in a number of books[1] and reviews.[2,3]

B. Biological Aspects

Carotenoids occur in virtually all living organisms. Their abundance is particularly remarkable in algae, bacteria, higher plants, and invertebrates. Biosynthetically, they are derived from the acyclic lycopene via hydrogenation, dehydrogenation, cyclization, and oxygenation reactions.[4,5] So far, there are no examples of animals biosynthesizing carotenoids *de novo*.[6] Hence, all animal carotenoids are ultimately of plant origin, and assimilated through

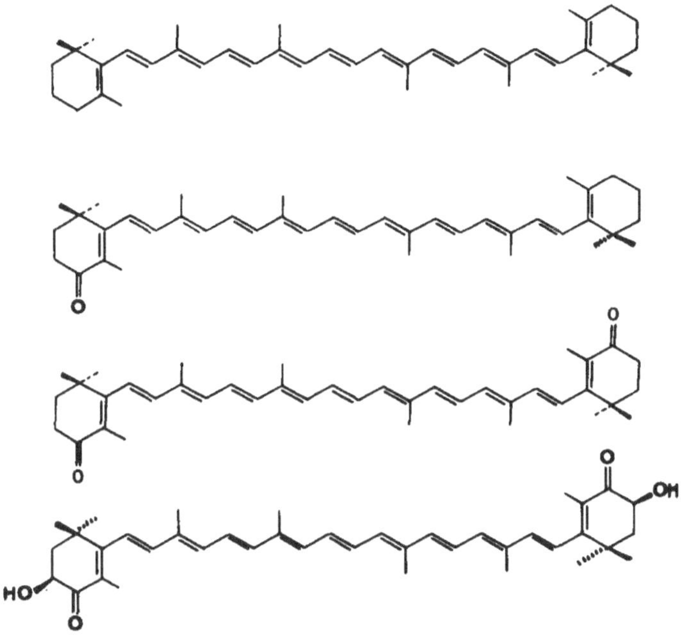

1

2

3

4

FIGURE 1. Structural formulas of carotenoids of interest. (1) β-Carotene; (2) echinenone; (3) canthaxanthin; (4) astaxanthin.

the diet, either with or without metabolic transformation. The biochemistry of carotenoids in plants and animals has been thoroughly surveyed by Goodwin in two authoritative books.[7,8]

The extended chromophore in the carotenoid structure points to a biological significance for these pigments, by suggesting photosensitive and antioxidant properties. In plants, carotenoids participate as secondary light-harvesting compounds in the photosynthetic process[9] and exert photoprotective effects[10] as well. As antioxidants they protect vulnerable tissues against the deleterious effects of free radicals.[11] These antioxidant properties are particularly pronounced at low oxygen partial pressures.[12]

Similar protective effects have been invoked in marine animals.[13,14] In addition to protection against harmful radiation, the characteristic colors of these organisms may also contribute to chromatic concealment.[13,15]

II. SCOPE

A. General Scope

Apart from the possible photochemical significance of carotenoids, their abundance in the reproductive system and eggs of Crustacea has led many researchers to assume a more direct role for these compounds in reproduction/embryonic development. The aim of this chapter is to examine the available evidence suggesting such a relationship in *Artemia* and other Crustacea. Data pertaining to Crustacea in general will be presented to put *Artemia* in a broader perspective. Such general background information concerning organisms belonging to the same phylum as *Artemia* may indeed shed light indirectly on the role of carotenoids in *Artemia* itself.

Following a critical survey of the literature on the fate of carotenoids in Crustacea, six criteria were selected as a basis for the evaluation. Unless the term "carotenoprotein" is specifically used, this evaluation deals with "free" carotenoids, i.e., irrespective of their association with macromolecules.

B. Criteria to be Evaluated

1. Is there an active mobilization of carotenoids in the reproductive organs and/or eggs?
2. Do the animals display sexual dimorphism as far as their pigment composition is concerned? Specifically, do males and females contain specific pigments (qualitative aspect) or are there significant differences in the abundance of certain carotenoids between sexes (quantitative aspect)?
3. Is there evidence for the presence, in the reproductive organs and eggs, of specific pigments which are absent or substantially less abundant in other tissues?
4. Do qualitative or quantitative changes in the pigment profile occur in the course of development; i.e., is there utilization, synthesis, or transformation of pigments?
5. Is there a pigment transfer or redistribution in connection with the developmental process?
6. Do the macromolecular carotenoid complexes (carotenoproteins) undergo changes in relation to development?

C. Hypotheses

Following careful consideration of the available evidence, some hypotheses regarding a function of carotenoids in reproduction/development will be presented. In this respect, the recent findings of *cis-trans* isomers in *Artemia* eggs (see Section IV.D.) have most likely provided a clue to their function.

III. *ARTEMIA* IN PERSPECTIVE: THE FATE OF CAROTENOIDS IN CRUSTACEA

A. Introduction

The pigment composition of representatives of virtually all subclasses of Crustacea has been studied.[16] However, unlike the isolation, identification, and metabolism of these pigments, the function of carotenoids, particularly in connection with reproduction/development has only occasionally been addressed. Organisms that have been investigated in this respect belong mainly to the subclasses of Branchiopoda (including *Artemia*) and of Malacostraca (Decapoda in particular).

B. Occurrence of Carotenoids in Crustacea

1. Overall Picture

Crustacea lack the ability to biosynthesize carotenoids *de novo*, but efficiently absorb these pigments from the diet.[6,17] They may be deposited unaltered in the tissues (a typical example is β-carotene) or they may be metabolically transformed to keto- and hydroxy derivatives[16-18] (see Section III.C). An exhaustive compilation of all derivatives so far reported in Crustacea has been conducted by Goodwin.[16] In characterizing specific carotenoids, special attention should be paid to their stereochemistry. Geometrical (*cis-trans*) and optical isomerism are theoretically possible. The absolute configuration of optical isomers of astaxanthin occurring in Crustacea has been established.[16] Until recently (see Section IV.D.) there were no reports on the presence of significant amounts of *cis* isomers in Crustacea. Physicochemically, carotenoids may occur in different states, i.e., "free", esterified, or bound to macromolecules.[16,17,19] Free and esterified carotenoids may accumulate as lipid dispersions, e.g., in the chromatophores[20] or the hepatopancreas.[16,19] Binding with structural elements, e.g., chitin in the carapace has been reported,[15] while keto-carotenoids are frequently noncovalently attached to proteins to form water soluble carotenoproteins.

2. Carotenoproteins

Carotenoproteins are particularly abundant in Crustacea. Several reviews have appeared on this subject.[13,21-25] Two types of carotenoproteins have been distinguished. In complexes of type I, the carotenoid is bound nonstoichiometrically to a lipo(glyco)protein, often through nonspecific association with the lipid portion, or, alternatively, via direct binding with the protein (e.g., through hydrophobic interactions). The prosthetic group may either be a carotene, a keto-carotenoid or another xanthophyll. Very often, the relative distribution of the pigments in the carotenoprotein is similar to the overall pigment pattern. Complexes of type I are very common in the blood, eggs, and ovaries of female Crustacea. Lipovitellins, the storage lipoglycoproteins in crustacean eggs, are typical examples of this kind of association. Type II complexes (e.g., Crustacyanin) constitute "real" carotenoproteins in which the carotenoid, nearly always astaxanthin or rarely canthaxanthin, is stoichiometrically attached at specific sites on a (glyco)protein. The carotenoid often maintains the tertiary and quaternary structure of the protein. The association commonly results in a substantial bathochromic shift in the absorption spectrum, presumably caused by polarization of the carotenoid by charged groups in its vicinity. Such bathochromic shift is also observed with type I carotenoproteins. "Real" carotenoproteins are frequently present on the external surfaces of Crustacea, where they account for the vivid coloration. Cleavage of the complex results in a color change caused by the liberation of the free carotenoid.

C. Metabolism

There is a wealth of literature about the metabolic transformations by Crustacea of dietary carotenoids.[6,16-18] Metabolism is reportedly under endocrine control, particularly from the eyestalk.[16,17] Considerable efforts have been devoted to elucidating the biochemical pathways and the intermediates involved. The most common reaction is oxidation of carotenes to hydroxy- and keto- carotenoids, with astaxanthin as the most oxidized end product. The latter may be reversibly esterified and de-esterified. Apart from reductive pathways operative in other marine organisms,[26] there is presently no evidence indicating other types of metabolic transformations, e.g., cleavage of the polyene chain to yield vitamin A or modification of end groups.[6] For a more detailed picture of the different pathways leading to canthaxanthin and astaxanthin, the reader is referred to three reviews.[16-18]

In the context of this chapter, carotenoid metabolism will primarily be considered in view of possible links with reproduction/development.

D. Participation of Carotenoids in Reproduction/Embryonic Development

1. Introduction

A critical survey of the literature in the light of the criteria established in Section II.B provides only indirect evidence suggesting a relationship between carotenoids and reproduction/development of Crustacea. A number of findings even seem to contradict a participation of carotenoids (see Section III.D.8.a).

2. Mobilization of Carotenoids in Reproductive Organs and Eggs

The observation that carotenoids are massively mobilized into the gonads of Crustacea at the onset of reproduction dates back decades ago and has been repeatedly confirmed ever since.[19] The well-established relationship between carotenoid fluctuations in the body and the sexual cycle led Cheesman et al. to state that "the large numbers of carotenoproteins that have been identified in the eggs and the ovaries of invertebrates suggest their participation in development."[13] The pigments are most likely mobilized from the hepatopancreas, transported in the hemolymph, presumably as carotenoproteins, deposited in the ovaries of the females, and further transferred to the eggs.[19] Establier-Torregrosa observed an increase in free and esterified astaxanthin from 20 to 40 μg/g in resting ovaries of *Plesiopenaeus edwardsianus* (Crustacea: Decapoda) to 430 to 518 μg/g in mature ovaries, with a concomitant disappearance of both forms from the hepatopancreas.[27] A similar mobilization takes place in *Emerita* (Crustacea: Decapoda), which "prepares" for reproduction by enriching its ovaries in carotenoids.[28]

3. Sexual Dimorphism

There are few examples of Crustacea in which the pigment composition is the determining factor of sexual dimorphism.[16] In *Emerita*, differences in the relative abundance of carotenes and astaxanthin occur between males and females, the latter containing relatively more carotenes and less astaxanthin.[28] Obviously, there is a difference in uptake and mobilization between both sexes. Similar relative differences in the levels of canthaxanthin and its metabolic precursor echinenone have been observed in several Branchiopoda, i.e., *Branchipus*, *Branchinecta*, and *Chirocephalus*.[29] There are, however, no examples of true "sex-specific" carotenoids in Crustacea. Real sexual dimorphism, e.g., in Branchiopoda is sometimes displayed as a result of the presence of specific carotenoproteins in the blood or tissues. For example, female *Branchinecta* contain a blue carotenolipoprotein in connective tissue storage cells, which is absent in males.[30] Colored carotenoid-lipoproteins have been detected in the blood of either male or female *Tanymastix*, at the same stage of growth and under identical environmental conditions, suggesting another function than merely protective coloration.[21]

4. Specific Carotenoid Patterns in Ovaries/Eggs

The only differences in pigment patterns between eggs and whole animals so far reported relate to esterification/de-esterification. For example, the eggs of the decapods *Orconectes*,[31] *Nephrops*,[32] and *Homarus*[32] contain only free astaxanthin and no esters, as opposed to other tissues in these decapods in which both forms occur.

In general, however, the distribution of carotenoids in the ovaries and the eggs of Crustacea is the same as in other tissues and merely reflects the availability of the component or its metabolic precursors in the food.[28]

5. Qualitative and Quantitative Changes in the Carotenoid Profile during Development

Most investigators believe that, except for esterification/de-esterification, no qualitative or quantitative changes in the ("free") carotenoids occur in the course of development.[19,28,32-34] In *Acanthephyra* (Crustacea: Decapoda) eggs, astaxanthin is progressively esterified towards the end of development, before transfer of the ester to the appearing chromatophores in the embryo.[35] There is no substantial evidence suggesting utilization or synthesis of pigments.[16,17,19,28,32-35] There has been a report about a reduction of the amount of carotenoids in eggs, which allegedly coincides with the biosynthesis of visual pigments in the nauplius.[36] Interconversion of carotenoids during development is exceptional. In one report, β-carotene has been found to be converted to astaxanthin in developing locust eggs to yield equal amounts of both pigments at the moment of hatching.[37] Color changes are normally caused by modification of the carotenoproteins[38] (see Section III.D.7), rather than of "free" carotenoids.

6. Pigment Transfer

Examples of transfer of carotenoids to the chromatophores[35] or to the naupliar eye[36] have been given above. In *Acanthephyra*, astaxanthin is liberated from an association with a lipoprotein as yolk consumption proceeds.[35] Free astaxanthin is subsequently esterified to be deposited into the chromatophores. This phenomenon led Herring and Morris[35] to hypothesize that the metabolism of carotenoids during development is primarily directed at the formation of chromatophores. In a broader sense, carotenoproteins could thus be regarded as mere carriers of the carotenoid to equip the young animals with chromatophores[35] or photosensitive pigments.[31,36]

7. Changes in Carotenoproteins

Two types of changes in carotenoproteins are common, i.e., structural changes in the course of development and disruption towards the end of development. Structural modifications often result in characteristic color changes, which may be correlated with particular stages of development.[39]

In some cases, such transformation of one form of carotenoprotein into another may even be a necessary step preceding its usage by the embryo.[28] As part of the hatching process, carotenoproteins are usually disrupted to yield the free carotenoid and a (lipoglyco)protein.[40] The carotenoid may subsequently be concentrated in the eye[36,41] or the chromatophores[35] of the nauplii or excreted in the gut,[42] whereas the (lipoglyco)protein satisfies growth and energy requirements during early larval development. A typical example in this regard is the breakdown of the yolk platelets in *Branchipus stagnalis*,[43] a freshwater anostracan closely related to *Artemia*. These yolk platelets consist of lipoglycoproteins associated with canthaxanthin, yielding blue complexes. Two to three hours after hatching, the blue yolk platelets are broken down and the liberated canthaxanthin is incorporated in (orange) lipid globules.[43] Similar phenomena have been demonstrated in Cladocera, e.g., *Daphnia*[34] and *Simocephalus*.[42]

8. Discussion

a. Is there a Function for Carotenoids?

Various investigators believe that the accumulation of carotenoids in Crustacea is purely fortuitous, that the pigment is merely a waste product and hence serves no function. A number of facts indeed seem to support this view. Most puzzling is the observation that nonpigmented animals, reared on carotenoid-free diets for prolonged periods, remain perfectly viable and produce viable eggs.[19,34] It was concluded by Herring that, in *Daphnia magna*, neither growth, viability, nor fertility of eggs or adults exposed to a variety of environmental conditions was affected by the presence of carotenoids.[34] While the mobilization of carotenoids in the ovaries and eggs would appear to suggest a function, it has been argued that their carotenoid pattern is the same as seen in whole animals and is strongly dependent on food abundance and feeding activity.[28] The pigment transfer from ovary to egg is purely passive and relative only to overall carotenoid uptake.[28] In addition, the formation of lipovitellin-carotenoid complexes may also be considered fortuitous since the nature of the carotenoid-protein linkage is rather nonspecific and rarely stoichiometric.[21] Again, the amount of carotenoid in the lipovitellin, available at vitellogenesis, will depend on dietary and environmental factors.[21] The spectral shift may have no significance but merely reflect the polarizability of the polyene.[21]

Alternatively, animals may derive some adaptive advantages from accumulating carotenoids. The key to a function in development, if any, clearly lies in the characteristic fate of carotenoproteins (formation, structural modifications, disruption at specific stages of development).[13] Basically, two opposite views exist about a possible role for carotenoids in developing embryos. Either the pigment exerts a function at an early stage, i.e., in the eggs, before hatching,[13,40] or the carotenoid satisfies larval requirements.[31,35,44]

b. Functions in the Eggs/Embryos

A metabolic function for carotenoids in developing embryos can be virtually ruled out because there is no utilization or transformation of pigment.[45] Some kind of structural[45] or protective role[13] in the embryos seems to be more likely. Irrespective of their association with macromolecules, the carotenoids may act as a light or heat shield to protect the exposed eggs.[13,40] However, in view of the phenomenon of carotenoprotein formation, protection of either of the components constituting the complex is also plausible.[13,44] Thus, the protein may protect the carotenoid from photooxidation, or, alternatively, the carotenoid may stabilize the protein and/or control its configuration.[13] Carotenoproteins may also, in analogy with vitamin A[46] be involved in the stabilization and permeability control of membranes by acting as a cross link between lipid and protein. In view of the partial mutual exclusivity of carotenoids and vitamin A in invertebrate eggs,[47] carotenoids may very well replace vitamin A in this regard.

Aside from these possible functions of "true" carotenoproteins, the complexes of carotenoids with storage lipoglycoproteins deserve a special note. By complexing lipovitellin with carotenoids, the animals are supposedly able to stabilize and protect it from enzymatic attack and thus hold it in reserve until needed.[13] Disruption of the carotenoprotein would make the free protein available for consumption by the nauplius.

c. *Functions in the Nauplii*

The alternative possibility is that carotenoproteins in the eggs merely have a temporary transport function, i.e. by providing a vehicle for transfer of the carotenoid to the nauplii.[31,35,44] As long as there is no feeding activity, the nauplii have to derive their carotenoids from the mother (through the eggs). As noted above, when the carotenoprotein is split the free carotenoid reappears in the naupliar eye[36,41] where it may serve a role as an accessory visual pigment, or in the chromatophores.[35] Furthermore, the pigment may confer some degree of photoprotection and/or chromatic concealment upon the animal.[13] In certain Copepoda for example, the degree of pigmentation is the result of two selective counteracting forces, i.e., protection from the harmful effects of sunlight and predator pressure.[48,49] Thus, the small nauplii which live near the surface of the water and are relatively invisible to predators are highly pigmented, as opposed to the more exposed mature animals which live in deeper water and are less pigmented. In general, the above hypothesis implies that the role of carotenoids in crustacean eggs is merely reduced to serving larval requirements in integumental and photosensitive pigments.

E. Conclusions

An investigation into the fate of carotenoids in Crustacea, including freshwater Anostraca, has revealed that these pigments most likely serve no metabolic function during development but rather exert a structural/protective role in the eggs or provide some adaptive advantages to the young nauplii, e.g., in terms of protective pigmentation. Unfortunately despite all recent research in this area, a somewhat pessimistic statement made in 1950 by Goodwin[50] remains very timely. He wrote "It will be apparent (from this article) that a considerable amount of information is available about the occurrence and distribution of carotenoids in reproductive and allied structures in both plants and animals. Very little is known, however, concerning their direct participation in reproduction. This lack of precise information occurs in spite of a great deal of circumstantial evidence pointing to, but never specifically defining, a sexual function." It appears that some key is missing as yet that would allow a definitive unraveling of this function. Section IV will examine how well *Artemia* fits into the overall picture outlined for Crustacea. Specifically, the question arises whether the research on *Artemia* has provided additional clues supporting some of the above hypotheses regarding the role of carotenoids in Crustacea in general.

IV. CAROTENOIDS IN *ARTEMIA*

A. Occurrence and Metabolism

1. *Historical Aspects*

Investigators have been intrigued by the orange pigment in *Artemia* since the 19th Century.[51] In 1930, Needham and Needham[52] claimed that it was synthesized during development. For some time, there was a dispute about the identity of the pigment. Gilchrist and Green were the first to conduct a thorough identification study and concluded that the major carotenoid in *Artemia* was esterified astaxanthin.[51] A few years later, however, Krinsky[53] and Czygan[54] independently identified this pigment as canthaxanthin. While the former stated that *Artemia* and related Branchiopoda do not have the ability to convert carotenoids beyond the oxidation level of canthaxanthin, the latter maintained that some strains, e.g., those of Canadian origin, contained astacene in addition to canthaxanthin. Gilchrist[29] later partly refuted her preliminary identity report, confirmed the data obtained by Czygan and

Krinsky, but continued to find small amounts of astaxanthin in addition to canthaxanthin. Other reports on the presence of astaxanthin in *Artemia* have appeared,[55] although it is now generally accepted that canthaxanthin constitutes the major pigment.[56-61] For some years, further studies were mainly directed towards elucidating the metabolic pathways involved[56-59] (see Section IV.A.2). Soejima et al. did some pioneering work in that they recognized the potential of the carotenoid content of cysts and nauplii as a parameter for the biochemical characterization of various *Artemia* geographical strains.[60] The years 1983-1984 saw the discovery, independently by two research groups, of *cis*-carotenoids in *Artemia* eggs. *Cis*-canthaxanthin was demonstrated by Zagalsky et al. on the basis of resonance Raman and circular dichroism (CD) evidence[62] and by the authors of this chapter using high performance liquid chromatography (HPLC) in conjunction with spectral techniques.[61] Since then, the latter have further elaborated on the significance of these unusual carotenoids.[63] It is hoped that this finding may eventually contribute to a better understanding of the relevance of carotenoids to *Artemia* development.

2. Metabolism

Artemia converts β-carotene, absorbed from the (algal) food, via echinenone to canthaxanthin. This metabolic pathway was unraveled by Hata and Hata,[57] Hsu et al.,[58] and Davies et al.[59] by feeding canthaxanthin precursors, algae, and labeled β-carotene. The hypothetical intermediates isocryptoxanthin, isozeaxanthin, and 4-hydroxy-4′-keto-carotene could not be detected. Therefore, the oxidation of β-carotene to canthaxanthin may formally be regarded as a two-stage process, involving direct insertion of oxygen and catalyzed by hydroxylase and dehydrogenase enzymes.[59] Nevertheless, at the enzymatic level, these intermediates probably occur as transient enzyme-bound compounds.[59] This overall metabolic scheme implies that, apart from the presence of other ingested and untransformed carotenoids (e.g., lutein or even astaxanthin), *Artemia* displays a relatively simple carotenoid pattern of only three derivatives of interest, i.e., canthaxanthin and its precursors echinenone and β-carotene. However, Hata and Hata detected a number of previously unrecognized keto-carotenoids,[57] which physicochemically closely resembled canthaxanthin. Unfortunately, these workers failed to identify them as *cis*-canthaxanthins, which were only rediscovered in 1983-1984.[61,62]

3. Distribution

Encysted embryos (cysts) are very rich in canthaxanthin but contain little echinenone and no β-carotene.[60] Upon hatching, the carotenoids are found again in the nauplii.[60] Adults contain pigments only when fed on an algal diet.[57] Yeast-fed *Artemia* are pale and produce white eggs.[57] In adult *Artemia* reared on *Spirulina* algae, the principal carotenoids were canthaxanthin, echinenone, β-carotene, and some xanthophylls including zeaxanthin or lutein.[64] All these pigments except canthaxanthin were already present in the *Spirulina*. The detailed occurrence and distribution of *cis*- and all *trans*-canthaxanthin are described below (see Section V).

4. Binding to Macromolecules

As early as in 1960, Gilchrist and Green[51] suggested that canthaxanthin in the cysts of *Artemia* was associated with a protein, a statement repeated by Hata and Hata.[57] Canthaxanthin is particularly abundant in the lipovitellin, present in the yolk platelets.[65-67] This is consistent with the occurrence of similar canthaxanthin-protein complexes in the eggs of other Anostraca.[21,29] In *Artemia*, as in *Branchinecta*,[30] this canthaxanthin-lipovitellin complex is orange-red[66] as opposed to the blue lipovitellin in *Branchipus*.[43] The biochemical composition of the *Artemia* lipovitellin as well as the mode of binding of the carotenoid to the lipoglycoprotein have been extensively studied.[62,66,67] Lipovitellin isolated from the yolk granules is basically a lipoglycoprotein containing 8.6% lipid and 3.3% carbohydrate.[66,67] Two apoproteins were found, having molecular weights of 190,000 and 68,000, respectively.[66,67] Lipovitellin contains a single canthaxanthin binding site at which the polyene can exist in three conformations of minimum energy, i.e., in a twisted all-*trans*, a twisted *cis*, and a twisted conjugation-ruptured form.[62] A similar picture is seen in the canthaxanthin-lipovitellin complex from *Branchipus stagnalis*.[68] Upon extraction with organic solvents, part of the canthaxanthin is liberated from the lipovitellin as a sterically unhindered *cis* form. After hatching, the *Artemia* lipovitellin, as in *Branchipus*[43] and other Crustacea[21] breaks down by the action of proteolytic enzymes to satisfy protein and energy requirements of the nauplii, with concomitant liberation of the canthaxanthin and its incorporation in orange lipid globules.[66]

B. Participation in Reproduction/Embryonic Development

Before the discovery of *cis*-canthaxanthin, only scarce data were available concerning the biological fate of carotenoids in relation to *Artemia* development. The following summary ignores the observations pertaining to *cis*-canthaxanthin, which will be discussed later. Based on the criteria set above, the following facts are of interest:

1. Canthaxanthin is mobilized in the cysts.[60] The transfer from the body tissues to the ovaries occurs at a particular stage of development.[57]
2. There is no real pigmentary sexual dimorphism except for the greater relative abundance of echinenone in males.[58]
3. The pigment patterns in eggs, nauplii, and adults are comparable in that canthaxanthin is the predominant carotenoid.[57-60] The levels of echinenone in cysts and nauplii are low and β-carotene is absent. Both metabolic precursors reappear as soon as the nauplii start feeding on algae.[69] Up to recently, there were no reports of specific pigments in cysts (see Section IV.D).
4. There is no indication of any qualitative or quantitative changes in the carotenoid profile in connection with development. Early reports of a pigment synthesis[52] have not been confirmed.
5. Canthaxanthin is associated with lipoglycoproteins in the eggs. Upon hatching, the carotenoprotein is broken down and the "free" carotenoid is deposited in fat globules.[66] A definite link between the characteristic fate of carotenoproteins and development was suggested by Dutrieu.[70,71] Her hypothesis was an attempt to present a unifying concept explaining a number of apparently related phenomena, including mode of reproduction, cyst formation, and diapause deactivation. The general premise was that the nature of the food (rich vs. poor in chlorophyll and carotenoids) as well as environmental factors (e.g. oxygen supply) determine the mode of reproduction of *Artemia* (ovoviviparity vs. oviparity) and the establishment of diapause. The cleavage of carotenoproteins in the embryo is assumed to play a decisive role in the developmental process. Thus, Dutrieu proposes that in a medium rich in oxygen and chlorophyll, the female animals do not secrete hemoglobin (or a derivative, no identification was done) in the shell glands and no brown shell is formed around the embryos. The gray-blue embryo contains intact carotenoproteins which, as development proceeds, are disrupted when the naupliar stage is reached. The animals are ovoviviparous and orange nauplii, containing free carotenoid are released into the medium. In contrast, under conditions of poor oxygen supply, the animal synthesizes hemoglobin and secretes it, via the shell gland, into the brown shell surrounding the embryo, resulting in oviparity (cyst formation). Development proceeds but, allegedly because of the impermeable shell, the carotenoproteins are not split at a particular stage. Consequently, further development is arrested and the newly laid cysts are in diapause. The disruption of the carotenoproteins can subsequently be brought about by desiccation, which leads to diapause deactivation and resumption of development. Furthermore, in the absence of

chlorophyll and carotenoids, hemoglobin is synthesized by the females but is not eliminated and the animals remain ovoviviparous. Experimental evidence supporting this oversimplified theory is lacking and several authors have refuted it over the years, by reporting conflicting results.[72-75] Yet, Dutrieu rightly pointed to the formation and cleavage of carotenoproteins, at particular stages, as important phenomena connected with embryonic development. In addition, recent evidence (see Sections V and VI) supports her view that the carotenoid possibly only plays an indirect role in development, by temporarily "locking up" some proteins that are required for development.

6. Eggs laid by animals fed on yeast are colorless and yet tend to hatch normally.[57,76]

C. Carotenoids and Anoxia

From the above considerations it must have become increasingly clear that, if carotenoids really play a role in *Artemia* development, they probably do so by virtue of their binding with proteins. Dutrieu was already thinking along these lines when she postulated a link between desiccation of cysts and cleavage of carotenoproteins. Similarly, Karmiol, in his 1981 Master's thesis,[77] experimentally studied the fate of the carotenoproteins and a few selected enzymes following exposure of *Artemia* cysts to anoxic conditions. His starting point was that cysts can tolerate anoxia owing to some adaptive mechanisms at the cellular level. Basically, cysts respond to anoxybiosis by becoming quiescent,[75,78] a state in which most enzymatic processes are arrested[79] and the structural integrity of vital cell constituents (enzymes, membranes) is maintained. The aim of the work was to get an insight into the biochemical mechanisms that underly this preservative action. According to Karmiol, carotenoids would be key components in this respect. In summary, he made the following intriguing observations:

1. Cysts contain two major carotenoids, a "red one and a yellow one" identified as all-*trans*-canthaxanthin and "echinenone", respectively. As it turns out, the latter yellow compound was misidentified, which was in part due to the lack of authentic echinenone as a reference component. As evidenced in Sections IV.D and V, the latter is in fact a mono *cis*-canthaxanthin. The spectral data presented by Karmiol were not consistent with echinenone. No explanation could be given at that time for the discrepancy between the observed absorption maximum (454 nm, petroleum ether) and the literature value for echinenone (458 nm), for the bathochromic shift in the absorption spectrum upon saponification (most peculiar for a keto-carotenoid) and, particularly, for the presence of a second absorption maximum around 360 to 370 nm, which is totally absent in the echinenone spectrum. *Cis*-canthaxanthins on the other hand do display a second maximum in this region, this "*cis* peak" being a most characteristic feature,[80,81] and their main absorption maximum (454 nm) is hypsochromically shifted by approximately 6 nm with reference to all-*trans*-canthaxanthin. The bathochromic shift following saponification is easily rationalized on the basis of the fact that upon heating or even upon standing at room temperature, part of the *cis*-canthaxanthin is back isomerized, to yield eventually an equilibrium mixture of *cis*- and *trans*- isomers.[80]
2. The total carotenoid concentration allegedly increases in anoxic embryos. This puzzling observation can obviously not be explained in terms of *de novo* synthesis or transformation of β-carotene. Instead, Karmiol postulates that a reversible hydrogenation/dehydrogenation of canthaxanthin occurs.[77] A similar mechanism involving in-chain epoxidation/de-epoxidation has been proposed by Karnaukhov[82] in an attempt to explain the tolerance of certain mollusks to environmental pollution. Other authors however failed to find experimental evidence for this theory.[83] In conditions of high oxygen tension, canthaxanthin is supposedly partly hydrogenated, which leads to disruption of the chromophore and, hence bleaching of the compound. When hypoxia ensues,

the chromophore is allegedly re-expressed by a reverse dehydrogenation reaction. This reversible process is believed to be coupled with the enzymes malate dehydrogenase and α-glycerophosphate dehydrogenase and to require NAD^+/ NADH as co-factors. However, solid experimental data supporting this hypothesis are lacking.

3. In yolk platelets, carotenoids are associated with lipoglycoproteins, but reportedly exist in different populations. The spectral characteristics of the carotenoproteins, both in native form and after urea treatment indicate different modes of binding, i.e., a rather firm hydrophobic binding with the protein and a more loose association with the lipid portion. Extraction with sodium chloride or digestion with trypsin yields different carotenoprotein moieties. As a reaction to anoxia, the carotenoproteins undergo a number of characteristic alterations. Their color changes from brown to olive-green. They become increasingly resistant to disruptive forces, e.g., elevated temperatures and solubilization with sodium chloride. Extraction of the carotenoids with organic solvents becomes more difficult and some changes occur in the spectral properties of the carotenoproteins. Most important, these changes are accompanied by a conversion of a substantial part of canthaxanthin to "echinenone," apparently indicating a reversal of the biosynthetic pathway. However, as shown in Section V this process would actually appear to be *trans* → *cis* conversion.
4. In anoxic cysts, still according to Karmiol, an apparent redistribution of macromolecules over the different subcellular fractions takes place. Specifically in the post mitochondrial supernatant (PMS) there is a shift to compounds with higher molecular weight. This is accompanied by an enhanced binding of carotenoids to proteins.[77]
5. Some enzymes, e.g., GTP:GTP guanylyltransferase supposedly become associated with carotenoids during anoxia. This association confers stability against structural damage and accounts for a decreased activity of the enzymes.

The most relevant data are summarized in Table 1. It shows the following biochemical effects of anoxia: (1) a subcellular redistribution of the carotenoids (an increase in the carotenoid/protein ratio in the postmitochondrial supernatant, PMS) and (2) the apparent increase in the total carotenoid content and the significant conversion of all-*trans*-canthaxanthin to "echinenone" (*cis*-canthaxanthin). Karmiol concludes that encysted *Artemia* embryos have acquired an adaptive mechanism to stabilize enzymes and membrane systems (yolk platelets) under anoxic conditions, by complexing them with carotenoids.[77] Consequently, their structural integrity is maintained and the activity of enzymes is suppressed. The implications of Karmiol's hypothesis for our overall understanding of the function of carotenoids in connection with *Artemia* development are discussed in Section VI, specifically in the light of the phenomenon of *cis-trans*-canthaxanthin isomerism.

D. *cis-trans*-Canthaxanthin Isomerism in *Artemia*

1. Demonstration and Identification of cis-Canthaxanthin[61]

In 1983, we reported as part of a new method for the quantitative determination of canthaxanthin and other carotenoids in *Artemia*, a nonaqueous reversed phase liquid chromatographic method.[64] The initial aim was to use this HPLC technique as a basis for a "routine" assay of carotenoids in *Artemia* cysts of different geographical origin. Preliminary application of the method to adult animals revealed the expected carotenoid profile consisting of canthaxanthin, echinenone, β-carotene, and some xanthophylls.[64] However, in chromatograms obtained on extracts of cysts, a significant unknown peak eluting very close to the peak of all-*trans*-canthaxanthin showed up[61] (Figure 2A). This peak proved composite upon rechromatography on silica and could be further differentiated into three individual components (Figure 2B). Newly hatched nauplii displayed a similar two-peak pattern in the reversed phase system except that the peak height ratio between the unknown and all-*trans*-canthaxanthin was considerably lower. The unidentified pigment was also found in repro-

Table 1
EFFECT OF ANOXIA ON THE CAROTENOID CONTENT OF VARIOUS *ARTEMIA* (GREAT SALT LAKE STRAIN) EMBRYO SUBCELLULAR FRACTIONS[77]

Sample	Period under anoxia	μg canthaxanthin per mg of protein	μg "echinenone" (*cis*-canthaxanthin) per mg of protein	μg total carotenoids per mg of protein	% canthaxanthin	% "echinenone" (*cis*-canthaxanthin)
C-TH	NIL	0.27	0.10	0.37	73.0	27.0
C-YP	NIL	0.64	0.41	1.05	61.0	39.0
C-PMS	NIL	0.63	0.23	0.86	73.3	26.7
C-NaCl	NIL	0.63	0.60	1.23	51.2	48.8
C-trypsin	NIL	0.86	0.64	1.50	57.3	42.7
E-TH	1 month	0.78	0.29	1.07	72.9	27.1
E-YP	1 month	0.53	0.70	1.23	43.1	56.9
E-PMS	1 month	0.94	0.32	1.26	74.6	25.4
E-NaCl	1 month	0.51	0.68	1.19	42.9	57.1
E-trypsin	1 month	0.75	0.73	1.48	50.7	49.3

C, control, without anoxia.
E, experimental (1 month anoxia).
TH, total homogenate; YP, yolk platelets; PMS, postmitochondrial supernatant; NaCl, carotenoprotein extracted with 1M NaCl; pH 7.4; trypsin, carotenoprotein obtained after digestion with trypsin (10 μg/mℓ).

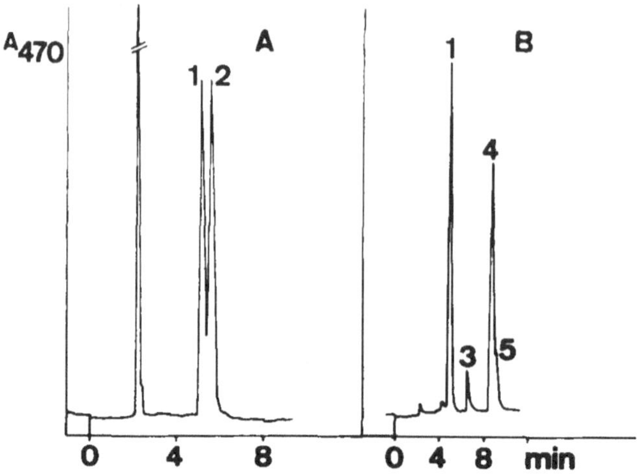

FIGURE 2. Carotenoid pattern of *Artemia* cysts. (A) Chromatogram of an extract of reference *Artemia* cysts[97] obtained on a 5 μm Zorbax ODS column (25 × 0.46 cm). Eluent, acetonitrile:methanol:dichloromethane (5:3:2, v/v); flow rate, 1.0 mℓ/min; detection, 470 nm. (B) Chromatogram of an extract of reference *Artemia* cysts[97] obtained on a 5 μm ROSIL column (20 × 0.46 cm). Eluent, dichloromethane:isopropyl alcohol (99.5:0.5, v/v). Flow rate, 1.0 mℓ/min; peak identification: 1, all-*trans*-canthaxanthin; 2 (composite), *cis*-canthaxanthins, further differentiated in peaks 3, 4, and 5 (peak 5, shoulder peak), on the (B) silica column.

ductively active females but was virtually absent in mature males. It was identified as a mixture of *cis*-canthaxanthins by absorption and mass spectrometry and by comparison with synthetic *cis*-canthaxanthins, prepared by stereomutation of all-*trans*-canthaxanthin.[80] The molecular ion and the fragment ions in the electron impact (EI) mass spectrum of the major unknown component isolated by semipreparative liquid chromatography (LC) strongly suggested the canthaxanthin structure. In addition, the same spectrum was observed on the major synthetic *cis*-canthaxanthin, which was purified from the stereomutation mixture. Additional evidence for the identity was obtained from the mass spectra of the unknown and the reference component under chemical ionization conditions. The mono-*cis* configuration was assigned to the compound on the basis of its highly characteristic absorption spectrum. The 8-nm hypsochromic shift of the main maximum (λ_{max} = 468 nm in the chromatographic solvent) with reference to all-*trans*-canthaxanthin (λ_{max} = 476 nm) as well as the appearance of the intense "*cis* peak" at 363 nm absent in the spectrum of all-*trans*-canthaxanthin, are consistent with a mono-*cis* configuration. A number of observations indicated the nonartefactual character of the *cis*-canthaxanthins. Their distribution and occurrence witness a remarkable biological selectivity (see Section V). For example, their selective localization in ovaries and cysts, their disappearance in nauplii, and their virtual absence in immature animals and adult males all provide important evidence for their non-artefactual character, particularly since all samples were analyzed under identical conditions. *Trans*⇄*cis* isomerization could theoretically occur during sample pretreatment and lead to an equilibrium *trans-cis* mixture. However, the *cis-trans* ratios found in cysts and in subcellular fractions dramatically exceed this equilibrium ratio and, thus cannot be rationalized in terms of stereomutation. In addition, supplementing cysts with all-*trans*-canthaxanthin prior to analysis did not lead to "extra" *cis*-canthaxanthin formation. For reasons of convenience, the *cis*-canthaxanthins, being a mixture in which one isomer predominates, will further be treated as a single component and designated as "*cis*-canthaxanthin".

2. *Quantitative Determination of Canthaxanthins in Artemia*

To study the biological significance of *cis*- and all-*trans*-canthaxanthin, a method was required to quantitate both forms. The previously described nonaqueous reversed phase system was modified to optimize the separation between the all-*trans*-canthaxanthin peak and the "total *cis*" peak, consisting of three mono-*cis* isomers.[84] In view of the instability of the *cis*-canthaxanthin, sample pretreatment was kept to an absolute minimum. Basically, it consists of a mechanical homogenization step in a methanol-formic acid medium, followed by direct injection of the extract, without further clean-up, on the chromatographic column. Later, milder approaches were considered as alternatives to the rather drastic formic acid treatment, but their evaluation is still in progress. Quantitation is based on the determination of peak height ratios (all-*trans*- or *cis*-canthaxanthin vs. the internal standard, i.e., β-*apo*-8′-carotenoic acid ethyl ester or β-*apo*-8′-carotenal) and comparison with a calibration curve. To construct this curve, known amounts of authentic all-*trans*- or *cis*-canthaxanthin (the latter purified from stereomutation mixtures), together with the internal standard were analyzed as the unknowns and the resulting peak height ratios plotted vs. the corresponding canthaxanthin concentration. When necessary, the carotenoid determination was combined with an assay of total protein. The latter was performed on the residue left after extraction of the carotenoids. This residue was hydrolyzed using barium hydroxide[85] and the amino acids were determined after reaction with ninhydrin.[86] Results were then expressed as amount of canthaxanthin (μg) per mg of protein. This particular approach was indicated in cases where the small sample size prevented accurate weighing of the material.

V. SIGNIFICANCE OF *cis*-CANTHAXANTHIN IN *ARTEMIA*

A. General

Further research on the biological fate of *cis*-canthaxanthin in *Artemia* seemed to be warranted for a number of reasons. First, it was felt that there must be some rationale for the animal to make the thermodynamically unfavorable *cis*-isomers rather than the physicochemically stable all-*trans*-canthaxanthin. Two phenomena may conceivably underly this formation: either, an enzyme is involved catalyzing the *trans* → *cis* rearrangement (an energy requiring process) or, alternatively preferential binding of the *cis* form to a macromolecule progressively shifts the *cis-trans* equilibrium in favor of the former. In both cases, some kind of more or less specific mechanism would seem to be operative, suggesting that the *cis*-canthaxanthin formation is not merely fortuitous, but is rather governed by a biological necessity. Second, preliminary observations showed that the *cis*-canthaxanthin appears and disappears at particular stages of the *Artemia* life cycle and, unlike the all-*trans*-canthaxanthin is distributed throughout the animals in a remarkably selective fashion. Therefore, this peculiar "biological selectivity" formed the subject of a detailed study. Specifically, the characteristic fate of the *cis*-canthaxanthin in the course of *Artemia* development as well as its distribution in tissues and subcellular fractions were investigated. Following HPLC analysis, results were expressed either in terms of *cis-trans* ratios (peak height or concentration ratios), as μg/g dry weight, or as μg/mg of protein.

B. Sexual Dimorphism

The presence or virtual absence of *cis*-canthaxanthin reflects sexual dimorphism in mature, i.e., reproductively active animals. Females fed on algae were found to contain 0.23 ± 0.01 μg/mg of protein ($n = 3$), whereas the content in males was only 0.03 ± 0.01 μg/mg ($n = 3$). *cis-trans* Peak height ratios in females were in the order of 0.3 and nearly unmeasurable in males (values <0.1). The corresponding all-*trans*-canthaxanthin levels in mature females and males amounted to 0.52 ± 0.095 ($n = 3$) and 0.11 ± 0.02 ($n = 3$) μg/mg of protein, respectively. This is also in keeping with the well known fact that female crustaceans usually contain a higher total amount of carotenoids than their male counterparts.[16] It seems to be justified to designate *cis*-canthaxanthin as a ''female-specific' pigment, especially in view of its very pronounced affinity for the ovaries and the cysts (see Section V.D).

C. Subcellular Localization

A crude subcellular fractionation of cysts was carried out by homogenizing the material in buffer or water, followed by differential centrifugation.[61] Fractions enriched in yolk platelets, mitochondria, and PMS were obtained. Yolk platelets were further disrupted in buffered 2 *M* sodium chloride to yield lipovitellin and a pellet enriched in yolk platelet membranes. All fractions were quantitatively analyzed as described above.

Preliminary experiments, not involving quantitation, had shown that *cis*-canthaxanthin was preferably localized in the yolk platelets, whereas all-*trans*-canthaxanthin predominated in the fractions enriched in mitochondria, the PMS, and in some flotating lipids.[61] Following lysis of the yolk platelets, *cis*-canthaxanthin was apparently mainly deposited in the insoluble fraction, enriched in membranes, and to a lesser extent in the soluble fraction containing the lipovitellin.[61] Quantitative analysis confirmed the overall picture, particularly with respect to the strong affinity of *cis*-canthaxanthin for the yolk platelets and its lower abundance in the PMS (Table 2). However, after disruption of the yolk platelets an enrichment of the insoluble fraction in *cis*-canthaxanthin[61] was no longer observed, which prompts a dismissal of the hypothesis about the compound's possible membrane-specific localization. The reason for this discrepancy of results most probably lies in the higher extraction efficiency of the lipovitellin from yolk platelets obtained with a modification (lower ratios of protein to extraction buffer) of the original extraction approach.

D. Organ and Tissue Distribution

The *cis-trans*-canthaxanthin distribution in mature female *Artemia* was examined by analyzing various dissected organs and tissues, including ovaries, ripening eggs, the gut, the hemolymph, the shell gland, and the remaining carcass. The isolated tissues of 10 to 20 animals were pooled and carotenoids and proteins assayed as described above. As shown in Figures 3 and 4, *cis*-canthaxanthin was particularly abundant in the oocytes and in mature but not in immature ovaries. *cis*-Canthaxanthin levels in hemolymph, although not impressive in absolute terms, exceeded those of all-*trans*-canthaxanthin. In contrast, all-*trans*-canthaxanthin largely predominated in the carcass and was the only canthaxanthin in the gut. The metabolic precursor echinenone was mainly found in the carcass and the gut and β-carotene (and lutein as well) was nearly exclusively confined to the gut. The presence of the latter two compounds merely reflects the algal diet.

E. *Artemia* Life Cycle

To study their fate in the course of the *Artemia* life cycle, *cis*- and all-*trans*-canthaxanthin were quantitatively determined at various stages of embryonic development (starting from hydrated cysts), in freshly hatched nauplii as well as in growing animals, up to the mature stage. The animals were fed on *Dunaliella tertiolecta* algae, except in a separate experiment with starved nauplii. The major difficulty in this particular experiment was to find a suitable ''reference point'' so as to correlate the results obtained on diverse matrices (cysts, nauplii, adults). Dry weight was dismissed because of the often very small sample sizes involved. Alternatively, the carotenoid levels were referred on the total amount of protein in the sample, a value which however, does not remain constant during growth either. In addition, the residues left after extraction of the carotenoids showed variable resistance to solubilization prior to the protein determination. A solution was found by using an assay involving total alkaline hydrolysis (and, hence, complete solubilization) of the proteins[85] in conjunction with a ninhydrin reaction.[86] This method was, however, subject to considerable variability.

Table 2
DISTRIBUTION OF *CIS*-AND ALL-*TRANS*-CANTHAXANTHIN IN SUBCELLULAR FRACTIONS PREPARED FROM AEROBIC AND ANOXIC *ARTEMIA* CYSTS (SAN FRANCISCO BAY STRAIN, BATCH 288/2606)

	Canthaxanthin concentration (μg/mg of protein)			
Subcellular fraction	All-*trans*	*cis*	Total	*cis*-*trans* ratio
A. Aerobic Cysts (Control for B)				
Total cysts[a]	0.35	0.53	0.88	1.51
Total homogenate[b]	0.33	0.46	0.79	1.39
PMS[c]	0.47	0.32	0.79	0.68
Yolk platelets	0.24	0.86	1.10	3.58
Lysed yolk platelets				
Insoluble fraction[d]	0.27	0.59	0.86	2.19
Lipovitellin	0.20	0.83	1.03	4.15
B. Anoxic Cysts (6 Weeks of Anoxia)				
Total cysts[a]	0.38	0.54	0.92	1.42
Total homogenate[b]	0.38	0.49	0.87	1.29
PMS[c]	0.60	0.35	0.95	0.58
Yolk platelets	0.33	0.93	1.26	2.82
Lysed yolk platelets				
Insoluble fraction[d]	0.37	0.66	1.03	1.78
Lipovitellin	0.23	0.79	1.02	3.43
C. Aerobic Cysts (Control for D)				
Total homogenate[b]	0.45	0.62	1.07	1.38
Yolk platelets	0.30	0.90	1.20	3.00
PMS[c]	0.68	0.39	1.07	0.57
D. Anoxic Cysts (1 1/2 Years Anoxia)				
Total homogenate[b]	0.53	0.83	1.36	1.57
Yolk platelets	0.26	1.07	1.33	4.12
PMS[c]	0.66	0.41	1.07	0.62

[a] Analyzed directly.
[b] Homogenate in water, filtered over Miracloth.
[c] PMS: postmitochondrial supernatant.
[d] Contains yolk platelet membranes and undisrupted yolk platelets. Disruption was carried out in buffered 2 *M* NaCl.

Another factor of irreproducibility lay in the rather drastic methanol-formic acid extraction, which causes partial degradation of canthaxanthin. Meanwhile, alternatives for both the carotenoid extraction and the protein determination have been developed. Hence, the "life cycle experiment" awaits repetition. The lack of refinement of the analytical technique accounts for a certain scattering of the results obtained in this first experiment. Yet, some trends became evident. The data are summarized in Table 3 and Figure 5. For interpretation, results were divided arbitrarily into four groups. During cyst incubation under standard

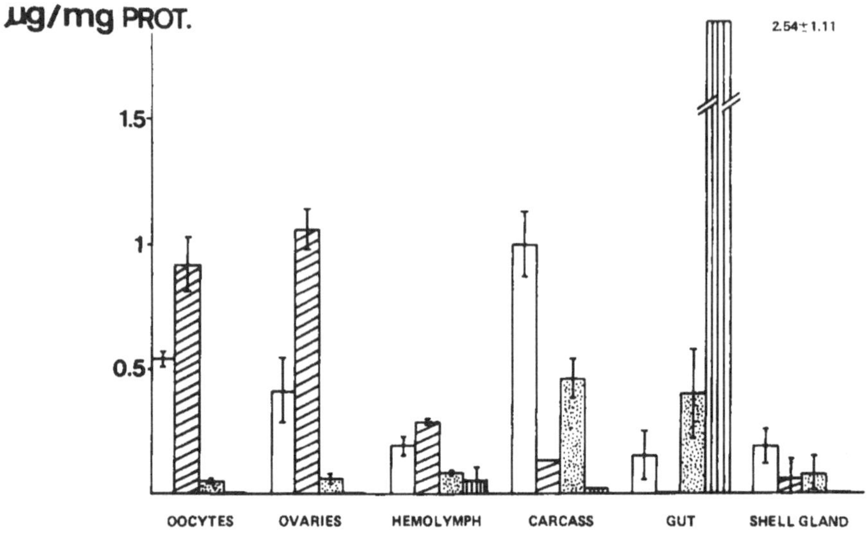

FIGURE 3. Organ and tissue distribution of all *trans*-canthaxanthin, *cis*-canthaxanthin, echinenone, and β-carotene in vitellogenic female *Artemia* (San Francisco Bay strain). Results are expressed as µg/mg of protein. For each sample, the first block (white) represents all *trans*-canthaxanthin, the second (oblique stripes) *cis*-canthaxanthin, the third (dots) and fourth (vertical stripes) echinenone and β-carotene, respectively.

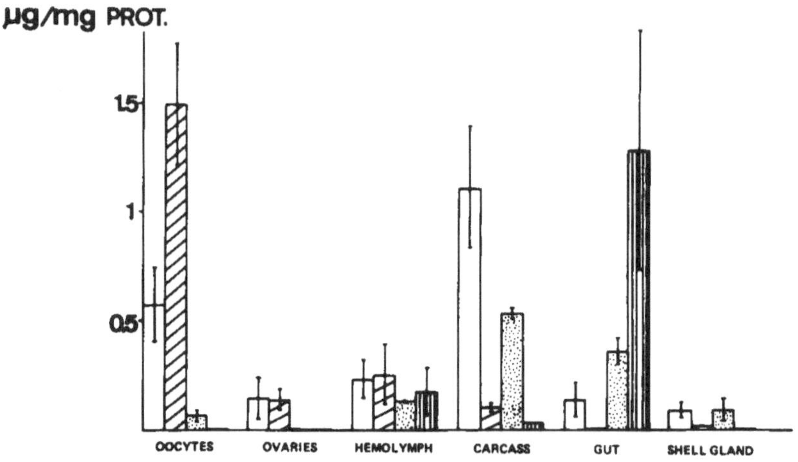

FIGURE 4. Organ and tissue distribution of all *trans*-canthaxanthin, *cis*-canthaxanthin, echinenone, and β-carotene in nonvitellogenic female *Artemia* (San Francisco Bay strain). For details see Figure 3.

conditions in sea water,[87] the *cis-trans*-canthaxanthin ratio remained virtually constant (1.24 ± 0.12, relative standard deviation (RSD) = 9.7%, n = 9). Although hatching was observed initially after 15 hr incubation, sufficient freshly hatched nauplii could only be collected after 16 hr. The second group of data spans the 17 to 27 hr period, during which the *cis-trans*-concentration ratio dropped from 1.00 to 0.65 (mean 0.84 ± 0.14, n = 7). In the course of the third period (32 to 39 hr), the ratios changed very little (0.37 ± 0.01, n = 5), whereas in starved nauplii (2 to 5 days old), a further decline down to a mean ratio of 0.21 ± 0.01 (n = 4) or 0.16 ± 0.02 (n = 4) was observed. The average absolute concentrations of all-*trans*- and *cis*-canthaxanthin over this time period are given in Table

Table 3
CONTENTS OF *CIS*- AND ALL-*TRANS*-CANTHAXANTHIN DURING EARLY *ARTEMIA* DEVELOPMENT (SAN FRANCISCO BAY STRAIN)

Stage of development (in hr after incubation)	Concentrations (μg/mg protein) All-*trans*-canthaxanthin	*cis*-canthaxanthin	Total	*cis-trans* ratio
1—16 (n = 9) hr	0.41 ± 0.09	0.50 ± 0.09	0.91 ± 0.18	1.24 ± 0.12
17—27 (n = 7) hr	0.65 ± 0.11	0.50 ± 0.11	1.14 ± 0.18	0.84 ± 0.14
32—39 (n = 5) hr	0.85 ± 0.08	0.31 ± 0.03	1.16 ± 0.11	0.37 ± 0.01
2—5 days old	0.74 ± 0.05	0.15 ± 0.02	0.89 ± 0.06	0.21 ± 0.01
starved nauplii[a] (n = 4)	0.66 ± 0.07	0.10 ± 0.01	0.76 ± 0.07	0.16 ± 0.02

[a] Analyzed in duplicate on different days.

3. The picture is somewhat complicated by the concomitant disappearance of proteins, which affects the absolute values (μg/mg of protein), unlike the *cis-trans* ratios, which only give an idea about the relative distribution of both isomers. The above observations can be summarized and interpreted as follows:

1. During pre-emergence development,[88] the levels of *cis*- and all-*trans*-canthaxanthin show little, if any, change. No *cis-trans* isomerization takes place.
2. After hatching, there is a progressive decrease of the levels of *cis*-canthaxanthin with a concomitant increase in the all-*trans*-canthaxanthin concentration, resulting in a dramatic drop in *cis-trans* ratios (from 1.24 to 0.16 to 0.21) (Figure 5).
3. The total amount of canthaxanthin (*cis* + all-*trans*) per mg of protein remains virtually constant during early naupliar development.

These data suggest that after hatching *cis*-canthaxanthin is progressively converted to the all-*trans* form. In addition, the catabolism of carotenoids apparently keeps equal pace with the disappearance of proteins (yolk consumption): since there was no feeding activity, no new protein was formed.

In a parallel experiment, the hatched nauplii were fed on *Dunaliella tertiolecta* algae. As a first sign of feeding activity, the canthaxanthin precursor echinenone appeared in naupliar extracts. However, the formation of new canthaxanthin is apparently superseded by protein synthesis, as evidenced from the further decline of the all-*trans*-canthaxanthin/protein ratio (from 0.65 to 0.35 μg/mg during the first day), down to a point where the ratio tends to remain constant in further growing immature animals (Figure 6A). The levels of *cis*-canthaxanthin rapidly decrease, become unmeasurable after 6 days, and remain so until the onset of reproductive activity when they sharply rise in females (from 0 to 0.23 μg/mg of protein) but not in mature males (0.03 μg/mg) (Figure 6A). At the same time, an increase in the levels of all-*trans*-canthaxanthin is observed as well (from 0.17 to 0.52 μg/mg of protein in females). In Figure 6B the concentrations of echinenone and β-carotene are also plotted vs. time. Since the latter is mainly present in the gut (see Section V.D) the fluctuations of its concentrations probably only reflect feeding activity and food abundance. Suprisingly, however, changes of the echinenone levels seem to follow a similar pattern. This would suggest echinenone formation in the gut. Chromatograms of extracts of *Dunaliella tertiolecta* and mature female *Artemia* are compared in Figure 7. The algae are a very rich source of β-carotene, but do not contain echinenone or canthaxanthin (elution positions indicated by arrows). The polar xanthophylls (peaks 1 to 3) are absorbed by the animal but not further metabolized.

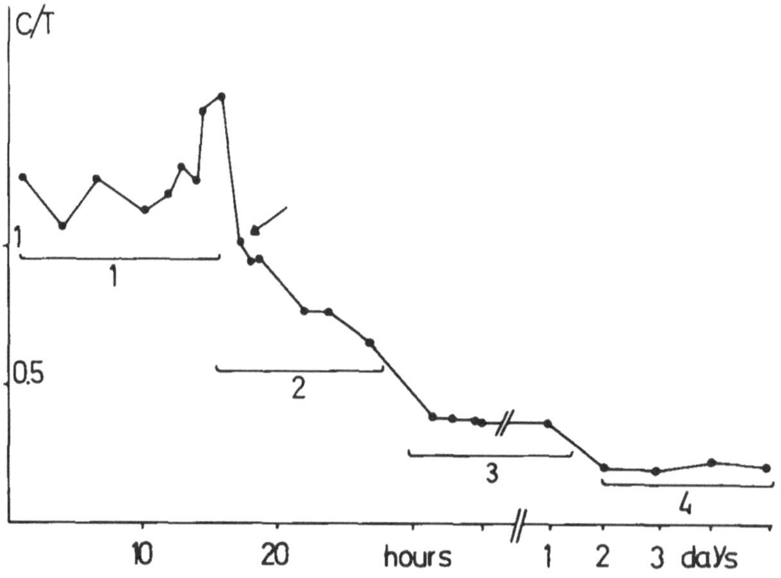

FIGURE 5. Concentration ratios (C/T) of *cis*- to all *trans*-canthaxanthin observed in the course of early *Artemia* development (San Francisco Bay strain). The arrow indicates the time of hatching. The numbers 1, 2, 3, 4 refer to the 4 arbitrarily chosen groups (Table 3).

F. Dehydrated/Hydrated Cysts

In the course of a study of *Artemia* cysts from different geographical strains, the degree of hydration was found to affect the relative *cis-trans*-canthaxanthin abundance.[63] It was observed that, upon hydration of cysts, the *cis* form was favored, while the opposite was true in the case of subsequent dehydration. The *cis-trans* ratio progressively increased as a function of the hydration time (Figure 8), and the reversible *cis-trans* "cycle" could be repeatedly reproduced in the course of successive dehydration/hydration treatments (Figure 9). To investigate the significance of this phenomenon, both isomers were quantitatively determined in dehydrated as well as fully hydrated cysts from 17 geographical strains.[89] Dehydrated cysts were prepared by drying 10-mg samples in an oven at 40°C for 24 hr, followed by cooling in a desiccator. Hydration was achieved by soaking dehydrated cysts in 35 ppt sea water for 5 hr at room temperature, using small plastic filter cups (Beckman). The experimental all-*trans*- and *cis*-canthaxanthin levels in dehydrated and fully hydrated cysts from 17 geographical strains (18 batches, Table 4) of *Artemia* are summarized in Figures 10 and 11. The previously observed reversible *cis-trans* conversion associated with hydration/dehydration was confirmed in all strains studied, though to different extents. For example, upon hydration the *cis*-canthaxanthin concentration increases by approximately 28% in San Francisco Bay cysts, whereas in Argentinian cysts this increase is only 6%. Soejima et al. also observed differences in the (total) canthaxanthin content of cysts from nine geographical strains,[60] although their values tend to be somewhat higher. This is probably due to the lower specificity of their analytical technique, which failed to differentiate between all-*trans*- and *cis*-canthaxanthin.

Differences in total canthaxanthin content are likely to be influenced by environmental factors (e.g. composition of food available to the parental population) and, hence, may be of limited relevance. However, the observed differences in relative abundance of all-*trans*- and *cis*-canthaxanthin may have a more fundamental significance, e.g., in connection with the cryptobiotic process (see Section VI). On the other hand, the *cis-trans*-canthaxanthin ratios, possibly in combination with the absolute canthaxanthin levels, may be a valuable parameter for the identification of strains or even specific batches of cysts.

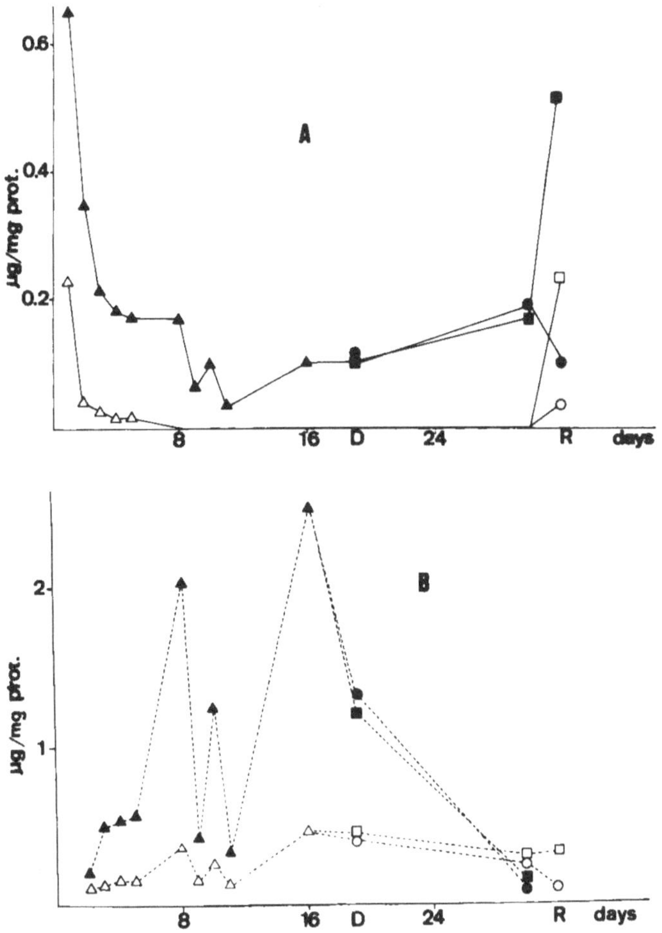

FIGURE 6. Levels of (A) all-*trans*- and *cis*-canthaxanthin and (B) echinenone and β-carotene during the *Artemia* life cycle (San Francisco Bay strain fed on *Dunaliella tertiolecta*). Mean results (n = 3) are expressed as µg/mg of protein. (A) all-*trans*-Canthaxanthin in undifferentiated animals (▲—▲), males (●—●), and females (■—■); *cis*-canthaxanthin in undifferentiated animals (△—△), males (○—○), and females (□—□). (B) Echinenone in undifferentiated animals (△- -△), males (○- -○), and females (□- -□); β-carotene in undifferentiated animals (▲- -▲), males (●- -●), and females (■- -■). D indicates the point at which differentiated immature males and females could be distinguished and separated, R indicates reproductively active animals.

G. Anoxia

In view of Karmiol's pioneering work,[11] it looked straightforward to apply the analytical technique for *cis-trans*-canthaxanthin determination to anoxic cysts. His data indeed seemed to suggest a similar *cis-trans* "cycle" during anoxybiosis as the one observed in connection with hydration/dehydration (Table 1).

Basically, two phenomena needed verification, i.e., the alleged increase in the total carotenoid content per mg of protein in anoxic cysts and the apparent all-*trans*- → *cis*-canthaxanthin ("echinenone") conversion following anoxybiosis. As shown in Table 2, there may be a slight trend towards such increase (10.1 and 27.1% in total homogenates) but certainly not to the extent as reported by Karmiol (189%). In addition, no evidence

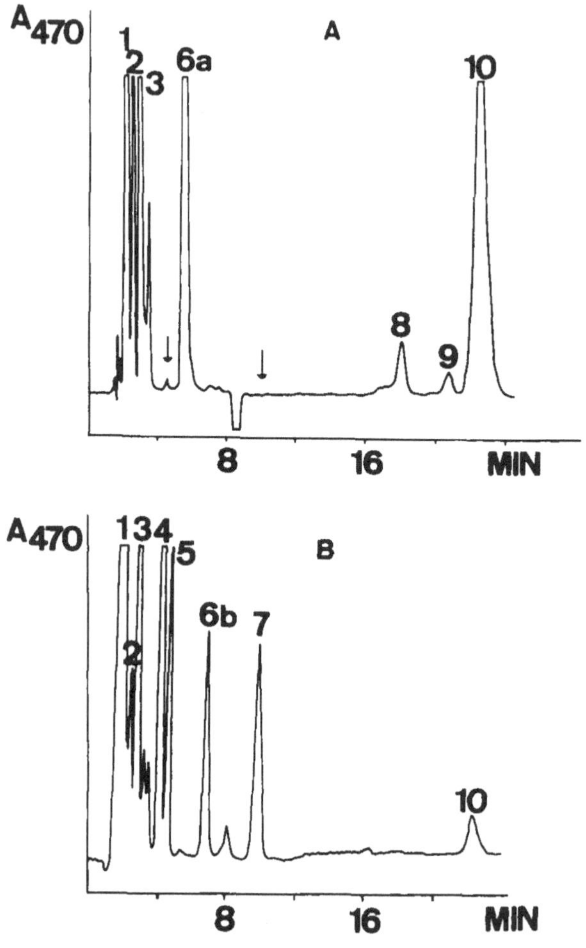

FIGURE 7. Chromatograms of (A) an extract of *Dunaliella tertiolecta* and (B) an extract of mature female *Artemia* (San Francisco Bay strain). Chromatographic conditions: column, 5 μm Zorbax ODS, 15 × 0.46 cm: eluent, methanol:acetonitrile:dichloromethane (5:4.1:0.9, v/v); flow rate, 1 mℓ/min; detection, 470 nm. Peak identification: 1, neoxanthin + violaxanthin; 2, antheraxanthin; 3, lutein; 4, all-*trans*-canthaxanthin; 5, *cis*-canthaxanthin; 6 a and b, unidentified (not carotenoids) (not necessarily identical); 7, echinenone; 8, γ-carotene (tentative); 9, α-carotene (tentative); 10, β-carotene. The arrows in (A) indicate the elution positions of canthaxanthin and echinenone, respectively.

could be produced indicating a *trans* → *cis* conversion in yolk platelets in response to anoxia (Table 5). Only minor differences were observed: in two experiments there was a drop in the relative content of *cis*-canthaxanthin by 5.6% (sample 1) and 1.0% (sample 3), respectively, whereas in two others increases by 2.1% (sample 2) and 6.5% (sample 4), respectively, occurred. These figures stand in sharp contrast to Karmiol's findings,[77] suggesting an increase in the relative *cis*-canthaxanthin ("echinenone") concentration in anoxic yolk platelets by 45.9% compared to the controls (Table 1). Hence, at this point neither of the two claims can be substantiated. It is felt that the slight increase, if any, in total carotenoid content is merely an apparent effect, to be rationalized as yet. The pronounced discrepancy between the results may be partly due to the fundamentally different extraction techniques used. While Karmiol's analyses relied on a Bligh and Dyer technique (involving an evaporation

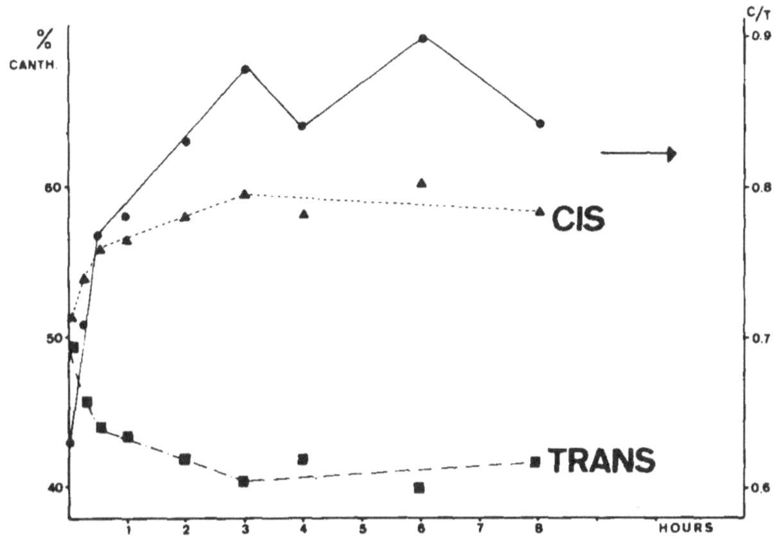

FIGURE 8. Influence of hydration time (hydration in 35 ppt sea water at 20°C) on the relative distribution of *cis-* and all-*trans*-canthaxanthin in reference *Artemia* cysts.[97] Upper trace (●—●), peak height ratios (C/T) (*cis-trans*); middle trace (▲- -▲), % *cis-*; and lower trace (■· - ·■), % all-*trans*-canthaxanthin in the extract.

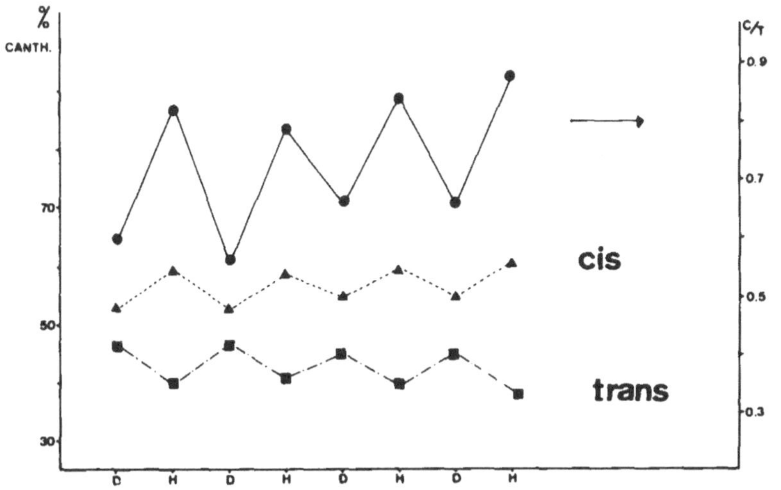

FIGURE 9. Influence of repeated dehydration (D) (40°C) and hydration (H) (35 ppt sea water, 20°C) on the relative distribution of *cis-* and all-*trans*-canthaxanthin in reference *Artemia* cysts.[97] Upper trace (●—●), peak height ratios (C/T) (*cis-trans*); middle trace (▲- -▲), % *cis*; and lower trace (■- · - ·■), % all-*trans*-canthaxanthin in the extract.

step during which *cis* ⇄ *trans* rearrangement may likely occur) the present authors used an exhaustive extraction using methanol and sodium deoxycholate, with direct injection of the extract into the liquid chromatograph, thus minimizing the risk of artefactual *cis* ⇄ *trans* conversion. It is remarkable that the latter approach seemed to yield higher *cis-trans*-canthaxanthin ratios (e.g., 4.15 in lipovitellin from aerobic cysts vs. 0.95 in Karmiol's study). However, part of this difference will obviously be due to the different geographical strain used, i.e., Great Salt Lake (Karmiol) vs. San Francisco Bay (this work).

Table 4
GEOGRAPHICAL STRAINS STUDIED

Strain	Origin	Abbreviation	Batch No.	Year
Artemia franciscana				
Manaure	Colombia	MAN	2[a]	1983
Macau	Brazil	MAC1	200[a]	1978
Macau	Brazil	MAC2	971051	1979
Chaplin Lake	Canada	CHL	241[a]	1979
San Francisco Bay	U.S.	SFB	288—2596	1976
San Pablo Bay	U.S.	SPB	1628	1978
Great Salt Lake	U.S.	GSL	285[a]	1978
Reference Artemia Cysts	—	RAC	—	1979
Artemia parthenogenetica				
Shark Bay	Australia	SB	299[a]	1980
Tientsin	China	TIE	242[a]	1978
Lavalduc	France	LVD	—	1980
Tuticorin	India	TUT	215[a]	1978
Izmir	Turkey	IZM	209[a]	1979
Altai-Siberia	U.S.S.R.	SIB	—	1983
Artemia persimilis				
Buenos Aires	Argentina	BA	223[a]	1979
Artemia urmiana				
Lake Urmia	Iran	URM	338[a]	1982
Artemia tunisiana				
Larnaca	Cyprus	LAR	305[a]	1980
Sfax	Tunisia	SFA	362[a]	1983

[a] Numbers referring to the collection at the *Artemia* Reference Center.

VI. DISCUSSION AND PERSPECTIVES FOR FUTURE RESEARCH

A. *Artemia:* a Representative Example Among Crustacea?

Before the demonstration of *cis*-canthaxanthin in *Artemia*, nothing suggested that this organism differed from related Crustacea as far as the fate of its carotenoids in relation to development was concerned. Like other Crustacea, *Artemia* actively mobilizes carotenoids in the gonads and the eggs. Pigmentary sexual dimorphism was supposedly nonexistent, no specific carotenoids were known to occur in the reproductive system and no metabolic changes had become apparent during development, except for cleavage of canthaxanthin-lipovitellin complexes in connection with hatching of cysts. The overall picture is particularly obscured by the documented claim that carotenoid-free *Artemia* retains its viability and produces viable and apparently healthy offspring. However, it appears from research in progress that these unpigmented animals show functional defects, which may not be immediately recognizable.[76] It seems more plausible indeed that in the course of evolution, *Artemia* and possibly other Crustacea, have acquired some adaptive advantages from accumulating carotenoids in their tissues, and their cysts or eggs in particular. The experimental evidence available for Crustacea in general does however not permit one to decide conclusively in favor of a photoprotective function, another protective role, or a more direct participation in reproduction/development. It even has proven difficult to distinguish between a role at an early stage (in the cysts, containing the intact carotenoproteins) or at a later stage (when, after disruption of the carotenoproteins, the free carotenoid shows up in the nauplii). Neither did the early literature on *Artemia* add further to the understanding of such function. However, the finding of *cis*-canthaxanthin opened up new perspectives, putting *Artemia*, for the time being, in a unique position among the Crustacea.

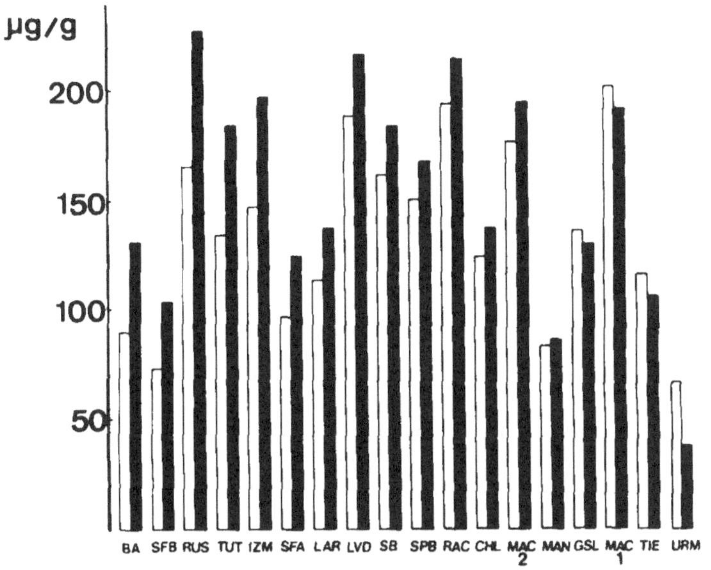

FIGURE 10. Levels (µg/g) of all-*trans*- and *cis*-canthaxanthin in dehydrated *Artemia* cysts of different geographical origin. White and black blocks represent the all-*trans*- and *cis*-canthaxanthin concentrations, respectively. For explanation of the abbreviations see Table 4.

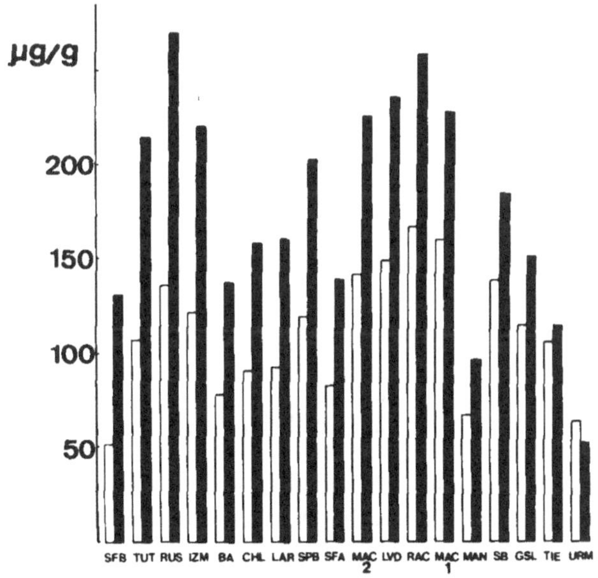

FIGURE 11. Levels (µg/g) of all-*trans*-and *cis*-canthaxanthin in hydrated *Artemia* cysts of different geographical origin. See Figure 10 for details.

B. The Biological Fate of *cis*-Canthaxanthin

For the first time a specific carotenoid with a most characteristic biological fate and potential relevance to developmental processes has been demonstrated in a crustacean. In summary, the following facts await further study:

1. A carotenoid with an unusual *cis* configuration, never demonstrated before in marine

Table 5
RELATIVE DISTRIBUTION (%) OF ALL-*TRANS*- AND *CIS* CANTHAXANTHIN IN YOLK PLATELETS (YP) OF ANOXIC AND AEROBIC CYSTS

	Anoxic cysts		Aerobic cysts	
Sample no.	All-*trans*	*cis*	All-*trans*	*cis*
1	26.4	73.6	22.0	78.0
2	23.3[a]	76.7[a]	24.9[b]	75.1[b]
3	24.1[b]	75.9[b]	23.3[b]	76.7[b]
4	19.2[b]	80.8[b]	24.1[b]	75.9

Note: Samples 1 and 2 (San Francisco Bay strain, batch 288/2606), 6 weeks of anoxia; samples 3 and 4, 1½ years of anoxia (San Francisco Bay strain, same batch).

[a] Mean of 3 samples.
[b] Mean of 2 samples.

animals, accumulates in mature female *Artemia*. It is virtually absent in immature animals and mature males, both of which, however, contain the all-*trans* form. In the females, the *cis* isomer is *grosso modo* confined to the ovaries and the cysts.

2. In encysted embryos, *cis*-canthaxanthin is particularly abundant in the yolk platelets. Reversible conformational changes (*cis*⇄ *trans* isomerization) occur in close connection with hydration/dehydration, probably paralleled by configurational changes of the (caroteno)proteins themselves. Breakdown of yolk lipovitellin in hatched nauplii is associated with liberation of canthaxanthin and concomitant irreversible isomerization of the *cis* to the more stable all-*trans*-canthaxanthin. Canthaxanthin disappears from the nauplii at a rate comparable to the breakdown of the yolk protein.

C. *cis*-Canthaxanthin: a Clue to the Function of Carotenoids in *Artemia?*

In view of the peculiar nature and biological fate of *cis*-canthaxanthin, the hope seems to be justified that this finding may eventually clear up some of the existing confusion about the role of carotenoids in *Artemia* and possibly in Crustacea in general. Specifically, a number of questions need to be answered.

1. What makes *Artemia* unique among the Crustacea in that this animal is incited to make an unusual, thermodynamically unfavorable pigment, apparently absent in related organisms?
2. Has the finding of *cis*-canthaxanthin contributed to a clearer insight into the developmental stage, i.e., cysts or nauplii, at which carotenoids are supposed to play a role?
3. Does the characteristic biological fate of *cis*-canthaxanthin provide experimental evidence supporting some of the hypotheses outlined for Crustacea in general, while ruling out alternative possibilities?

D. The "Uniqueness" of *Artemia?*

One of the differences between higher Crustacea (e.g., Decapoda) and Anostraca, including *Artemia*, lies in the ability of the latter to produce resistant encapsulated embryos. In addition, all Anostraca so far investigated contain canthaxanthin, and not astaxanthin, as the most oxidized carotenoid.[29] Hence, the question arises whether *cis*-canthaxanthin formation is typical for *Artemia* only or whether it is a universal feature of Anostraca or even Branchiopoda in general. Preliminary analytical data obtained on *Branchipus* and *Branchinecta* indeed revealed the presence of *cis*-canthaxanthin in their cysts. In the same study, *cis*- unlike all-*trans*-canthaxanthin was found to be virtually absent from *Daphnia* epphipia. However, this single observation should be further verified in other *Daphnia* samples. In case this preliminary finding is confirmed, other Cladocera should be taken into account as well. The presence of (astaxanthin) *cis* isomers in Decapoda has not been investigated as yet. Such investigation will however be hampered by the difficulty to separate astaxanthin geometrical isomers.[90] The nonaqueous reversed phase and normal phase systems successfully used for canthaxanthin isomer separation[84] indeed totally fail to resolve *cis-trans*-astaxanthins.

Hence, at this point it is premature to conclude that *cis*-carotenoid formation is confined to Anostraca, although considerable evidence seems to suggest a link with cyst production. It is remarkable in this respect that, when *Artemia* switches to ovoviviparous reproduction, the newly laid nauplii contain little *cis*-canthaxanthin.[69]

E. The Site of Action of *cis*-Canthaxanthin: Cysts or Nauplii?

The fact that conformational changes of bound canthaxanthin specifically occur in cysts as a response to such events as hydration and dehydration is consistent with a hypothesized function in cysts. So is the selective localization of *cis*-canthaxanthin in yolk platelets. The alternative possibility of a role for *cis*-canthaxanthin in the nauplii can be dismissed with a reasonable degree of certainty. It would indeed make little sense for the embryos to be equipped with a component that has nothing to do but serve only larval requirements, when this particular component immediately starts to disappear, at a fast rate, at the onset of hatching. Theoretically, *cis*-canthaxanthin may be a precursor of a metabolite which itself could be the biologically active form in nauplii. However, *cis*-canthaxanthin is most probably converted to the all *trans* form, rather than to some other unknown substance.

F. Towards the Definition of a Function of *cis*-Canthaxanthin in *Artemia* Cysts

The *cis* $\rightleftarrows$ *trans*-canthaxanthin interconversion in cysts following hydration or dehydration seems to point to some kind of relevant mechanism. This *cis-trans* "cycle" may be accompanied by simultaneous conformational changes in the (caroteno)proteins. These changes are however not necessarily biologically relevant. At present, apart from some data in Karmiol's work[77] and the studies on lipovitellin[66,67] there is little information about the mode of binding of *cis*- and *trans*-canthaxanthin to yolk platelets. *cis*-Canthaxanthin is apparently more tightly bound than the all-*trans* form, as evidenced from its greater resistance to solvent extraction. Whether, as Karmiol suggests[77] there are two distinct carotenoid pools in *Artemia* cysts, one associated with lipids and the other with proteins, is speculative as yet. It is tempting to hypothesize that *cis*-canthaxanthin would be the "active" component which supposedly provides a "lock" on certain proteins.[13] This association may either temporarily protect the protein from usage by the developing embryo, confer some general stability upon vital proteins, or even maintain them in particular conformations.[13] Further to the first possibility, the *cis*-canthaxanthin "lock" has subsequently to be removed from the protein to make it available for consumption by the nauplius. *cis*-Canthaxanthin, once in free form, would then be spontaneously converted to the more stable all-*trans* form. The gradual disappearance of *cis*-canthaxanthin from nauplii would thus reflect the progressive deblocking of the protein in the course of its digestion by the animal.

Alternatively, the binding of canthaxanthin to proteins may be part of the cyst's adaptive mechanisms to protect vital structures (membranes, enzymes) against such environmental stress situations, as anhydrobiosis, cryobiosis, or anoxybiosis. This is in keeping with Karmiol's suggestion[77] that the carotenoids (or, in this case possibly the *cis*-canthaxanthin), are involved in maintaining the structural integrity and the overall architecture of the yolk platelets during anoxybiosis. However, so far the existence of a *cis*⇄*trans* "cycle" in connection with anoxybiosis, as suggested by Karmiol's findings, could not be experimentally confirmed. It may be speculated that *cis-trans* conformational changes following hydration/dehydration of cysts form a link in the chain of biochemical events underlying diapause deactivation. As an echo of Dutrieu's work,[70,71] it is conceivable that the *cis*-canthaxanthin temporarily locks up a specific protein, until, by the action of desiccation, part of the *cis* is converted to the all-*trans* form, possibly in close connection with a conformational change in the protein, so that the protein becomes available for whatever action is required. If so, however, it is presently not understood why the process is reversed upon hydration of cysts. This mere reversibility of the phenomenon seems to argue against a participation in the process of diapause deactivation. Rather, the *cis-trans* conformational changes appear to coincide with the onset of metabolic activity in hydrated cysts, which is a reversible process, influenced by such factors as the hydration level,[91-93] temperature,[92,94,95] light,[94,96] etc.

A unifying hypothesis explaining all observations on the fate of *cis*- and all-*trans*-canthaxanthin is still lacking. Particularly, it remains obscure to what extent the *cis* and the all-*trans* form have different significance. It is likely that *cis*- and/or all-*trans*-canthaxanthin are involved in some kind of protection of proteins, perhaps as a mechanism to keep proteins temporarily in reserve or to stabilize them during exposure to environmental stress situations (anhydrobiosis and possibly anoxybiosis or cryobiosis). Future research should therefore be directed towards characterizing the macromolecular complexes containing the *cis*- and all-*trans*-canthaxanthin, particularly those located in the yolk platelets. In addition, the reversibility of the *cis-trans* isomerization should be reinvestigated in the light of some of the reversible processes triggering the resumption of metabolic activity in the cysts.

ACKNOWLEDGMENTS

This work was supported by F.K.F.O. Contract 2.0012.82 and F.G.W.O. Contract 3.0011.81. H.J.N., L.M., and P.S. acknowledge financial support (permanent position of Research Associate) from the Belgian Foundation for Scientific Research (Nationaal Fonds voor Wetenschappelijk Onderzoek).

NOTE ADDED IN PROOF

An updating of the work on *cis*-canthaxanthin can be found in the April issue of the *Journal of Lipid Research,* 1988.

REFERENCES

1. **Isler, O., Ed.,** *Carotenoids,* Birkhaüser, Basel, 1971.
2. **Moss, G. P. and Weedon, B. C. L,** Chemistry of the carotenoids, in *Chemistry and Biochemistry of Plant Pigments,* Vol. 1, 2nd ed., Goodwin, T. W., Ed., Academic Press, New York, 1976, chap. 3.
3. **Davies, B. H.,** Carotenoids, in *Chemistry and Biochemistry of Plant Pigments,* Vol. 2, 2nd ed., Goodwin, T. W., Ed., Academic Press, New York, 1976, chap. 19.
4. **Britton, G.,** Biosynthesis of carotenoids, in *Chemistry and Biochemistry of Plant Pigments,* Vol. 1, 2nd ed., Goodwin T. W., Ed., Academic Press, New York, 1976, chap. 5.
5. **Spurgeon, S. L. and Porter J. W.,** Biosynthesis of carotenoids, in *Biosynthesis of Isoprenoid Compounds,* Vol. 2, Porter, J. W. and Spurgeon, S. L., Eds., John Wiley & Sons, New York, 1983, 1.
6. **Davies, B. H.,** Carotenoid metabolism in animals: a biochemist's view, *Pure Appl. Chem.,* 57, 679, 1985.
7. **Goodwin, T. W.,** *The Biochemistry of the Carotenoids,* Vol. 1, 2nd ed., Chapman & Hall, London, 1980.
8. **Goodwin, T. W.,** *The Biochemistry of the Carotenoids,* Vol. 2, 2nd ed., Chapman & Hall, London, 1984.

9. **Mathis, P. and Schenck, C. C.,** The functions of carotenoids in photosynthesis, in *Carotenoid Chemistry and Biochemistry,* Britton, G. and Goodwin, T. W., Eds., Pergamon Press, Oxford, 1982, 339.
10. **Burnett, J. H.,** Functions of carotenoids other than in photosynthesis, in *Chemistry and Biochemistry of Plant Pigments,* Vol. 1, 2nd ed., Goodwin, T. W., Ed., Academic Press, New York, 1976, chap. 14.
11. **Krinsky, N. I.,** Carotenoid protection against oxidation, *Pure Appl. Chem.,* 51, 649, 1979.
12. **Burton, G. W. and Ingold, K. U.,** β-Carotene: an unusual type of lipid antioxidant, *Science,* 224, 569, 1984.
13. **Cheesman, D. F., Lee. W. L., and Zagalsky, P. F.,** Carotenoproteins in invertebrates, *Biol. Rev.,* 42, 131, 1967.
14. **Hairston, N. G.,** Photoprotection by carotenoid pigments in the copepod *Diaptomus nevadensis, Proc. Natl. Acad. Sci. U. S. A.,* 73, 971, 1976.
15. **Ghidalia, W.,** Pigments with a chromatic function, in *The Biology of Crustacea,* Vol. 9, Bliss, D. E. and Mantel. L. H., Eds., Academic Press, New York, 1985, 347.
16. **Goodwin, T. W.,** Crustacea, in *The Biochemistry of the Carotenoids,* Vol. 2, 2nd ed., Chapman & Hall, London, 1984, chap. 6.
17. **Castillo, R., Nègre-Sadargues, G., and Lenel, R.,** General survey of the carotenoids in Crustacea, in *Carotenoid Chemistry and Biochemistry,* Britton, G. and Goodwin, T. W., Eds., Pergamon Press, Oxford, 1982, 211.
18. **Nègre-Sadargues, G.,** Sur les transformations métaboliques des pigments caroténoìdes chez les Crustacés: synthèse bibliographique, *Ann. Biol.,* 17, 415, 1978.
19. **Lenel, R., Nègre-Sadargues, G., and Castillo, R,** Les pigments caroténoìd chez les Crustacés, *Arch. Zool. Exp. Gen.,* 119, 297, 1978.
20. **Fingerman, M.,** Chromatophores, *Physiol. Rev.,* 45, 296, 1965.
21. **Zagalsky, P. F.,** Carotenoid-protein complexes, *Pure Appl. Chem.,* 47, 103, 1976.
22. **Britton, G., Armitt, G. M., Lau, S. Y. M., Patel, A. K., and Shone, C. C.,** Carotenoproteins, in *Carotenoid Chemistry and Biochemistry,* Britton, G. and Goodwin, T. W., Eds., Pergamon Press, Oxford, 1982, 237.
23. **Zagalsky, P. F.,** Carotenoid-protein complexes in marine organisms, *Oceanis,* 9, 73, 1983.
24. **Goodwin, T. W.,** Carotenoid-protein complexes, in *The Biochemistry of the Carotenoids,* Vol. 2, 2nd ed., Chapman & Hall, London, 1984, chap. 1.
25. **Zagalsky, P. F.,** Invertebrate carotenoproteins, *Methods Enzymol.,* 111, 216, 1985.
26. **Matsuno, T., Katsuyama, M., Maoka, T., Hirono, T., and Komori, T.,** Reductive metabolic pathways of carotenoids in fish (3S, 3′S)-astaxanthin to tunaxanthin A, B and C, *Comp. Biochem. Physiol.,* 80B, 779, 1985.
27. **Establier-Torregrosa, R.,** Carotenoids of plants and marine animals. II. Variations in the carotenoid content of the ovaries and hepatopancreas of the crustacean *Plesiopenaeus edwardsianus* during the sexual cycle, *Invest. Pesquera,* 30, 223, 1966; *Chem. Abstr.,* 65, 19041f, 1966.
28. **Gilchrist, B. M. and Lee, W. L.,** Carotenoid pigments and their possible role in reproduction in the sand crab, *Emerita analoga* (Stimpson, 1857), *Comp. Biochem. Physiol.,* 42B, 263, 1972.
29. **Gilchrist, B. M.,** Distribution and relative abundance of carotenoid pigments in Anostraca (Crustacea: Branchiopoda), *Comp. Biochem. Physiol.,* 24, 123, 1968.
30. **Gilchrist, B. M. and Zalgalsky, P. F.,** Isolation of a blue canthaxanthin-protein from connective tissue storage cells in *Branchinecta packardi* Pearse (Crustacea: Anostraca) and its possible role in vitellogenesis, *Comp. Biochem. Physiol.,* 76B, 885, 1983.
31. **Wolfe, D. A. and Cornwell, D. G.,** Composition and tissue distribution of carotenoids in crayfish, *Comp. Biochem. Physiol.,* 16, 205, 1965.
32. **Goodwin, T. W. and Srisukh, S.,** Some observations on astaxanthin distribution in marine Crustacea, *Biochem. J.,* 45, 268, 1949.
33. **Goodwin, T, W.,** Carotenoid metabolism during development of lobster eggs, *Nature (London),* 167, 559, 1951.
34. **Herring, P. J.,** The carotenoid pigments of *Daphnia magna* Straus. II. Aspects of pigmentary metabolism, *Comp. Biochem. Physiol.,* 24, 205, 1968.
35. **Herring, P. J. and Morris, R. J.,** Embryonic metabolism of carotenoid pigments and lipid in species of *Acanthephyra* (Crustacea: Decapoda), *Proc. 9th Eur. Mar. Biol. Symp.,* Barnes, H., Ed., Aberdeen University Press, Aberdeen, 1975, 299.
36. **Ceccaldi, H. J.,** Recherches sur la biologie des associations entre protéines et caroténoïdes chez les Crustacés Décapodes: aspects métaboliques et moléculaires, thèse de doctorat en Sciences Naturelles, Aix-Marseille, CNRS n° AO 2530, 1968.
37. **Goodwin, T. W. and Srisukh, S.,** The biochemistry of locusts. I. The carotenoids of the integument of two locust species *(Locusta migratoria migratorioides* R. & F. and *Schistocerca Gregaria* Forsk.), *Biochem. J.,* 45, 263, 1949.
38. **Lenel, R., Castillo, R., and Nègre-Sadargues, G.,** Les pigments caroténoïdes, intérêt de leur étude dans la physiologie des Crustacés, *Oceanis,* 9, 61, 1983.

39. **Ceccaldi, H. J.**, Evolution des oeufs et cycle de réproduction chez *Plesionika Edwardsi* (Brandt), *Rec. Trav. Stn. Mar. Endoume-Marseille Fasc. Hors Ser. Suppl.*, 44, 403, 1968.
40. **Green, J.**, Chemical embryology of the Crustacea, *Biol. Rev.*, 40, 580, 1965.
41. **Lwoff, A.**, Le cycle du pigment carotinoïde chez *Idya furcata* (Baird), *Bull. Biol. Fr. Belg.*, 61, 193, 1927.
42. **Green, J.**, Variation in carotenoid pigmentation of *Simocephalus vetulus* (Crustacea: Cladocera), *J. Zool. (London)*, 149, 174, 1966.
43. **Zagalsky, P. F. and Gilchrist, B. M.**, Isolation of a blue canthaxanthin-lipovitellin from the yolk platelets of *Branchipus stagnalis* (L.) (Crustacea: Anostraca), *Comp. Biochem. Physiol.*, 55B, 195, 1976.
44. **Herring, P. J.**, The pigments of plankton at the sea surface, *Symp. Zool. Soc. London*, 19, 215, 1967.
45. **Griffiths, M.**, The carotenoids of the eggs and embryos of the sea urchin *Strongylocentrotus purpuratus*, *Dev. Biol.*, 13, 296, 1966.
46. **Lucy, J. A. and Dingle, J. T.**, Vitamin A and membrane systems. II. Membrane stability and protein-vitamin A-lipid interactions, *Biochem. J.*, 84, 76P, 1962.
47. **Plack, P. A. and Kon, S. K.**, A comparative survey of the distribution of vitamin A aldehyde in eggs, *Biochem. J.*, 81, 561, 1961.
48. **Hairston, N. G.**, The adaptive significance of color polymorphism in two species of *Diaptomus* (Copepoda), *Limnol. Oceanogr.*, 24, 15, 1979.
49. **Hairston, N. G.**, The relationship between pigmentation and reproduction in two species of *Diaptomus* (Copepoda), *Limnol. Oceanogr.*, 24, 38, 1979.
50. **Goodwin, T. W.**, Carotenoids and reproduction, *Biol. Rev.*, 25, 391, 1950.
51. **Gilchrist, B. M. and Green, J.**, The pigments of *Artemia*, *Proc. R. Soc. London, Ser. B*, 152, 118, 1960.
52. **Needham, J. and Needham, D. M.**, On phosphorus metabolism in embryonic life. I. Invertebrate eggs, *J. Exp. Biol.*, 7, 317, 1930.
53. **Krinsky, N. I.**, The carotenoids of the brine shrimp *Artemia salina*, *Comp. Biochem. Physiol.*, 16, 181, 1965.
54. **Czygan, F.-C,.** Über den Stoffwechsel von Keto-carotinoiden in niederen Krebsen, *Z. Naturforsch.*, 21B, 801, 1966.
55. **Czeczuga, B.**, Studies on the carotenoids in *Artemia salina* L. eggs, *Comp. Biochem. Physiol.*, 40B, 47, 1971.
56. **Czygan, F.-C.**, On the metabolism of carotenoids in the crustacean *Artemia salina*, *Z. Naturforsch.*, 23B, 1367, 1968.
57. **Hata, M. and Hata, M.**, Carotenoid metabolism in *Artemia salina L.*, *Comp. Biochem. Physiol.*, 29, 985, 1969.
58. **Hsu, W.-J., Chichester, C. O., and Davies, B. H.**, The metabolism of β-carotene and other carotenoids in the brine shrimp, *Artemia salina* L. (Crustacea: Branchiopoda), *Comp. Biochem. Physiol.*, 32, 69, 1970.
59. **Davies, B. H., Hsu, W.-J., and Chichester, C. O.**, The mechanism of the conversion of β-carotene into canthaxanthin by the brine shrimp, *Artemia salina* L. (Crustacea: Branchiopoda), *Comp. Biochem. Physiol.*, 33, 601, 1970.
60. **Soejima, T., Katayama, T., and Simpson, K. L.**, International study on *Artemia*. XII. The carotenoid composition of eight geographical strains of *Artemia* and the effect of diet on the carotenoid composition of *Artemia*, in *The Brine Shrimp Artemia*, Vol. 2., Persoone, G., Sorgeloos, P., Roels, O., and Jaspers, E., Eds., Universa Press, Wetteren, Belgium, 1980, 613.
61. **Nelis, H. J. C. F., Lavens, P., Moens, L., Sorgeloos, P., Jonckheere, J. A., Criel, G. R., and De Leenheer, A. P.**, *Cis*-canthaxanthins: unusual carotenoids in the eggs and the reproductive system of female brine shrimp *Artemia*, *J. Biol. Chem.*, 259, 6063, 1984.
62. **Zagalsky, P. F., Gilchrist, B. M., Clark, R. J. H., and Fairclough, D. P.**, The canthaxanthin — lipovitellin of the brine shrimp, *Artemia* (Crustacea: Anostraca): A resonance Raman and circular dichroism study, *Comp. Biochem. Physiol.*, 74B, 647, 1983.
63. **Nelis, H. J., Lavens, P., Moens, L., Sorgeloos, P., Criel, G. R., and De Leenheer, A. P.**, Occurrence, concentration and function of *cis*-carotenoids (*cis*-canthaxanthins) in the brine shrimp *Artemia*, Abstracts of the 7th Int. Symp. on Carotenoids, München, August 27, 1984, 26.
64. **Nelis, H. J. C. F. and De Leenheer, A. P.**, Isocratic nonaqueous reversed-phase liquid chromatography of carotenoids, *Anal. Chem.*, 55, 270, 183.
65. **Warner, A. H., Puodziukas, J. G., and Finamore, F. J.**, Yolk platelets in brine shrimp embryos, *Exp. Cell. Res.* 70, 365, 1972.
66. **de Chaffoy, D., Heip, J., Moens, L., and Kondo, M.**, *Artemia* lipovitellin, in *The Brine Shrimp Artemia*, Vol. 2, Persoone, G., Sorgeloos, P., Roels O., and Jaspers E., Eds., Universa Press, Wetteren, Belgium, 1980, 379.
67. **de Chaffoy de Courcelles, D. and Kondo, M.**, Lipovitellin from the crustacean *Artemia salina:* biochemical analysis of lipovitellin complex from the yolk granules, *J. Biol. Chem.*, 255, 6727, 1980.

68. **Zagalsky, P. F., Gilchrist, B. M., Clark R. J. H., and Fairclough, D. P.**, The canthaxanthin-lipovitellin of *Branchipus stagnalis* (L.) (Crustacea: Anostraca): a resonance Raman and circular dichroism study, *Comp. Biochem. Physiol.*, 75B, 163, 1983.
69. **Nelis, H. J. C. F. and De Leenheer, A. P.**, unpublished results.
70. **Dutrieu, J.**, Observations biochimiques et physiologiques sur le développement d' *Artemia salina* Leach, *Arch. Zool. Exp. Gen.*, 99, 1, 1960.
71. **Dutrieu, J.**, Incidence du type alimentaire sur le mode de réproduction d'un Crustacé *(Artemia salina), Ann. Nutr. Aliment.*, 16, 227, 1962.
72. **Lochhead, J. H.**, Oviparity versus ovoviviparity in the brine shrimp, *Artemia, Biol. Bull.*, 121, 396, 1961.
73. **Versichele, D. and Sorgeloos, P.**, Controlled production of *Artemia* cysts in batch cultures, in *The Brine Shrimp Artemia*, Vol. 3, Persoone, G., Sorgeloos, P., Roels, O., and Jaspers, E., Eds., Universa Press, Wetteren, Belgium, 1980, 231.
74. **Lavens, P. and Sorgeloos, P.**, Controlled production of *Artemia* cysts under standard conditions in a recirculating culture system, *Aquacult. Eng.*, 3, 221, 1984.
75. **Lavens, P. and Sorgeloos, P.**, The crytobiotic state of *Artemia* cysts and its diapause deactivation and hatching: a review, in *Artemia Research and its Applications*, Vol. 3, Sorgeloos, P., Bengston, D. A., Decleir, W., and Jaspers, E., Eds., Universa Press, Wetteren, Belgium, 1987, 27.
76. **Lavens, P. and Sorgeloos, P.**, unpublished results.
77. **Karmiol, S.**, Cellular Responses of *Artemia salina* to Anaerobic Conditions, M.S. thesis, University of Windsor, Ontario, Canada, 1981.
78. **Crowe, J. H., Crowe, L. M., Drinkwater, L., and Busa, W. B.**, Intracellular pH and anhydrobiosis in *Artemia* cysts, in *Artemia Research and its Applications*, Vol. 2, Moens, L., Slegers, H., Decleir, W., Sorgeloos, P., and Jaspers, E., Eds., Universa Press, Wetteren, Belgium, 1987, 19.
79. **Ewing, R. D. and Clegg, J. S.**, Lactate dehydrogenase activity and anaerobic metabolism during embryonic development in *Artemia salina, Comp. Biochem. Physiol.*, 31, 297, 1969.
80. **Gansser, C. and Zechmeister, L.**, Uber einige *cis*-Formen des Canthaxanthins, *Helv. Chim. Acta*, 40, 1757, 1957.
81. **Zechmeister, L.**, *Cis-trans Isomeric Carotenoids, Vitamins A and Arylpolyenes*, Springer-Verlag, Berlin, 1962, chap. 4.
82. **Karnaukhov, V. N., Milovidova, N. Y., and Kargopolova, I. N.**, On a role of carotenoids in tolerance of sea molluscs to enviromental pollution, *Comp. Biochem. Physiol.*, 56A, 189, 1977.
83. **Raines, P. S. and Davies, B. H.**, Carotenoids of the common mussel, *Mytilus edulis:* their hypothetical role in ATP production during anoxia, Abstracts of the 7th Int. Symp. on Carotenoids, München, August 27, 1984, 15.
84. **Nelis, H. J. C. F., Van Steenberge, M. M. Z., Lefevere, M. F., and De Leenheer, A. P.**, Selectivity effects in non-aqueous reversed-phase and normal-phase chromatography of geometric canthaxanthin isomers, *J. Chromatogr.*, 353, 295, 1986.
85. **McGrath, R.**, Protein measurement by ninhydrin determination of amino acids released by alkaline hydrolysis, *Anal. Biochem.*, 49, 95, 1972.
86. **Moore, S.**, Amino acid analysis: aqueous dimethyl sulfoxide as solvent for the ninhydrin reaction, *J. Biol. Chem.*, 243, 6281, 1968.
87. **Sorgeloos, P., Bossuyt, E., Lavens, P., Léger, P., Vanhaecke, P., and Versichele, D.**, The use of brine shrimp *Artemia* in crustacean hatcheries and nurseries, in *CRC Handbook of Mariculture*, Vol. 1, McVey, J. P., Ed., CRC Press, Boca Raton, Fla., 1983, 71.
88. **Clegg, J. S. and Conte, F. P.**, A review of the cellular and developmental biology of *Artemia*, in *The Brine Shrimp Artemia*, Vol. 2, Persoone, G., Sorgeloos, P., Roels, O., and Jaspers, E., Eds., Universa Press, Wetteren, Belgium, 1980, 11.
89. **Nelis, H. J., Lavens, P., Sorgeloos, P., Van Steenberge, M., and De Leenheer, A. P.**, *Cis*- and *trans*-canthaxanthin levels in *Artemia* cysts of different geographical origin, in *Artemia Research and its Applications*, Vol. 2, Moens, L., Slegers, H., Decleir, W., Sorgeloos, P., and Jaspers, E., Eds., Universa Press, Wetteren, Belgium, 1987, 99.
90. **Englert, G. and Vecchi, M.**, *Trans/cis* isomerization of astaxanthin diacetate/isolation by HPLC and identification by ^{1}H NMR spectroscopy of three mono-*cis* and six di-*cis*-isomers, *Helv. Chim. Acta*, 63, 1711, 1980.
91. **Morris, J. E.**, Hydration, its reversibility and the beginning of development in the brine shrimp, *Artemia salina, Comp. Biochem. Physiol.*, 39A, 843, 1971.
92. **Sorgeloos, P.**, De invloed van Abiotische en Biotische Faktoren op de Levenscyclus van het Pekelkreeftje, *Artemia salina L.*, Ph.D. thesis, State University of Ghent, Ghent, Belgium, 1975.
93. **Benijts, F., Vandeputte, G., and Sorgeloos, P.**, Energetic aspects of the metabolism of hydrated *Artemia* cysts, in *Fundamental and Applied Research on the Brine Shrimp Artemia salina* (L.) *in Belgium*, EMS Spec. Publ. No. 2, Jaspers, E. and Persoone, G., Eds., Inst. Mar. Sci. Res., Bredene, Belgium, 1977, 79.

94. **Sorgeloos, P., Baeza-Mesa, M., Benijts, F., and Persoone, G.,** Current research on the culturing of the brine shrimp *Artemia salina* L. at the State University of Ghent, Belgium, in *Proc. 10th European Symposium on Marine Biology,* Vol. 1, Persoone, G. and Jaspers, E., Eds., Universa Press, Wetteren, Belgium, 1976, 473.
95. **Vallejo, C. G., De Luchi, F., Laynez, J., and Marco, R.,** The role of cytochrome oxidase in the resumption of the development of *Artemia* dormant cysts, in *The Brine Shrimp Artemia,* Vol. 2, Persoone, G., Sorgeloos, P., Roels, O., and Jaspers, E., Eds., Universa Press, Wetteren, Belgium, 1980, 185.
96. **Sorgeloos, P.,** First report on the triggering effect of light on the hatching mechanism of *Artemia salina* dry cysts, *Mar. Biol.,* 22, 75, 1973.
97. **Sorgeloos. P.,** Availability of reference *Artemia* cysts, *Marine Ecol. Prog. Ser.,* 3, 363, 1980.

Chapter 8

PHYSICOCHEMICAL CHARACTERIZATION OF RIBOSOMAL PARTICLES FROM *ARTEMIA*

Julius Clauwaert and Paul Nieuwenhuysen

TABLE OF CONTENTS

I. INTRODUCTION

A. Ribosomes

The biosynthesis of proteins is a process in which many complex molecules take part. It occurs on ribosomes, subcellular organelles present in all organisms, and is essential for the translation of the genetic information encoded in messenger RNA (mRNA) into proteins. The ribosomes of prokaryotic and eukaryotic cells are structurally and functionally similar, but ribosomes from eukaryotic cells are larger and more complex. Therefore, most structural and functional studies have been performed on the ribosomes from prokaryotes, particularly from the bacterial species *Escherichia coli*. All ribosomes consist of two dissimilar subunits and these must associate with mRNA before protein synthesis can take place. Each ribosome can only be engaged in the synthesis of one protein at a time, but a messenger may be translated simultaneously by several ribosomes. The complex which includes mRNA and attached ribosomes is termed a polyribosome or polysome.

Elucidating the complex structure of the ribosome is necessary to understand in detail its involvement in protein synthesis. However, since each of its two subunits consists of RNA and many different protein molecules, any single experimental approach to a structural determination of this organelle (such as by X-ray crystallography) would be inadequate.

Research into ribosome structure can be divided roughly into three categories. The first category consists of the general characterization of ribosomes and their subparticles including the determination of the RNA and protein content, and of physical properties such as molecular (or particle) mass, size, shape, and solvation. The second category consists of more detailed structure determinations of the isolated ribosomal components including the analysis of their primary structure (i.e., the amino acid and nucleotide sequence of the protein and RNA molecules, respectively), and the determination of their secondary structure and their shape in solution. The third category is that of ribosomal "topography" or how the various individual ribosomal components are arranged relative to each other in space and how they interact.[1,2]

B. *Artemia*

Our attention has focused on the ribosomal particles of the brine shrimp *Artemia* for the following reasons:

1. The desiccated, cryptobiotic, undeveloped *Artemia* embryos (cysts) are available in large quantities and they can easily be stored.
2. The cysts contain a large supply of free, unprogrammed ribosomes, devoid of mRNA, and few if any polysomes.[3]
3. It is possible to isolate ribosomes and polysomes from *Artemia* at various stages of development, such as the free-swimming nauplii, which allows a comparison of the physical-chemical properties of ribosomes isolated from different stages of development.
4. The unprogrammed ribosomes are particularly useful for in vitro translation of exogeneous mRNA.[4]

5. *Artemia* is widely used to study the role of protein biosynthesis in executing the developmental program.[5]
6. The structure of the dried ribosomal particles of *Artemia* and *Escherichia coli* has been compared by electron microscopy.[6]
7. There has been an extensive interest in studying different aspects of the molecular biology of *Artemia* development, as described in the contributions of two of our colleagues (Dr. L. Moens and Dr. H. Slegers) elsewhere in this book.

II. ISOLATION OF RIBOSOMAL PARTICLES

A. Solvent Conditions

The following solvent conditions were used:

1. For ribosomes: 20 m*M* 4-(2-hydroxyethyl)-1-piperazine-ethanesulfonic acid (Hepes)-KOH, pH 7.5, 100 m*M* KCl, 9 m*M* Mg acetate, and 1 m*M* dithiothreitol.
2. For dissociating the ribosomes to subunits: 20 m*M* Hepes-KOH, pH 7.5, 400 m*M* KCl, 11 m*M* Mg acetate, and 1 m*M* dithiothreitol.
3. For ribosomal subunits: 20 m*M* Hepes-KOH or triethanolamine, pH 7.5, 250 m*M* KCl, 11 m*M* Mg acetate, and 1 m*M* dithiothreitol.

B. Isolation of Ribosomal Particles

Cryptobiotic, undeveloped embryos (cysts) of *Artemia* (San Francisco Bay brand) were stirred in a saturated NaCl solution and subsequently washed with distilled water and 150 m*M* sucrose in ribosome buffer. The cysts were dried by suction and homogenized in a mortar or with a motorized grinder at 4°C. The homogenate was centrifuged in a Beckman JA10 rotor at 6000 rpm at 4°C for 20 min. The supernatant was filtered through cheesecloth and centrifuged twice in a JA20 rotor at 18,000 rpm at 4°C for 30 min. This procedure yielded orange-tinted supernatants from which intact (complete) ribosomes or their subunits were isolated.[7]

For the isolation of intact ribosomes, the preparation was continued with the following procedures: overnight ultracentrifugation of the supernatants in a high density sucrose gradient, precipitation of the ribosomes from the top fractions with polyethylene glycol 6000, and rate-zonal sucrose density gradient sedimentation of resuspended ribosomes.

For the isolation of ribosomal subunits, the supernatants were layered on 20-mℓ cushions of 50% sucrose in ribosome buffer for overnight centrifugation in a Ti-60 rotor at 50,000 rpm at 4°C. This yielded clear colorless ribosome pellets which were then suspended in dissociation buffer (see above). The large and small subunits were separated by rate-zonal sedimentation in a 15 to 50% sucrose gradient in dissociation buffer using a Ti-40 zonal rotor.[8]

To obtain nauplii, the cysts were first incubated in a medium containing artificial sea salt at 30°C until hatching. In the preparation of polysomes and ribosomes from these nauplii different procedures were followed which yielded preparations of different degrees of purity.[9] However, pelleting of the ribosomal particles from the postmitochondrial supernatant (PMS) with high centrifugal force was always avoided.

C. Stability Conditions for the Ribosomal Particles

Freshly prepared solutions of complete ribosomes after dialysis against the ribosome buffer were intact and free of aggregates as judged by analytical boundary sedimentation. Initially, ribosomes were dissociated and their subunits separated in 20 m*M* Hepes-KOH buffer pH 7.5, containing 1 m*M* dithiothreitol, 700 m*M* KCl, and 11 m*M* Mg acetate because these ionic concentrations were used in the study of *Artemia* ribosomes by other groups. However, the absorbance profile of the gradient after centrifugation revealed that sometimes a part of small subunits sedimented at a slower rate.[8] Furthermore, after overnight centrifugation in a Beckman Ti-60 rotor at 40,000 rpm to pellet the subunits, the small ribosomal subunits did not redissolve as well as the large subunits or the complete ribosomes. The small subunits were not intact as evidenced by their high A_{260}/A_{280} ratio, and they had a higher buoyant density and a 20% lower sedimentation and diffusion coefficient than the intact subunits. When the KCl concentration of the dissociation buffer was lowered to 400 m*M*, the absorbance profile after zonal centrifugation also showed complete ribosomal dissociation, but the zone between the small and large subunits was closer than when the dissociation buffer contained 700 m*M* KCl.

In contrast to the complete ribosomes, and as is often the case with eukaryotic ribosomal subunits, the solutions containing the subunits could not be kept completely free of aggregates. Since this problem hinders the application of light, X-ray, and neutron scattering, we have tried to avoid conditions which lead to aggregation as much as possible. While lowering the KCl concentration induces aggregation, lowering the K^+/Mg^{2+} ratio stabilizes the isolated subunits. As a compromise, we have used a buffer containing 250 m*M* KCl and 11 m*M* Mg acetate in order to minimize the aggregation problem.

D. Pressure-Induced Dissociation of Ribosomes

Protein synthesis in vivo and in vitro is barosensitive and the ribosome seems to be the primary site of action of hydrostatic pressure. The dissociation of ribosome subunits is important for their function in protein biosynthesis, and the dissociation-association reaction of isolated subunits in solution is in dynamic equilibrium as shown below.

$$\text{ribosomes} \leftrightarrow \text{large} + \text{small subunit}$$

The equilibrium constant depends on the ionic concentrations and hydrostatic pressure; high K^+ concentrations promote dissociation, while high Mg^{2+} concentrations favor association. The hydrostatic pressure generated during ultracentrifugation can also induce dissociation of ribosomes when they are sedimented in a density gradient. This effect could complicate the analysis of the sedimentation patterns obtained after zonal centrifugation of ribosomal particles, such as when procedures for the preparation of ribosomal particles are being tested.

We have studied the hydrostatic pressure-induced dissociation of ribosomes and polysomes by monitoring the light scattering of solutions containing ribosomes or polysomes under different conditions of hydrostatic pressure and at specified solvent conditions. Light scattering data allowed us to determine the equilibrium dissociation constant as follows:

$$K = \frac{[\text{small subunits}] \times [\text{large subunits}]}{[\text{ribosomes}]}$$

Using the equation $\log K = \log K_{P=0} - P\Delta V/RT$ the volume change (ΔV) accompanying the dissociation of the complete ribosome to the subunits has been determined to be -250 ± 50 mℓ/mol.[10]

We have also studied the influence of the varying ionic conditions on the dissociation constant (K). The equation above then allowed us to estimate the influence of the hydrostatic pressure (P) on K under varying ionic conditions. Using the equation of Baierlein and Infante,[11] we have calculated that the pressures generated during boundary sedimentation in the analytical ultracentrifuge or during band sedimentation in a sucrose gradient in the ribosome solvent conditions were lower than 150 atm at the highest rotor speed used (36,000 rpm). We concluded that our results with ribosomes in ribosomal solvent conditions were not disturbed by the increase, during centrifugation, of the dissociation equilibrium constant K.

The application of high pressures on polysome solutions also caused a decrease of the light scattering which was nonexponential and slower than in the case for ribosomes. Only small decreases in light scattering were found for ribosomes attached to mRNA obtained by incubation of polysomes with pancreatic RNAase. Fixation of the ribosomes and polysomes with formaldehyde eliminated changes in light scattering with increasing hydrostatic pressure.

III. RNA AND PROTEIN CONTENT OF RIBOSOMES AND RIBOSOMAL SUBUNITS

A. Determination by Chemical Methods

The amount of RNA per A_{260} unit of ribosomal particles has been derived from phosphorus content determinations.[12] Taking into account the base composition of ribosomal subunits from *Artemia*,[13] the phosphorus content of the free acid of RNA in the large and small ribosome subunits was calculated to be 9.62 and 9.64% (by weight), respectively. Solutions of $NaHPO_4 \cdot 12H_2O$, their concentration measured accurately by acid-base titration before each series of determinations, were used as standards in the phosphorus determination.

The amount of protein per A_{260} unit of ribosomal particles has been determined using a modification of the Lowry and the Biuret methods.[14] Because of interference with the Lowry assay,[15] Hepes was replaced by phosphate buffer and dithiothreitol was omitted from the solutions. Bovine serum albumin (BSA) was used as the protein standard accepting an absorbance of 6.6 at 280 nm (1 cm path length) for a 1% solution.[8]

Based on the above standards the large and small ribosomal subunits of *Artemia* ribosomes contain (40.8 $\pm$ 1) μg of RNA/A_{260} unit. The same value has been found for particles with different ratios of RNA to protein content, because the contribution of the proteins to the absorbance at 260 nm is negligible,[8] The same value can be used for the complete ribosomes. Thus, for RNA in the ribosomal particles, the absorbance (A_{260}) of a 1 mg/mℓ solution is 24.5 $\pm$ 0.5. Different colorimetric determinations have shown that complete ribosomes contain (42 $\pm$ 2) μg of protein per A_{260} unit.[14]

In conclusion, *Artemia* ribosomes contain 49 $\pm$ 2% RNA and 51 $\pm$ 2% protein by weight. These results are consistent with published data.[16]

B. CsCl Isodensity Gradient Equilibrium Centrifugation

From a survey of data in literature, it is clear that the RNA/protein content of ribosomal particles is related to their buoyant density in CsCl.[16] However, variations in experimental procedures such as fixation conditions, solvents, density gradients, speed and duration of centrifugation, and method of density measurements may cause variations in the value of the buoyant density finally obtained. Therefore, we have used equilibrium sedimentation in CsCl density gradients comparatively, under identical (constant) experimental conditions to determine the differences between the RNA/protein content of the ribosomes and their subunits.

Solutions of ribosomal particles were diluted to 1 A_{260} unit/mℓ to avoid aggregation and were fixed for 20 hr with 1% freshly neutralized formaldehyde. The ribosomal particles were then dialyzed (together) against triethanolamine buffer to remove sucrose and formaldehyde and to ensure that the various ribosomal preparations had an identical composition. CsCl was added to the dialyzed solutions to bring the density to 1.6 g/mℓ, and the preparations were centrifuged in a Beckman SW40 rotor at 25,000 rpm at 4°C for 65 hr. During fractionation of the density gradients, the absorbance at 260 nm was recorded continuously. The exact position of the ribosomal particles was confirmed by measuring the absorbance

Table 1
PHYSICOCHEMICAL PROPERTIES OF RIBOSOMAL PARTICLES FROM *ARTEMIA*

	Complete ribosomes	Large subunits	Small subunits
Buoyant density in CsCl (g/mℓ)	1.551 ± 0.004	1.567 ± 0.004	1.519 ± 0.004
RNA content (% by weight)	49 ± 2	52 ± 2	43 ± 2
Partial specific volume (mℓ/g)	0.63 ± 0.02	0.62 ± 0.02	0.65 ± 0.02
$A^{1\ mg/ml}_{260,1\ cm}$	12.1 ± 0.4	12.8 ± 0.4	10.5 ± 0.4
A_{260}/A_{280}	1.92 ± 0.02	1.94 ± 0.02	1.87 ± 0.02
Sedimentation coefficient, ($s^{o}_{20,w} \times 10^{-13}$ sec)	81 ± 2	59 ± 2	38 ± 1
Diffusion coefficient, ($D^{o}_{20,w} \times 10^{-7}$ cm^2 sec^{-1})	1.41 ± 0.02	1.61 ± 0.03	1.80 ± 0.04
Molar mass, *M* (g/mole $\times 10^{+6}$)	3.85 ± 0.2	2.45 ± 0.2	1.45 ± 0.1
Total molecular mass of RNA $\times 10^6$ Da	1.9 ± 0.1	1.27 ± 0.08	0.62 ± 0.04
Total molecular mass of proteins $\times 10^6$ Da	2.0 ± 0.1	1.2 ± 0.1	0.83 ± 0.08
Hydrodynamic radius (R_h nm)	15.2 ± 0.2	13.3 ± 0.3	11.9 ± 0.3
Maximal hydrodynamic volume ($V_{h,max}$) $\times 10^3$ nm^3	14.7 ± 0.6	9.9 ± 0.6	7.1 ± 0.5
Dry particle volume (V_d) $\times 10^3$ nm^3	4.1 ± 0.3	2.5 ± 0.2	1.6 ± 0.2
Dry particle radius (R_d nm)	9.9 ± 0.3	8.4 ± 0.3	7.2 ± 0.3
Frictional ratio (R_h / R_d)	1.54 ± 0.04	1.58 ± 0.05	1.58 ± 0.06
Maximal hydrodynamic solvation			
δ_{max} mℓ solvent/g ribosome	1.7 ± 0.1	1.9 ± 0.2	1.7 ± 0.2
F_{max} mℓ solvent/mℓ ribosome	0.72 ± 0.03	0.75 ± 0.03	0.72 ± 0.03

of the individual fractions. The refractive index of each fraction was measured, from which the density of each fraction (at 4°C) was calculated using the relationship for CsCl in water at 25°C given in Reference 17. The calculated densities were confirmed by weighing aliquots of the gradient samples with calibrated constriction pipettes.[8] The length of the centrifugation tubes was exploited by using a low centrifugation speed to distribute the particles at sedimentation equilibrium in a broad band and shallow density gradient to reduce the distortions due to the fractionation. Samples of various ribosomal particles were centrifuged in different tubes in the same rotor, and they always distributed in a single, symmetrical band with a half width of 0.01 g/mℓ. Equal bandwidths were found with ribosomes of *Artemia, Euglena gracilis,* and *Tetrahymena pyriformis.*[18] These data indicated that the various ribosomal particles have similar densities and that the large subunits were not contaminated with dimers of the small subunits. In an earlier study, the large ribosomal subunits of rat liver showed a broader density distribution on similar gradients.[19]

The buoyant densities and other physicochemical properties for *Artemia* ribosomes and subunits are listed in Table 1. Based on a linear relationship between the inverse of the buoyant density and the RNA/protein content, the differences in RNA/protein content between the ribosomal particles were calculated. These values were then combined with values of the chemically determined RNA and protein content of the complete ribosomes (see above) to yield the composition of the subunits (Table 1). Small subparticles obtained by ribosomal dissociation in high KCl solutions had a higher density which indicates the loss of some proteins. The values of RNA and protein content were combined with the fact that 1 A_{260} unit of ribosomal particles contains 40.8 ± 1 μg of RNA (see above) to calculate the absorbances at 260 nm of 1 mg/mℓ solutions, $A^{1\ mg/m\ell}_{260,\ 1\ cm}$.

C. Determination of the Density Increment and Partial Specific Volume

Solutions of ribosomal particles were brought to thermodynamic equilibrium by exhaustive dialysis against the appropriate buffer using Visking dialysis membranes, centrifuged briefly, and their densities were measured with a resonant oscillator densimeter DMA O2C (Anton

Paar, Graz, Austria).[20] The temperature of the oscillator was kept constant within 0.005°C at 25.0°C. Dry air and twice-distilled water made dust free by filtration through a 0.1-μm Millipore filter and by centrifugation were used for calibration. After dialysis, the densities of the buffers inside and outside the dialysis bag were equal within experimental error so that no corrections had to be made for eventual membrane asymmetry.

The determination of the density increment and partial specific volume by measurement of the density as a function of the absorbance of ribosomal solutions proved to be sensitive to the loss of proteins from the ribosomal particles. Since the accuracy of the density measurements increases with concentration, purified solutions were concentrated by centrifugation. However, this caused an increase in the A_{260}/A_{280} absorbance ratio of the small subunits, reflecting the loss of some ribosomal proteins which remained in the supernatant. Repeated cycles of centrifugation leading to a gradual loss of proteins from slime mold ribosomal particles have been reported.[21]

To show that the values obtained for the density increment are reasonable, they have been plotted together with analogous values for other ribonucleoproteins, as a function of the RNA and protein content of the particles (Figure 1). The partial specific volume (υ°) is clearly correlated with the RNA and protein content. The differences between the respective υ° values of the complete ribosomes and their large and small subunits are meaningful, although they have the same magnitude as the estimated uncertainty in υ°. The statistical errors are small compared to the systematic errors which are equal and cancel each other out when density differences between various ribosomal particles are considered.

IV. PHYSICAL CHARACTERIZATION OF RIBOSOMES AND RIBOSOMAL SUBUNITS: MASS, VOLUME, FRICTIONAL COEFFICIENT RATIOS, SHAPE, AND SOLVATION

A. Experimental Methods

1. Analytical Boundary Sedimentation

The sedimentation coefficient of the ribosomal particles, their aggregation, homogeneity, and integrity were determined by boundary sedimentation in an MSE analytical ultracentrifuge equipped with a photoelectric scanner. Double sector cells were used in a six-hole rotor. Their temperature was kept at 20.0°C ± 0.1°C.[9]

2. Photon Correlation Spectroscopy of Scattered Light

The diffusion coefficient of the various ribosomal particles was determined from their quasi-elastic light scattering. In particular, single-clipped photon-count autocorrelation spectroscopy was used for the efficient and accurate analysis of the intensity fluctuations of laser light scattered by a solution of the macromolecules in Brownian movement. In the case of a dilute and monodisperse solution of macromolecules which are small compared to the wavelength of light, the measured correlation function G(iT) decays exponentially with a decay time proportional to the searched diffusion coefficient D. Thus,

$$G(iT) = A + B \exp(-2DK^2 iT)$$

where i is the channel number, T is the chosen sample time, K is the known modulus of the scattering vector, and A and B are constants. We have previously described our setup and experimental procedures, test experiments, and methods of data analysis.[9] Figure 2 shows a general block diagram of an intensity fluctuation spectrometer.

Solutions of ribosome subunits contain not only monomers but also aggregates and quasi-elastic light scattering would only yield their z-average diffusion coefficient from which no accurate values for the diffusion coefficient of the monomers can be derived. Therefore,

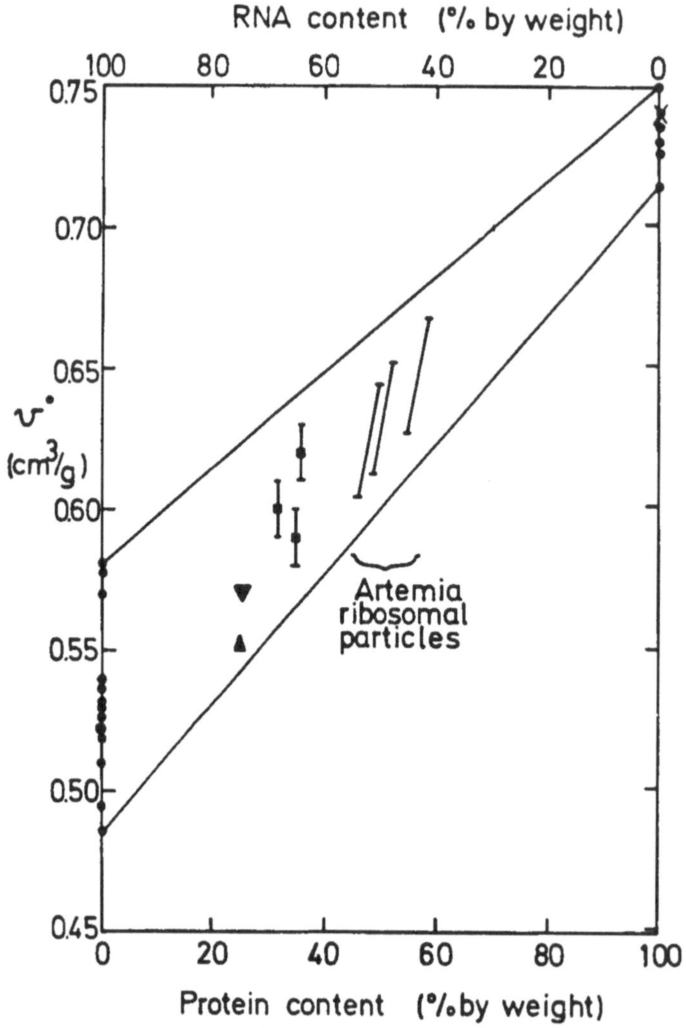

FIGURE 1. Correlation of the partial specific volume (υ°) of ribonucleoproteins with the RNA/protein content. Results for *Artemia* ribosomal particles have been plotted together with published data; ■, *E. coli* ribosomal particles; ▼, first reconstitution intermediate from the 50 S ribosomal subunit of *E. coli*; ▲, first reconstitution intermediate from the 30 S ribosomal subunit of *E. coli*. The points at 100% RNA indicate published values for υ° of various ribonucleic acids. At 100% protein, the points indicate published values for υ° of different ribosomal proteins from *E. coli*; x, the value for the preparation of *E. coli* ribosomal proteins denatured in 8 *M* urea. The same value has been derived from neutron scattering experiments for the proteins incorporated in the large subunits of *E. coli* ribosomes.[8] The straight lines connect the maximal and minimal υ° values of ribosomal proteins with the extreme values of ribonucleic acids.

photon correlation spectroscopy was performed on laser light scatterd exclusively by the monomeric subunits directly after they were separated from their aggregates and dust particles by band sedimentation in a CsCl or ^2H_2O density gradient in a transparent centrifugation tube.[22,23] The transparent centrifugation tube can then serve as the scattering cell. The monomers can be located by observing the intensity of the light which they scatter from the laser beam. Photon correlation spectroscopy then yields their diffusion coefficient.

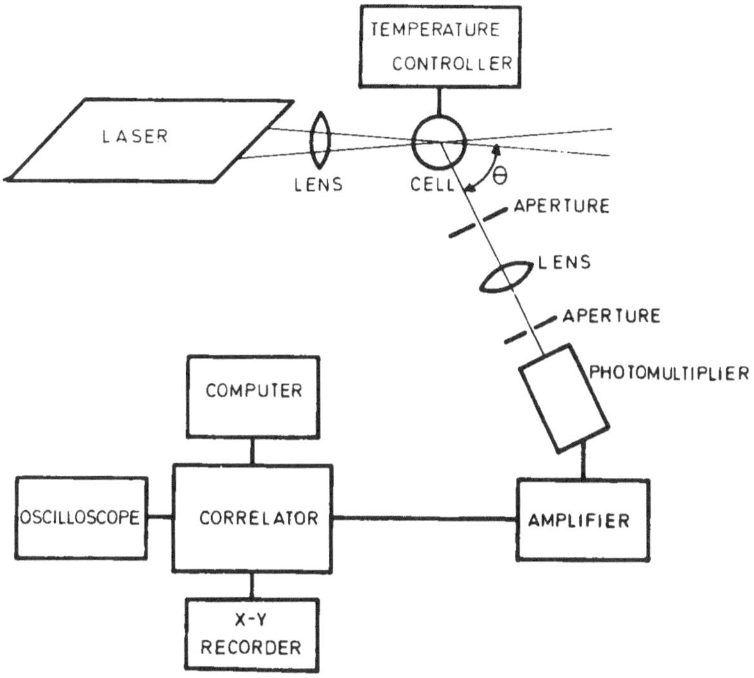

FIGURE 2. General block diagram for intensity fluctuation spectroscopy.

CsCl as the density gradient medium offers several advantages. Below room temperature the viscosity of aqueous CsCl solutions decreases with concentration allowing a good separation by sedimentation in the density gradient. Near room temperature, the viscosity is almost independent of concentration, and thus the uncertainty in the reduction of the measured diffusion coefficient to standard conditions is very small.

B. Sedimentation and Diffusion Coefficients and Molar Masses

The standard sedimentation coefficients $s^{o}_{20,w}$, of the ribosomal particles are listed in Table 1. When the dissociation buffer contained 700 m*M* KC1 instead of 400 m*M* KC1, the sedimentation behavior of the large subunits was nearly unchanged, but the $s^{o}_{20,w}$ of the small subunits was lowered to about 30 S for reasons discussed in Section II.C.

Photon correlation spectroscopy of laser light, scattered by monomeric ribosomal particles, yielded standard diffusion coefficients at infinite dilution $D^{o}_{20,w}$ and these are listed in Table 1.

The molar masses (M) listed in Table 1 have been calculated from the measured values of the density increment and the sedimentation and the diffusion coefficient using the Svedberg relation:

$$M = [s^{o}_{20,w}] \cdot [R.293\ K]/[D^{o}_{20,w}] \cdot [(\partial\rho/\partial c)^{o}_{20,w}]$$

where R is the gas constant and $(\partial\rho/\partial c)^{o}_{20,w}$ is the density increment. As expected, the sum of the molar masses of the large and small subunits were equal to the molar mass of the complete ribosomes within experimental error.

A combination of the values for the total molar mass of the various particles with their respective RNA and protein content yields the total molar masses (1) of the set of RNAs and (2) of the set of proteins incorporated in the ribosomal particles (see Table 1). The sum of the values for the molar masses of the two subunits and for the complete ribosome was equal within experimental error. These values are compared below with data obtained by relative measuring methods applied to the isolated RNAs and proteins, and for the RNAs only, to the values obtained directly by sequencing the RNAs.

Gel electrophoresis of *Artemia* ribosomal RNAs in denaturing conditions yielded estimates of their relative molecular mass by using ribosomal RNAs of *Escherichia coli* as standards. The relative molecular masses of the RNAs in the large and small ribosomal subunits of *Artemia* ribosomes were reported to be $(1.38 \pm 0.05) \times 10^6$ and $(0.66 \pm 0.05) \times 10^6$, respectively.[24] Direct sequencing yielded a molecular mass of 0.61×10^6 daltons for the RNA of the small subunit.[25] Our value $(0.62 \pm 0.04) \times 10^6$ daltons for the small rRNA fits the latter value nicely.

An extensive characterization of the ribosomal proteins of *Artemia* has been performed.[26,27] The total molecular mass of the proteins associated with the small ribosomal subunit has been estimated to be 0.56×10^6 daltons by sodium dodecyl sulfate-gel electrophoresis of the isolated proteins, assuming each protein to be present as one copy per ribosome.[26] Our value of $0.78 \pm 0.08 \times 10^6$ daltons is considerably higher. Furthermore, the reports on the RNAs and proteins considered together indicate that the small subunits contain much more RNA than proteins. However, most determinations of the RNA and protein composition of eukaryotic ribosomes, including ours, indicate the opposite.[28,29] This suggests that the total molecular mass of proteins associated with the small subunits is an underestimate. Perhaps this can be explained by a loss of proteins during the isolation (see above), by anomalous electrophorectic behavior of proteins, or by the presence of more than one protein copy per ribosome for one or more proteins as in *E. coli* ribosomes.[30]

C. Size, Shape, and Solvation of the Ribosomal Particles

From the diffusion coefficient of the ribosomal particles, the radius R_h of their hydrodynamic equivalent spheres can be calculated, using the Stokes-Einstein equation:

$$R_h = [k.293\ K]/[6\pi\eta_{20,w} \cdot D^o_{20,w}]$$

where k is the Boltzmann constant and η the viscosity of the solvent. These spheres have volumes $V_{h,max} = [4\pi R_h^3]/3$. Since the shapes of the ribosomal particles deviate from a sphere, these are maximal values for their real hydrodynamic volumes (V_h).

Dry particle volumes can be calculated from the equation:

$$V_d = [\upsilon^o M]/N_a$$

where N_a is Avogadro's number. These volumes can be packed into spheres with radii R_d.

The values calculated for R_h, $V_{h,max}$, V_d and R_d are given in Table 1. The sum of the dry particle volumes of the subunits is equal, within experimental error, to the dry particle volume of the complete ribosome.

The sizes of ribosomal particles in solution (see Table 1) are larger than the size of the dried particles determined by electron microscopy (even though the sandwich technique was used in the latter case, so that the particles appear to be larger than when they are deposited by the conventional technique on a single support).[6] The shrinking of ribosomes is hard to avoid because of their high degree of solvation.

The frictional coefficient ratio (f/f_{min}) of each ribosomal particle is the ratio of its hydrodynamic radius (R_h), to its dry particle radius (R_d) (see Table 1). This quantity depends on the solvation of the ribosomal particle and on the degree to which its shape deviates from a sphere. Thus, it follows that

$$f/f_{min} = R_h/R_d = [V_{h,max}/V_d]^{1/3} = [V_{h,max}/V_h]^{1/3} \cdot [V_h/V_d]^{1/3} > 1$$

In the last expression above, the first factor $[V_{h,max}/V_h]^{1/3}$ takes into account the shape effect and the second factor $[V_h/V_d]^{1/3}$ takes into account the effect of solvation; both factors are greater than 1. The degree of hydrodynamic solvation can be expressed as the fraction of the volume of the solvated particle which is occupied by solvent. This is

$$F = [V_h - V_d]/V_d$$

or as the volume of solvent per unit mass of "dry" particle

$$\delta = [V_h - V_d]/[M_r/N_A]$$

Both F and δ are maximal when $V_h = V_{h,max}$, i.e., when the ribosomal particles are spherical. These maximal values (F_{max} and δ_{max}) are listed in Table 1.

As the shape of *Artemia* ribosomal particles is visualized by electron microscopy,[6] one can estimate the extent to which the shape factor contributes to f/f_{min}. For the small ribosomal subunits, this factor must be somewhat higher than 1.05, which is that of a prolate ellipsoid of revolution with an elongation of 2. For the large ribosomal subunits and the complete ribosomes, this factor will be even closer to 1 since they are more spheroidal. The other factor contributing to f/f_{min} which is due to solvation, is somewhat lower than 1.54 ± 0.04, 1.58 ± 0.05, and 1.58 ± 0.06 for the complete ribosomes and their large and small subunits, respectively. From this, it can be concluded that

1. The contribution of solvation of the ribosomal particles to their f/f_{min} is much larger than the contribution due to deviation of their shape from a sphere.
2. Within experimental error, the solvation is equal for the various particles. These are 0.72 mℓ of solvent/mℓ of ribosome or 1.7 mℓ of solvent/g of ribosome.

The latter conclusion contrasts with the few reports on the solvation of eukaryotic ribosomes. In the case of ribosomes from rat liver[32] and yeast,[33] the small subunit seemed to have a much lower solvation than the large subunit. As an explanation, it was proposed that there is a deep groove or channel in the large subunit that contains the excess water.[33] However, the values of f/f_{min} obtained in our study are similar to the values obtained for *E. coli*. Thus, it may not be true that the small subunits of all eukaryotic ribosomes have a lower solvation or a higher compactness than the large subunits.

As mentioned above, electron microscopy has been used to study the shape and overall detailed structure of prokaryotic and eukaryotic ribosomes, including those from *Artemia*.[2,6] These studies have allowed us to build detailed models of prokaryotic and eukaryotic ribosomes and of their large and small subunits. The main criticism of this technique is the 10 to 20% shrinkage in ribosome size. This problem raises the question of whether the ribosome structures observed in the electron microscope are relevant to the situation in vivo.

X-ray and neutron small angle scattering offer researchers the opportunity to study ribosomes (native) in solution in much the same way as provided by the hydrodynamic methods. These techniques also allow the researcher to obtain more detailed structural information than from electron microscopy. However, some important uncertainty about these methods exists since several model molecules may result in scattering curves which fit the experimental data equally well, so that it is difficult to make a choice of a best-fitting model. Initially there was a problem of incompatibility between models deduced from small angle scattering data and electron micrographs of *Escherichia coli* ribosomal particles.[34] However, special handling of neutron scattering experiments, performed at different scattering backgrounds of the solvent by changing the ratio $H_2O/^2H_2O$, allowed deduction of the shape of the 50 S *E. coli* subunit.[35] This resulted in a unique model which showed great similarity to structures observed with the electron microscope.

Small angle X-ray scattering of rat liver ribosomes yielded dimensions which are quite close to the hydrodynamic dimensions determined by us and as discussed above. The diameter of the ribosome was determined to be 37 nm and for the large subunit a mean diameter of 28 nm was found.[36]

V. COMPARISON OF RIBOSOMAL PARTICLES FROM *E. COLI*, *ARTEMIA*, AND OTHER EUKARYOTES

A. Comparison of *E. coli* and *Artemia* Ribosome Structure

A combination of techniques including electron microscopy, small angle X-ray and neutron scattering, and chemical cross-linking has resulted in a detailed consensus model for the ribosomal particles of *E. coli*.[1,37] The ribosomal particles are asymmetric structures. The smaller subunit is divided into two unequal parts. The upper one third is called the head and the lower two thirds contain the base. A region called the platform extends from the base of the small subunit and forms a cleft between it and the head. The head and platform are part of the translational domain of the ribosome.

The large subunit consists of a central protuberance or "head" and two (lateral) protrusions inclined 50° to either side of the central protuberance. The latter structures are called the L7/L12 and the L1 protrusions in accordance with the proteins localized in these parts of the ribosomes. Several other proteins involved in the translation process have been localized in the upper protuberance and protrusions of the large subunit. Some proteins localized in the lower part are involved in the secretion of the growing peptide chain and they are part of the secretory domain.

In the complete ribosome, the small subunit is positioned asymmetrically on the large subunit. This arrangement permits the platform of the small subunit to contact the large subunit so that the head of the 30 S subunit contacts the L1 protrusion. The upper parts of both subunits are involved in the translation of the mRNA into a polypeptide chain and they form the translational domain. The base of the large subunit is involved in the secretion of the polypeptide chain and it forms the secretory domain.

There is substantial similarity in the overall shape and orientation of the subunits of the ribosome of *E. coli* and *Artemia*.[6] The small subunit of *Artemia* is oriented toward the large subunit in the same way as in *E. coli* ribosomes. Structural asymmetry is also a characteristic feature of the *Artemia* ribosomal particles, but aside from these features, there are clear differences in the fine structure of the small and large subunit between the two ribosome types. The *Artemia* 40 S particle contains three distinct structural domains; the upper third has a beak-like protrusion and it is more complex than the corresponding *E. coli* subunit. The *Artemia* large subunit is less homologous, having a stalk-like protrusion which has not been observed on any other ribosome, and it also contains a knob-like protrusion on its base.[6]

These structural features of *Artemia* ribosomes and their subunits must be reconciled with the following points:

1. A difference in molar mass and size between the *E. coli* and *Artemia* ribosomal particles: a ratio of 1.65 for the molar masses of the respective particles and of 1.25 for the ratio of the hydrodynamic radii;[16] a similar difference in dimensions has also been observed with the electron microscope.[6]
2. A difference in chemical composition: for the small subunits, the RNA protein ratio is 1.7:1 for the *E. coli* and 0.75:1 for *Artemia*; for the larger subunit the ratio's are 1.7:1 for the *E. coli* and 1.08:1 for *Artemia*, respectively.[16]
3. The similar frictional ratios $f{:}\ f_{min}$ of the ribosomal particles from *Artemia* and *E. coli* which reflect their similarity in solvation and overall shape in solution.[8]

B. Comparison of Ribosomal Particles from Various Other Eukaryotes

1. Comparison of Ribosomal RNAs and Proteins

A concensus is growing to support the existence of four separate lines of cellular descent: namely archaebacteria, eubacteria, eocytes, and eukaryotes. Each line has a characteristic set of basic structures such as ribosomes and ribosomal biomolecules (e.g. 5S rRNA[38]) which are used to study evolutionary relationships.

The size of the ribosomal RNAs has been used as a first parameter to characterize the ribosomal particle and to compare ribosomes from different species. This is due principally to the simplicity and sensitivity of the techniques involved since rRNAs can be isolated easily and fractionated by gel electrophoresis in denaturing or nondenaturing conditions. These analyses show that the molecular mass of the ribosomal RNA of the small subunit (S-rRNA) of all deuterostomes, from sea urchin to man, has remained constant at 0.65×10^5 daltons, whereas the molecular mass of the large RNA of the large ribosomal subunit (L-rRNA) has been conserved at 1.35×10^6 to 1.38×10^6 daltons in the lower members of the group but has increased to 1.65×10^8 daltons in warm-blooded animals.[24] The importance of the size of rRNAs in relation to evolution became less evident as it became clear that the change in mass of rRNA varies among the main eukaryotic phyla. For example the size of the S-rRNA is almost constant in plants and deuterostomes but changes from 0.65 to 0.95×10^6 daltons in molluska. The L-rRNA is constant in plants but varies from 1.35 to 1.65×10^6 daltons in deuterostomes and from 1.37 to 1.90×10^6 daltons in molluska.[39]

Due to the rapid development of nucleic acid sequencing methods, sequences of large RNA chains such as rRNAs are appearing at an increasingly rapid pace, and thus comparison betv'¬en ribosomal RNA sequences can be made. For S-rRNA, about 70 sequences have beeı. published from a wide range of eukaryotes, archaebacteria, and eubacteria.[40] These data show that although the size and primary structure of the S-rRNAs are quite different, the secondary structure is rather well conserved. Also the different lengths of the RNA strands from different species appear to be caused by deletions and insertions of structural elements while secondary structural patterns remain constant.

A detailed comparison of the ribosomal proteins has been more difficult due to the large number of different proteins, the different nomenclature used by different investigators, and the ambiguity in defining ribosomal proteins. It is highly probable that some ribosomal proteins are artefacts caused by the procedures of isolation and characterization while other ribosomal proteins are missing.[41] The introduction of a uniform nomenclature of ribosomal proteins has clarified the situation to a great extent[42] and a detailed comparison of ribosomal proteins from different sources such as *Artemia* and rat liver has been made.[27]

2. Stability, Density, Mass, Size, and Structure of Cytoplasmic Ribosomes in Solution from Tetrahymena, Artemia, and *Euglena*

Due to variations in the number and size of the ribosomal components, differences are expected in the mass of the cytoplasmic ribosomes of various eukaryotes and in their size in solution. However, direct evidence for the latter has been lacking. It is also surprising that ribosomal particles isolated from widely different eukaryotes show no clear difference in their sedimentation coefficient. Therefore, we began a study of three eukaryotes to reveal any differences in the stability, the ratio of RNA to protein content, the mass, and the size of the ribosomes in solution.[18] Our experimental approach allowed us to detect differences in the structure of the various ribosomes in solution. The organisms included in this study were

1. *Tetrahymena pyriformis*, a ciliated protozoan and one of the most primitive of eukaryotes.[43]

2. *Artemia*, a more highly evolved eukaryote.
3. *Euglena gracilis* (two strains), a flagellated alga for which it is well documented that their ribosomes have a sedimentation coefficient higher than 80 S which is usually assigned to cytoplasmic ribosomes of other eukaryotes.[44]

The experimental procedures have been described above in Section II.A (for the solvent conditions of the different ribosomal particles), Section III.B (for the isodensity centrifugation in CsCl), section III.C (for the determination of the partial specific volume), and sections IV.A, B, and C (for the physical characterization of the ribosomal particles).

a. Stability and Aggregation of the Ribosomal Particles

Freshly prepared solutions of whole ribosomes from different organisms were free of aggregates as judged by analytical boundary sedimentation, equilibrium centrifugation in CsC1 density gradients, and photon correlation spectroscopy.

The isolation of intact ribosomal subunits from *Artemia* has been discussed above (see Section II.C). Applying the same procedure with *Tetrahymena* yielded satisfactory results. Unfortunately, the isolation of intact and stable ribosomal subunits of *Euglena* is more difficult. The dissociation of *Euglena* ribosomes by centrifugation in a sucrose gradient (in dissociation buffer) yielded similar absorbance profiles to ribosomes from *Artemia* and *Tetrahymena*. However, an immediate fixation of the preparation with formaldehyde and analysis by centrifugation in a CsCl density gradient revealed heterogeneous material banding at higher densities than the intact subunits, indicating the partial loss of ribosomal proteins. This problem was avoided by lowering the concentration of KCl in the preparation to 350 m*M* KC1 (from 400 m*M*), and with a low concentration of ribosomes (<1 mg/mℓ), a dissociation absorbance profile was obtained. Fractions from the centrifuge tubes were dialyzed against 20 m*M* triethylethanolamine/HCl buffer (pH 7.5) containing 200 m*M* KC1, 10 m*M* Mg acetate, and 1 m*M* dithiothreitol. Analytical boundary centrifugation of the dialyzed preparations showed that the large subunits were partly converted to slower sedimenting particles, but that the small subunits obtained in this way were stable. No improvement was achieved when the ratio Mg^{2+} to K^+ (in m*M*) was changed during the dissociation from 10/350 as above to 20/500 or to 1/250. Clearly, the cytoplasmic ribosomal subunits of *Euglena* are less stable than the corresponding particles from *Artemia* and *Tetrahymena*.

The ribosomal subunits isolated from different organisms could not be kept completely free of aggregates. Therefore their diffusion coefficient has been determined by analyzing the intensity fluctuations of laser light scattered by the monomeric subunits directly in a density gradient after separating them from aggregates by rate-zonal sedimentation (see Section IV.A.2).

b. Sedimentation Coefficient

The sedimentation coefficients, $s^o_{20,w}$, of the ribosomal particles are listed in Table 2. No differences were found in $s^o_{20,w}$ of the ribosomal particles from *Artemia* and *Tetrahymena*, while larger values of $s^o_{20,w}$ were measured for *Euglena* ribosomal particles.[45]

c. Buoyant Densities in CsC1, RNA, and Protein Content and Partial Specific Volumes

The ratio of RNA to protein content of ribosomal particles is related to their buoyant density in CsC1. However, variations in experimental procedures may cause variation in the buoyant density finally obtained. Therefore we have used equilibrium sedimentation in CsC1 isodensity gradients comparatively, under identical experimental conditions, to determine differences in the ratio of RNA to protein content of the various ribosomal particles.

The buoyant densities of the corresponding ribosomal particles from the various species were found to be equal within experimental error and no differences were detected in the

Table 2
SUMMARY OF THE PHYSICAL PROPERTIES OF RIBOSOMES FROM *ARTEMIA, TETRAHYMENA PYRIFORMIS,* AND *EUGLENA GRACILIS*

	Artemia	*Tetrahymena pyriformis*	*Euglena gracilis*
Sedimentation coefficient, ($s^{o}_{20,w} \times 10^{-13}$ sec)	81 ± 1	81 ± 1	86 ± 1
Diffusion coefficient, ($D^{o}_{20,w} \times 10^{-7}$ cm^2 sec^{-1})	1.42 ± 0.02	1.51 ± 0.02.	1.29 ± 0.02
Buoyant density in CsCl (g/mℓ)	1.551 ± 0.004	1.551 ± 0.004	1.551 ± 0.004
RNA content (% by weight)	49 ± 2	49 ± 2	49 ± 2
Partial specific volume (mℓ/g)	0.63 ± 0.02	0.63 ± 0.02	0.63 ± 0.02
Molar mass, *M* (g/mole $\times 10^{+6}$)	3.85 ± 0.2	3.50 ± 0.2	4.40 ± 0.1
Hydrodynamic radius (R_h nm)	15.2 ± 0.2	14.2 ± 0.2	16.6 ± 0.3
Frictional ratio (R_h / R_d)	1.54 ± 0.04	1.48 ± 0.04	1.61 ± 0.04
GC/AU base ratio of total rRNA	1.09	0.50	1.12

ratio of their RNA to protein. Furthermore, the partial specific volume of ribonucleoproteins depends primarily on the ratio of their RNA to protein content. Thus, the values determined for both the RNA content and partial specific volume at vanishing concentration (υ^{o}), of *Artemia* ribosomal particles have been assigned to the corresponding particles of *Tetrahymena* and *Euglena*.

d. Diffusion Coefficients

Photon correlation spectroscopy of laser light scattered by the monomeric ribosomal particles, has yielded standard diffusion coefficients at infinite dilution ($D^{o}_{20,w}$) for various ribosomes (see Table 2). As expected the measured diffusion coefficients were independent of the ribosome concentration (<0.5 mg/mℓ).

e. Size in Solution

From the diffusion coefficient of the ribosomal particles, the radius (R_h) of their hydrodynamic equivalent spheres can be calculated (see section IV. C.). The resulting values are listed in Table 2. Although no differences were detected in their sedimentation coefficient, the hydrodynamic radius R_h of whole ribosomes of *Tetrahymena* is 6% smaller than the R_h of the *Artemia* ribosomes. On the other hand, our data indicate that the R_h of *Euglena* whole ribosomes is 10 and 16% larger than the R_h of *Artemia* and *Tetrahymena* ribosomes respectively, which correlates well with the exceptionally high sedimentation coefficients of the *Euglena* ribosomal particles. This demonstration that ribosomes from *Tetrahymena* have a smaller size than *Euglena* ribosomes partly explains why the former migrates faster than the latter during electrophoresis in polyacrylamide gels.[45] In contrast, no significant differences have been measured by electron microscopy in the size of the dried and negatively stained ribosomes.[18]

f. Molar Mass of the Ribosomal Particles

The molar mass (M) of the ribosomal particles has been calculated from the hydrodynamic parameters (as explained Section IV. B) and the resulting values are listed in Table 2. *Tetrahymena* ribosomes have a 6% smaller mass than *Artemia* ribosomes, whereas ribosomes of *Euglena* have a considerably larger mass than both *Artemia* and *Tetrahymena* ribosomal species.

Having determined the ratio of RNA to protein content as well as the molar mass of the ribosomes (Table 2), we can now discuss the molecular mass of their RNA and their proteins. The relative molecular masses of the Na salts of *Artemia* rRNA have been estimated by gel electrophoresis under denaturing conditions using *E. coli* rRNA as the standard.[39,46] In the interpretation of these results, we took into account the contribution of Na to the molecular masses, and we assigned more realistic molecular masses, based on their nucleotide sequences, to the standard *E. coli* rRNAs.[47] When this was done the data agreed, within experimental error, with our values for the total molecular mass and for the RNA content of *Artemia* ribosomal particles.

Using rRNA from *E. coli* as the standard, the molecular masses of *Tetrahymena* rRNA had been estimated previously, using relative measuring methods, such as by electrophoresis in nondenaturing or denaturing conditions,[24] and by sucrose-gradient centrifugation.[48] Unfortunately, such relative measurements, and in particular those in non-denaturing conditions, are obscured by variations in the conformation of the various rRNAs and concentrations of polyacrylamide gels used in electrophoresis. In one of these studies the size of the rRNA from *Tetrahymena* and *Artemia* was estimated by gel electrophoresis under identical denaturing conditions.[24] The data indicated that the total mass of RNA in a *Tetrahymena* ribosome may be a few percent less than the RNA mass in a ribosome of *Artemia*, which is consistent with the data reported here.

In the case of *Euglena* ribosomes, our values for the RNA content and molecular mass (Table 2) are consistent with published data if the correct values for the reference prokaryotic rRNAs are used.[24,47] Combining the values for the molecular masses of the *Euglena* rRNA with our estimate of the RNA content of the whole ribosome gives a molecular mass for whole ribosomes of *Euglena* of about 4.2×10^6 daltons. This is very close to 4.4×10^6 daltons as determined by us (see Table 2) and the earlier rough estimate of 5.5×10^6 daltons now seems too high.[49]

The differences in the size of the rRNA among the eukaryotes studied seem to be matched by approximately equal differences in the amount of ribosomal protein. Indeed, the ratio of the RNA to protein is similar for corresponding ribosomal particles of various species, while their total molar masses vary by up to 25% (Table 2). Thus, the number and/or size of the ribosomal proteins are expected to be quite variable among the eukaryotes which is relevant for a more detailed comparison of the ribosomal proteins of the various species.

g. Frictional Ratio, Solvation, and Compactness of the Ribosomal Particles

The frictional ratio (f/f_{min}) has been calculated for the different ribosomal particles. These data are listed in Table 2. The contribution of the shape factor to this ratio can be estimated from observation by electron microscopy of ribosome particles of *Tetrahymena*, *Euglena*, and *Artemia*.[31,50] For the small subunits, this factor will be about equal to that of an ellipsoid of revolution with an elongation of 2, which is close to 1.04. For the large subunits and the whole ribosomes, this factor will be closer to 1, since these particles are more spherical. Thus, in the case of ribosomal particles, the major contribution to f/f_{min} is the solvation factor. The differences in f/f_{min} are attributed to differences in the amount of solvent entrapped by the ribosomes (1.7 g solvent/g of "dry" ribosome in the case of *Artemia*, vs. 1.4 and 2.0 in the case of *Tetrahymena* and *Euglena*, respectively).

For the ribosomes of the three species mentioned, their solvation or degree of compactness, as reflected by the f/f_{min} correlates well with their size in solution and their molecular mass (see Figure 3). The larger the ribosome, the less compact is its structure. This may partly explain why no difference in size of the dried, whole ribosomes of *Tetrahymena* and *Euglena* has been measured by electron microscopy.[50] Indeed, when ribosomes are dried for electron microscopy, the larger *Euglena* ribosomes are expected to shrink more since they have a less compact structure than *Tetrahymena* ribosomes. Thus, differences in size and solvation of ribosomes may partly compensate each other when ribosome dimensions are estimated by electron microscopy.

The frictional ratio (f/f_{min}) of the ribosomes from the three species also increases with the GC/AU base ratio of their total rRNA (Table 2).[51] This may be relevant, since this guanosine,

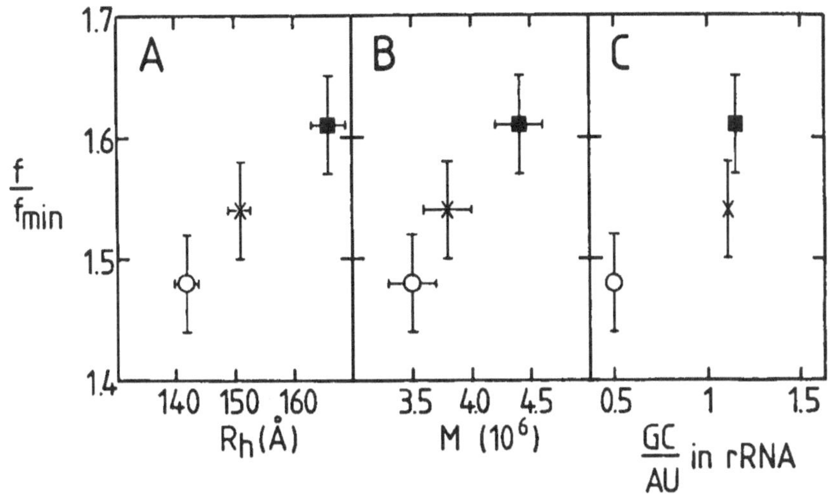

FIGURE 3. Correlation between the frictional coefficient ratio (f/f_{min}) and other physical properties of complete ribosomes of *Tetrahymena* (○), *Artemia* (x), and *Euglena* (■). (A) the hydrodynamic radius R_h; (B) the molar mass; (C) the GC/AU base ratio.

cytosine/adenine, uracil (GC/AU) ratio in the rRNA of both ribosomal subunits increases with the evolutionary level of the organism. Gel electrophoresis experiments on a large variety of rRNA have been interpreted as indicating that insect rRNA species with a low GC content (about 42%) have a less extended spatial distribution of mass than most other rRNA species.[39] Such variations in the structure of rRNA of various species may be a cause of differences in the degree of compactness between the whole ribosomes of various species such as described above.

VI. SIZE AND CONFORMATION OF RIBOSOMAL RNAs IN SOLUTION

The conformation of the RNA molecule free in solution must be known in order to understand the assembly of the RNA polymer into biological structures such as ribosomes. In this respect, the rRNA molecules of *E. coli* have been studied extensively. The degree of unfolding of the *E. coli* RNA depends on the ionic concentrations of the solution and in particular the Mg^{2+} concentrations which induce a tightening of the RNA.[52,53] However, data reported by various groups differ even when the same RNA was studied under identical conditions.[54] Thus, conflicting answers have been given to the questions concerning the extent to which the free RNA is unfolded and to the size of the conformational change of the RNA during its assembly into the ribosomal subunit.

In view of this controversial situation, we have compared the size and conformation in solution of the RNA of the small subunit (S-rRNA) and the large RNA isolated from the large ribosomal subunit (L-rRNA) of *Artemia*. Knowing the size of both RNAs allowed then a comparison to be made with the known size of the corresponding total ribosomal subunits.

A. Preparation of Ribosomal RNA

Ribosomal subunits were isolated from the cryptobiotic embryos of *Artemia* as described in Section II.B, and concentrated by centrifugation. Two alternative methods were used for the isolation of the rRNAs.

In the first method, a pellet of ribosomal subunits was dissolved in 10 m*M* Hepes-KOH buffer (pH 7.5) containing 50 m*M* KCl and 1 m*M* EDTA.[55] A 20% (w/v) solution of sodium dodecyl sulfate was added to give a final concentration of 2%. After 15 min at room temperature, the samples were chilled in an ice bath, and 1 volume of redistilled phenol/chloroform (1:1, v/v) was added. The samples were centrifuged at 5000 *g* for 30 min for phase separation, and the aqueous phase was re-extracted with the phenol/chloroform mixture. The RNA was recovered by precipitation overnight with 0.1 volume of 20% (w/v) potassium acetate and 2 volumes of ethanol at −20°C. The RNA was dissolved in the Hepes-KCl-EDTA buffer, and purified by rate-zonal centrifugation on a 10 to 30% (w/v) linear gradient of sucrose in the same buffer at 67,000 *g* for 16 hr in a Beckman SW40 rotor at 4°C. The fractions containing rRNA were dialyzed overnight at 4°C against the appropriate buffers.

In the second method, rRNA was extracted from the subunit by the sodium dodecyl sulfate procedure[56] which allows for the isolation of RNA without using EDTA. The pellet of ribosomal subunits was dissolved in 10 m*M* Hepes-KOH (pH 7.5) containing 50 m*M* KCl and 1 m*M* EDTA, or in 20 m*M* Tris-HCl (pH 7.5) containing 5 m*M* $MgCl_2$ and 50 m*M* KCl or 50 m*M* NH_4Cl. Sodium dodecyl sulfate (SDS) solution (20%) was then added to give a final concentration of 2% SDS. The mixture was kept at 37°C for 2 min and the rRNA was further purified by rate zonal centrifugation as described above. The rRNA fractions were dissolved and dialyzed against a buffer containing 10 m*M* Hepes-KOH (pH 7.5) and various concentrations of Mg and K ions as specified below. The experimental methods used for studying the size and conformation in solution, namely determination of the density increment, analytical boundary sedimentation, and photon correlation spectroscopy of scattered laser light, have been described above (see Sections III. C, IV. A.1, and IV. A.2).

B. Size and Conformation of the S-rRNA and L-rRNA in Solution

The influence of Mg^{2+} ions on the size of the rRNAs was studied (see Figure 4). At low Mg^{2+} concentration (<3 m*M*) the S-rRNA is much more expanded than the L-rRNA under the same conditions. The hydrodynamic radius (R_h) of the S-rRNA in a buffer lacking Mg^{2+} is relatively large, indicating that under these conditions the S-rRNA is very expanded. Such a "jump" in the R_h does not appear for the L-rRNA and suggests that this rRNA is not as highly expanded. This supports the model in which both ends of the 5.8 S rRNA are joined by H-bonding to different regions of the L-rRNA such that the L-rRNA does not unfold completely, even at low Mg^{2+} concentrations as long as the interaction with the 5.8 S rRNA prevails.

In order to obtain quantitative information about the degree of compactness and the size of the RNAs, which are reflected by the frictional coefficient ratio (f/f_{min}) and the hydrodynamic radius (R_h), the density increment or the partial specific volume was measured. For the L-rRNA a density increment ($\delta\rho/\delta c$) of 0.50 ± 0.01 was obtained. This value is rather low compared with other values published in one of our earlier papers.[16] Nevertheless, this value allowed us to calculate the correct molar mass of the rRNAs by using the Svedberg equation (see Section IV.B) giving us confidence in the accuracy of this measurement.

The size of the rRNAs has been compared with the size of the corresponding ribosomal subunit. The conformation of the free L-rRNA in solution is more extended than when it is incorporated in the ribosome as we have already reported.[57] This physical state of the rRNA is not due to the lack of bivalent or polyvalent cations, but it occurs also in the presence of Mg^{2+} and spermidine which cause a tightening of the RNA. The size of the S-rRNA resembles the size of the small ribosomal subunit. However, the size of both rRNAs free in solution is larger than the rRNA incorporated in the subunits when one takes into account the fact that 57% of the small and 48% of the large subunit is made up of proteins. Both rRNAs free in solution assume a looser conformation than when they are incorporated into the ribosomal subunits. Thus, the ribosomal proteins induce a tightening of the rRNA during the assembly of the ribosome.

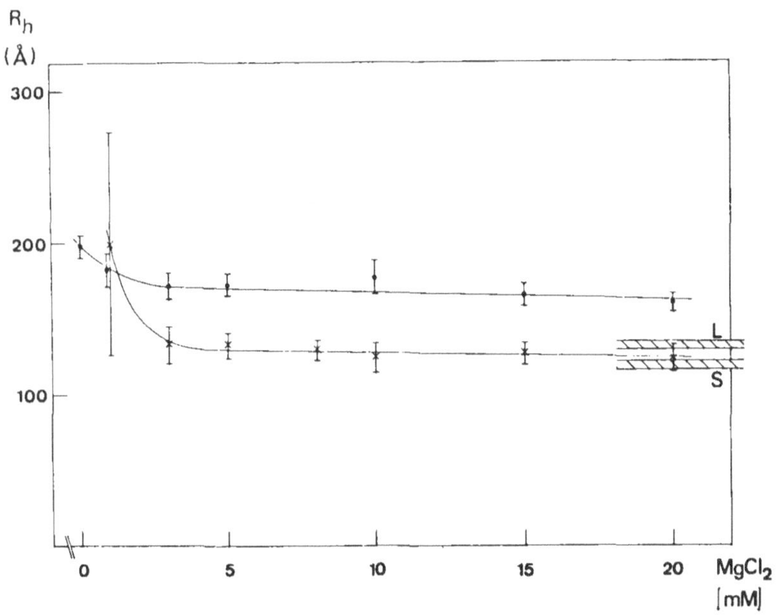

FIGURE 4. Hydrodynamic size of the *Artemia* ribosomal RNAs as a function of the Mg^{2+} concentration. (●) Large RNA of the large ribosome subunit; (×) RNA of the small ribosome subunit. The hydrodynamical radii of the larger subunit (L) and the small subunit (S) are indicated by the shaded regions.

ACKNOWLEDGMENTS

We thank the I.W.O.N.L./I.R.S.I.A. (Belgium) and the Belgian National Fund for Scientific Research for fellowships (N.P.). Our investigations were supported by grants from the Belgian National Fund for Scientific Research and from the F.K.F.O. (Belgium).

We thank T. Aerts, H. Backhovens, and D. Bauwens for their excellent technical assistance.

We thank Dr. M. Kondo for introducing us to *Artemia*.

REFERENCES

1. **Lake, J. A.,** Evolving ribosome structure: domains in archaebacteria, eubacteria, eocytes and eukaryotes, *Annu. Rev. Biochem.*, 54, 507, 1985.
2. **Liljas, A.,** Structural studies of ribosomes, *Prog. Biophys. Mol. Biol.*, 40, 161, 1982.
3. **Hultin, T. and Morris, J. E.,** The ribosomes of encysted embryos of *Artemia salina* during cryptobiosis and resumption of development, *Dev. Biol.*, 17, 143, 1968.
4. **Kramer, G. and Hardesty, B.,** Ribosomes from *Artemia salina* cysts in cell-free translation of eukaryotic mRNA, in *The Brine Shrimp Artemia*, Vol. 2, Persoone, G., Sorgeloos, P., Roels, O., and Jaspers, E., Eds., Universa Press, Wetteren, Belgium, 1980, 467.
5. **Bagshaw, J. C.,** Introduction, in *Biochemistry of Artemia Development*, Bagshaw J. C. and Warner, A. H., Eds., University Microfilms Int., Ann Arbor, Mich., 1979, 1.
6. **Boublik, M. and Hellmann, W.,** Comparison of *Artemia salina* and *Escherichia coli* ribosome structure by electron microscopy, *Proc. Natl. Acad. Sci. U.S.A.*, 75, 2829, 1978.
7. **Nieuwenhuysen, P. and Slegers, H.,** Purification of eukaryotic ribosomes by isopycnic centrifugation in sucrose, *Anal. Biochem.*, 89, 472, 1978.
8. **Nieuwenhuysen, P. and Clauwaert, J.,** Physicochemical characterization of ribosomal particles from the eukaryote *Artemia*, *J. Biol. Chem.*, 256, 9626, 1981.

9. **Nieuwenhuysen, P. and Clauwaert, J.,** Physical properties of *Artemia salina* ribosomes, *Biochemistry,* 17, 4261, 1978.
10. **Nieuwenhuysen, P., Heremans, K., and Clauwaert, J.,** High-pressure light scattering study of *Artemia salina* ribosomes and polysomes, *Biochim. Biophys. Acta,* 606, 292, 1980.
11. **Baierlein, R. and Infante, A. A.,** Pressure-induced dissociation of ribosomes, *Methods Enzymol.,* 30, 328, 1974.
12. **Ames, B.,** Assay of inorganic phosphate, total phosphate and phosphatases, *Methods Enzymol.,* 8, 115, 1966.
13. **Vandenberghe, A., Nelles, L., and De Wachter, R.,** High-pressure liquid chromatography analysis of oligo- and monoribonucleotide mixtures, with special reference to ribosomal RNA constituents, *Anal. Biochem.,* 107, 369, 1980.
14. **Nieuwenhuysen, P., De Herdt, E., and Clauwaert, J.,** Ribonucleic acid and protein content of eukaryotic ribosomes isolated from *Artemia salina, Anal. Biochem.,* 88, 532, 1978.
15. **Peterson, G. L.,** Review of the Folin phenol protein quantitation method of Lowry, Rosebrough, Farr and Randall, *Anal. Biochem.,* 100, 201, 1979.
16. **Nieuwenhuysen, P. and Clauwaert, J.,** Physical-chemical characterization of cytoplasmic ribosomal particles isolated from *Artemia,* in *The Brine Shrimp Artemia,* Vol. 2, Persoone, G., Sorgeloos, P., Roels, O., and Jespers, E., Eds., Universa Press, Wetteren, Belgium, 1980, 491.
17. **Ifft, J. B., Martin, W. R., and Kinzie, K.,** Density gradient proportionality constants for a number of aqueous binary solutions, *Biopolymers,* 9, 597, 1970.
18. **Nieuwenhuysen, P., Clauwaert, J., Ledoigt, G., and Curgy, J.,** Structural differences between ribosomes of various eukaryotes: stability, density, mass, size and structure in solution of cytoplasmic ribosomes from *Tetrahymena, Artemia,* and *Euglena, Int. J. Biol. Macromol.,* 5, 361, 1983.
19. **Hamilton, M. G., Cavalieri, L. F., and Petermann, M. L.,** Molecular weight, buoyant density and composition of active subunits of rat liver ribosomes, *Biochemistry,* 10, 3424, 1971.
20. **Kratky, O., Leopold, H., and Stabinger, H.,** The determination of the partial specific volume of proteins by the mechanical oscillator technique, *Methods Enzymol.,* 27, 98, 1973.
21. **Ramagopal, S. and Ennis, H. L.,** Studies on ribosomal proteins in the cellular slime mold *Dictyostelium discoideum, Eur. J. Biochem.,* 105, 245, 1980.
22. **Nieuwenhuysen. P. and Clauwaert, J.,** Quasi-elastic light scattering of *Artemia* ribosomes sedimented in a CsC1 density gradient, *J. Biochem. Biophys. Methods,* 5, 279, 1981.
23. **Nieuwenhuysen, P. and Clauwaert, J.,** Photon-correlation spectroscopy of centrifuged solutions. Application to *Artemia* ribosome subunits, in *Scattering Techniques Applied to Supramolecular and Nonequilibrium Systems,* Chen, S.-H., Chu, B., and Nosal, R., Eds., Plenum Press, New York, 1980, 845.
24. **Cammarano, P., Pons, S., and Londei, P.,** Discontinuity of the large ribosomal subunit RNA and rRNA molecular weights in eukaryot evolution, *Acta Biol. Med. Ger.,* 34, 1123, 1975.
25. **Nelles, L., Fang, B.-L., Volckaert, G., Vandenberghe, A., and De Wachter, R,** Nucleotide sequence of a crustacean 18 S ribosomal RNA gene and secondary structure of eukaryotic small subunit ribosomal RNAs. *Nucleic Acids Res.,* 12, 8749, 1984.
26. **Ting Shih, C.-Y., Toivonen, J. E., and Craven, G. R.,** Partial purification and characterization of the proteins from the 40 S ribosomes of *Artemia salina* and wheat germ, *Eur. J. Biochem.,* 97, 189, 1979.
27. **Kenmochi, N., Tsurugi, K., and Ogata, K.,** Analyses of 40 S and 60 S ribosomal proteins of *Artemia salina* with two- or "three-dimensional" acrylamide gel electrophoresis and comparison with rat liver 40 S and 60 S proteins. *J. Biochem. (Tokyo),* 89, 1293, 1981.
28. **Sacchi, A., Ferrini, U., Londei, P., Cammarano, P., and Maraldi, N.,** Mitochondrial and cytoplasmic ribosomes from mammalian tissues. (Further characterization of ribosomal subunits and validity of buoyant-density methods for determination of the chemical composition and partial specific volume of ribouncleo-protein particles), *Biochem. J.,* 168, 245, 1977.
29. **Hamilton, M. G.,** Isodensity equilibrium centrifugation of ribosomal particles: the calculation of the protein content of ribosomes and other ribonucleoproteins particles, *Methods Enzymol.,* 20, 512, 1971.
30. **Subramanian, A. R.,** Evidence for a repeated protein structure in the 30 S subunit of *Escherichia coli* ribosome, *J. Biol. Chem.,* 255, 6941, 1980.
31. **Boublik, M., Robakis, N., Hellmann, W., and Wall, J. S.,** Electron microscopy study of free and ribosome bound rRNAs from *E. coli, Eur. J. Cell Biol.,* 27, 177, 1982.
32. **Petermann, M. and Pavlovec, A.,** Effects of magnesium and formaldehyde on the sedimentation behavior of rat liver ribosomes, *Biopolymers,* 7, 73, 1969.
33. **Mazelis, A. G. and Petermann, M. L.,** Physical-chemical properties of stable yeast ribosomes and ribosomal subunits, *Biochim. Biophys. Acta,* 312, 111, 1973.
34. **Hill, W. E. and Fessenden, R. J.,** Structural studies on the 30 S ribosomal subunit from *Escherichia coli, J. Mol. Biol.,* 90, 719, 1974.
35. **Stuhrmann, H. B., Koch, M. H. J., Parfait, R., Haas, J., Ibel, K., and Crichton, R. R.,** Shape of the 50 S subunit of *Escherichia coli* ribosomes, *Proc. Natl. Acad. Sci. U.S.A.,* 74, 2316, 1977.

36. **Damaschun, G., Muller, J.J., Bielka, H., and Böttger, M.,** Studies on the structure of animal ribosomes. V. Studies on the size and structure of rat liver ribosomes by small angle X-ray scattering, *Acta Biol. Med. Ger.*, 33, 817, 1974.
37. **Giri, L., Hill, W. E., and Wittmann, H. G.,** Ribosomal proteins: their structure and spatial arrangement in prokaryotic ribosomes, *Adv. Protein Chem.*, 36, 1, 1984.
38. **De Wachter, R., Huysmans, E., and Vandenberghe, A.,** 5 S ribosomal RNA as a tool for studying evolution, in *Evolution of Prokaryotes*, Schleifer, K. H. and Stackebrandt, E., Eds., Academic Press, New York, 1985, 115.
39. **Cammarano, P., Londei, P., Mazzei, F., and Biagini, R.,** Size heterogeneity of ribosomal RNA in eukaryote evolution, *Comp. Biochem. Physiol.*, 73B, 435, 1982.
40. **Huysmans, E. and De Wachter, R.,** Compilation of small ribosomal subunit RNA sequences, *Nucleic Acids Res.*, 14, r74, 1986.
41. **Bielka, H.,** *The Eukaryotic Ribosome*, Springer-Verlag, Berlin, 1982, 61.
42. **Graves, M. C., Breitenberger, C. A., and Spremulli, L. L.,** *Euglena gracilis* chloroplast ribosomes: improved isolation procedure and comparison of elongation factor specificity with prokaryotic and eukaryotic ribosomes, *Arch. Biochem. Biophys.*, 204, 444, 1980.
43. **Svoboda, A. J. and McConkey, E. H.,** Identification of homologous ribosomal proteins in HeLa cells and in *Tetrahymena pyriformis*, *Eur. J. Biochem.*, 116, 379, 1981.
44. **Rawson. J. R. and Stutz, E. J.,** Characterization of *Euglena* cytoplasmic ribosomes and ribosomal RNA by zone velocity sedimentation in sucrose gradients, *J. Mol. Biol.*, 33, 309, 1968.
45. **Ledoigt, G. and Louvel, C.,** Study of the regulation of plastid development in *Euglena gracillis*. II. Functional localisation and synthesis of ribosomal chloroplast particles, *Biochim. Biophys. Acta*, 563, 432, 1979.
46. **Cruces, J., Sebastian, J., and Renart, J.,** Ribosomal ribonucleic acids from the crustacean *Artemia*, *FEBS Lett.*, 137, 108, 1982.
47. **Brosius, J., Palmer, M. L., Kennedy, P.J., and Noller, H. F.,** Complete nucleotide sequence of a 16 S RNA from *Escherichia coli* ribosome in reconstitution buffer, *J. Biol. Chem.*, 253, 8759, 1978.
48. **Chi, J. C. H. and Suyama, Y.,** Comparative studies on mitochondrial and cytoplasmic ribosomes of *Tetrahymena pyriformis*, *J. Mol. Biol.*, 53, 531, 1970.
49. **Rawson, J. R. and Stutz, E.,** Isolation and characterization of *Euglena gracilis* cytoplasmic and chloroplast ribosomes and their ribosomal components, *Biochim. Biophys. Acta*, 190, 368, 1969.
50. **Stevens, B. J., Curgy, J.-J., Ledoigt, G., and André, J.,** Comparison of *Tetrahymena* and *Euglena* ribosome structure by electron microscopy, in *Genetics and Biogenesis of Chloroplast and Mitochondria*, Bücher, T., Ed., Elsevier, Amsterdam, 1976, 731.
51. **Lava-Sanchez, P.A., Amaldi, F., and La Posta, A.,** Base composition of ribosomal RNA and evolution, *J. Mol. Evol.*, 2, 44, 1972.
52. **Tam, M. F., Dodd, J. A., and Hill, W. E.,** Physical characterization of 16 S rRNA under reconstitution conditions, *J. Biol. Chem.*, 256, 6430, 1981.
53. **Allen, S. H. and Wong, K. P.,** The hydrodynamic and spectroscopic properties of 16 S RNA from *Escherichia coli* ribosome in reconstitution buffer, *J. Biol. Chem.*, 253, 8759, 1978.
54. **Bogdanov, A. A., Zimmerman, R. A., Wang, C., and Ford, N. C.,** Conformational changes in 16 S ribosomal RNA induced by 30 S ribosomal subunit proteins from *Escherichia coli*, *Science*, 202, 999, 1978.
55. **Slegers, H. and Kondo, M.,** Messenger ribonucleoprotein complexes of cryptobiotic embryos of *Artemia salina*, *Nucl. Acids Res.*, 4, 625, 1977.
56. **Noll, H. and Stutz, E.,** The use of sodium and lithium dodecyl sulfate in nucleic acid isolation, *Methods Enzymol.*, 12, 129, 1968.
57. **Donceel, K., Nieuwenhuysen, P., and Clauwaert, J.,** The size and conformation of *Artemia* ribosomal RNA free in solution, *Biochem. J.*, 205, 495, 1982.

Chapter 9

ARTEMIA TUBULIN — A MODEL PROTEIN FOR THE STUDY OF GENE REGULATION AND ORGANELLE FORMATION DURING DEVELOPMENT

Thomas H. MacRae

TABLE OF CONTENTS

I. INTRODUCTION

The brine shrimp, with their unusual capacity among metazoan animals to withstand desiccation and to undergo extensive development in the absence of mitosis, have intrigued researchers for years.[1-3] The study of brine shrimp development is facilitated by a ready supply of animals arrested at the gastrula stage, the so-called cysts, and the relatively synchronous progression of the events which follow reinitiation of development. In spite of the advantages, the impermeability of the cyst shell prevents the monomeric building blocks of biologically important macromolecules from entering the cyst, placing experimental limitations on the system. In vivo labeling studies are difficult to perform, as is ultrastructural analysis of the cells during development.[4-6] This can be especially troublesome if one wishes to integrate biochemical or molecular data, obtained by the examination of components retrieved from cell homogenates, to the formation and/or function of specific cellular organelles during development. To alleviate such problems in the study of *Artemia*, two approaches are possible. First, one can study portions of the animal amenable to structural analysis at certain stages of development and attempt to relate the findings to biochemical or molecular data. Freeman (Chapter 10) is utilizing such an approach in his study of epidermal cells during segmentation of *Artemia*. It seems that Freeman's system will yield to the application of immunofluorescence microscopy and perhaps the use of radioactive isotopes to allow structural/functional integration of segmentation-related events.

A second way to circumvent the problems caused by cyst impermeability, if one is interested in developmental events in *Artemia* at the cellular or organellar level, is to choose for study a relatively simple orangelle which can be isolated and studied in vitro. One choice, as described by Clauwaert and Nieuwenhuysen (Chapter 8) is the ribosome and it has been the subject of extensive study during *Artemia* development.[7-13] Although ultrastructurally rather simple, the ribosome is complex at the molecular level, being composed of diverse proteins and RNA. A simpler organelle, manageable in vitro to the extent that its formation and breakdown under defined conditions are possible, is the microtubule[14,15] and it will be the subject of this chapter.

II. MICROTUBULES

A. Microtubule Proteins: Tubulin and Microtubule-Associated Proteins (MAP)

Microtubules are hollow, cylindrical organelles about 25 nm in diameter. They are constructed from one main structural protein termed tubulin, a heterodimer composed of the monomers α- and β-tubulin. The tubulin monomers each have molecular weights of 50,000 to 55,000 but they can be separated on one-dimensional SDS polyacrylamide gels. Both α- and β-tubulin are resolved into isoforms or isotubulins by isoelectric focusing or two-dimensional gel electrophoresis. The isotubulins may arise as the products of different genes[15-17] or they may result from post-transcriptional mechanisms including differential processing of mRNA or post-translational modifications such as acetylation, phosphorlyation, and tyrosination.[15,16,18-24]

Another group of molecules, MAP, adds a second level of complexity to microtubule structure. The MAP, first isolated from mammalian brain, are divided into high molecular weight MAP and tau proteins. Neural MAP synthesis is developmentally regulated and their distribution is restricted within neural cells.[25,26] MAP and their genes have now been isolated from several non-neural sources but controversy still exists as to whether these cells contain MAP typical of neural cells in addition to MAP specific to the individual cell types.[15,27-37] Equally uncertain is the function of MAP. It was initially thought that MAP are required for tubulin assembly but it is clear that tubulin, even from brain, can assemble in the absence of MAP. The MAP will, however, lower the critical concentration for tubulin assembly, this being the lowest tubulin concentration at which assembly of microtubules will occur. Tubulins from non-neural sources characteristically assemble more readily in the absence of MAP than will neural tubulin, even though their assembly is promoted by MAP from neural or non-neural sources.[38,39]

B. The Manipulation of Tubulin Assembly

Manipulation of microtubule assembly in vitro can be accomplished by adjusting the temperature of tubulin-containing solutions. Mircotubules disassemble into their molecular components when cooled. If guanosine triphosphate (GTP) is then added to 1 to 2 mM, care is taken to chelate free Ca^{2+} by addition of ethyleneglycol-*bis*-(β-aminoethyl ether) *N,N'*-tetraacetic acid (EGTA) and the solution rewarmed, assembly will occur. Assembly/disassembly accompanied by differential centrifugation is often exploited to purify the microtubule proteins, tubulin, and MAP. Further, tubulin-reactive drugs and chemicals which influence microtubule formation are available.[40] The drug, colchicine, inhibits tubulin assembly or leads to disassembly of microtubules. Vinblastine induces formation of tubulin polymers structurally different from microtubules,[41-43] whereas taxol, another assembly-promoting drug, causes formation of tubulin polymers similar, if not identical, to microtubules.[27,38] Drug-dependent disruption of microtubule formation in vivo allows study of tubulin function within the cell. Examination of drug binding to tubulin in vitro, especially when used in concert with tubulin-reactive chemicals[44] and determination of proteolytic enzyme effects on tubulin,[45-47] allows direct analysis of tubulin assembly and a comparison of specific amino acid domains within tubulins from different sources. Thus, from ultrastructural and biochemical perspectives, microtubules are relatively simple organelles, readily subject to experimental manipulation. As such they are, within the limits of the discussion to this point, promising candidates for study of the interplay of molecular and cellular events during *Artemia* development.

C. Tubulin Synthesis and Utilization

1. The Multitubulin Hypothesis

A further indication that microtubules are suitable organelles for the type of study proposed herein for *Artemia* is that tubulin synthesis and utilization are developmentally regulated. There are isotubulins specific to certain stages of development in *Drosophila*[17] and brain,[48] for example, and tubulins which occur predominantly, although not exclusively, in organelles such as flagella.[19,20] The occurrence of tubulin microheterogeneity led Fulton and Simpson[49] to formulate the multitubulin hypothesis, in which they proposed that isotubulins are functionally discrete. That is, if an organism possesses several different isotubulins, then each of the isotubulins will have a different function. The evidence concerning the multitubulin hypotheses is mixed but the proposal may be valid for flagella where a post-translationally modified α-tubulin is incorporated into the axoneme.[19,20] The function of the modified tubulin is unknown. On the other hand, there are situations where the multitubulin hypothesis does not seem appropriate. For example, in *Drosophila*,[17] a β-tubulin found only in testis is engaged in several functions while in *Crithidia fasciculata*,[50] a trypanosome, functionally different microtubules contain identical isotubulins. The myxamoebae stage of *Physarum*, has mitotic spindles, centrioles, and cytoplasmic microtubules but only a single species of α- and β-tubulin.[51] The *Physarum* plasmodium, on the other hand, has several α- and β-tubulins but forms only spindle microtubules. The possibilities are complex, ranging from situations in which an isotubulin is apparently limited to one function, to other situations where several isotubulins are present in a cell possessing only one microtubule-containing structure supposedly involved in only one activity.

2. Tubulin Genes and their Expression

Implied in the previous discussion is the presence of multiple tubulin genes in eukaryotic cells. The tubulin genes of several organisms have been characterized and they exist as relatively small families with members subject to differential regulation and expression.[16,17] The tubulin genes are generally rather widely dispersed among the chromosomes but in *Leishmania enrietti*[52] they occur as separate tandem repeats of α- and β-tubulin genes, whereas in *Trypanosoma brucei*[53] there is a tightly clustered family of alternating α- and β-tubulin genes. Some organisms, such as humans, have several pseudogenes within their tubulin gene families.[54]

An important consideration arising from the cloning and sequencing of the individual members of tubulin gene families is the relationship between amino acid sequence, isotubulin composition within a cell, and the function of the tubulins. This is partly a restatement of the concerns provoked by the multitubulin hypothesis[49] but it is now possible to ask the questions in more detail and to attempt to relate tubulin amino acid sequence to tubulin function. The chicken β-tubulin gene family is the best studied example of this approach to understanding how tubulin functions.[16,55,56] An unexpected observation was that upon comparison of the nucleotide sequences of the seven genes of the chicken β-tubulin family to one another there are regions of sequence divergence and other regions which are highly conserved. When compared to sequences of β-tubulins from other organisms it was found that the intraspecies conservation of the chicken β-tubulin genes was less than the interspecies conservation. That is, each β-tubulin gene in chicken is more similar to a β-tubulin gene from another source than it is to any of the other chicken β-tubulin genes. The suggestion arising from this observation is that the regions of intraspecies divergence, which are conserved at the interspecies level, determine isotype-specific tubulin function while the regions of intraspecies conservation are involved in tubulin functions common to all isotubulins, such as GTP binding. It is also possible that the regions of variability yielding distinct isotubulins result from gene duplication followed by genetic drift. Variability then resides in those regions of the tubulin molecule where its presence is not disruptive, such as in areas not required for specific functions, or to functional regions able to tolerate sequence divergence. A prime example is the carboxyl-terminal of tubulin which apparently depends on the maintenance of an overall negative charge, rather than a specific amino acid sequence per se, for MAP binding.[57]

D. Microtubule Organization

Microtubule function is determined by the spatial disposition of the microtubules within the cells, as well as by their isotubulin composition. The most dramatic change of organization is from the diffuse cytoplasmic network seen in interphase cells, to formation of the mitotic apparatus during mitosis. Equally as interesting are the organizational changes microtubules undergo upon formation of neural axons and the mechanisms whereby tubulin is assembled into the axoneme of the flagellum. Microtubule organization is influenced by microtubule-organizing centers (MTOC), cellular organelles often composed of one or more centrioles surrounded by an electron dense, amorphous material responsible for promotion of microtubule growth and the actual organization of the microtubules.[58,59] There is little information to indicate which of the MTOC molecules is required to organize the microtubules nor is it known how the control of spatial organization is accomplished. The MAP may cross-link microtubules or join microtubules to other cellular structures.[35,36] The isotubulin composition may influence spatial organization and thus exert a direct effect on microtubule function in this manner. Since the chicken erythrocyte contains a prominent β-tubulin not found in other chicken cells, this was proposed to be the case in formation of the marginal band, a bundle of microtubules running just under the plasma membrane of chicken and other nucleated erythrocytes. The proposal has been weakened, however, as brain tubulin will form a marginal band in vitro when mixed with chicken erythrocyte membranes.[61] Although the actual need for the β-tubulin is now in doubt, it may yet be proven to facilitate formation of the marginal band.

It is apparent from the preceding discussion that the properties of tubulin and MAP are such that microtubules are ideal organelles for the study of molecular and cellular integration and this has been in progress in my laboratory.

III. *ARTEMIA* MICROTUBULES

A. Purification of *Artemia* Tubulin and MAP

As there is sufficient tubulin in brain cell extracts to exceed the critical concentration for assembly, it is generally purified by assembly-disassembly or cycling of microtubules. Cycling of neural microtubules yields preparations of tubulin and MAP which can be separated by chromatography on ion-exchange columns. For example, tubulin in a low salt buffer will pass through a phosphocellulose P11 column while MAP are retained, but the MAP can be eluted with buffers containing increased concentrations of salt. Tubulin prepared in this manner is essentially homogeneous, whereas the MAP consist of a complex mixture of high molecular weight MAP and tau proteins. The MAP can be further subdivided or purified by heating and/or molecular sieve chromatography.

Attempts to purify tubulin from sources other than brain by direct assembly/disassembly in cell-free extracts have usually been unsuccessful, and *Artemia* is no exception to the general rule. The most common cause for lack of assembly is a low tubulin concentration in the cell-free supernatant but there are reports of inhibitors of tubulin assembly and of digestion of tubulin by proteolytic enzymes before purification is accomplished.[38,62] Inhibitors of tubulin assembly may, in reality, always be proteolytic enzymes for there are no reports of proteins which bind to tubulin in a stoichiometric relationship and prevent assembly.

To purify *Artemia* tubulin, which entailed circumventing the problem of low tubulin concentration, I have used two approaches. The first was to apply standard protein purification techniques to the preparation of tubulin.[39] Ion-exchange chromatography on phosphocellulose P11 and DEAE-cellulose, followed by $(NH_4)_2SO_4$ fractionation provided a preparation sufficiently enriched in tubulin to allow assembly to occur (Figure 1). The microtubules formed were composed only of tubulin with no associated proteins. The second method used to purify *Artemia* tubulin was to add the antileukemic drug, taxol, to cell-free *Artemia* supernatants (Figures 2 and 3).[38] Taxol is unusual among the tubulin-reactive drugs as it favors microtubule formation by lowering the critical concentration of tubulin. Taxol-induced *Artemia* microtubules, collected by centrifugation through sucrose cushions, are composed of tubulin and other proteins, of which some are MAP (Figure 2; unpublished data). Thus, a major advantage to using taxol as opposed to biochemical procedures is the copurification of MAP with tubulin from non-neural sources and this has been accomplished in experimental systems other than *Artemia*.[15,27,29,34-36] Another advantage related to the rapidity of tubulin preparation when taxol is used, is the lessening of proteolytic enzyme effects. For example, we were able to obtain tubulin from *Artemia* developed for 36 hr only by using taxol.[38] Under other conditions, the proteolytic enzymes formed as the organism develops digest the tubulin, even at 0°C and in the presence of proteolytic inhibitors, preventing large-scale tubulin preparation by chromatographic procedures (Figure 4). One disadvantage of this procedure is that taxol-induced *Artemia* microtubules are extremely stable and can only be depolymerized in the presence of high salt and low temperature. Therefore, it is difficult to use the taxol-purified tubulin for subsequent analysis, other than to determine the isotubulin composition by gel electrophoresis. Furthermore, unless purified homologous tubulin is available as a marker, or the experimenter has tubulin-specific antibodies, the copurifying proteins make it difficult to unambiguously identify the proteins appearing on two-dimensional gels as tubulins.

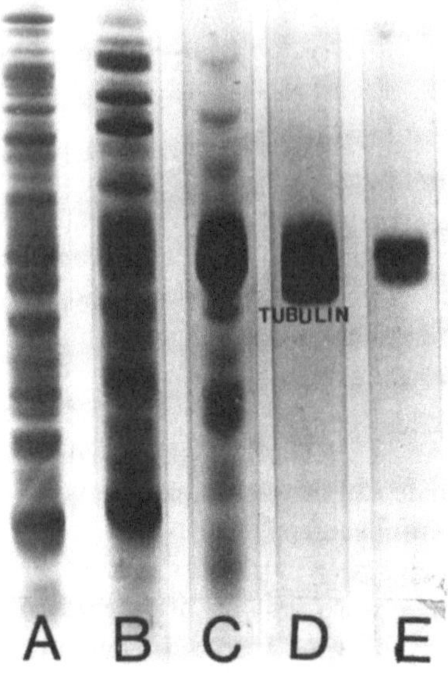

FIGURE 1. Purification of *Artemia* tubulin. Tubulin at various stages of purification was electrophoresed on SDS-polyacrylamide gels. (A) 100 μg of whole cell extract; (B) 100 μg of protein after fractionation on phosphocellulose P11; (C) 105 μg of protein after fractionation on DEAE-cellulose; (D) 80 μg of protein after $(NH_4)_2SO_4$ fractionation and assembly/disassembly; (E) 8 μg of bovine neural tubulin. The fraction shown in D is pure *Artemia* tubulin and does not contain any contaminating proteins visible by Coomassie blue-staining. (Reprinted by permission from MacRae, T. H. and Ludueña, R. F., *Biochem. J.*, 219, 137-148, copyright© 1984, The Biochemical Society, London.)

B. *Artemia* Isotubulins

1. Comparison of Artemia and Bovine Neural Isotubulins

Successful purification of *Artemia* tubulin has allowed extensive analysis of the protein in relation to development, assembly characteristics, and phylogenetic interrelationships. Comparative one-dimensional SDS polyacrylamide gel electrophoresis provided the first clue that *Artemia* tubulin differs from bovine neural tubulin. Specifically, the α- and β-subunits of *Artemia* tubulin were better resolved than the same subunits of bovine tubulin (Figure 5). This difference was reflected in the analysis by isoelectric focusing and two-dimensional gel electrophoresis where it was shown that bovine isotubulins differ in number and charge when compared to *Artemia* tubulin (Figure 6). That *Artemia* tubulin is less negatively charged than bovine neural tubulin has interesting implications for assembly, as will be discussed later.

2. Artemia Isotubulins during Development

It was particularly interesting to determine if the tubulin isoforms changed during differentiation in *Artemia* since tubulin is proposed to function in determination of cell morphology. Furthermore, since microtubules are integral components of the mitotic apparatus, it was possible that the onset of cell division at emergence is accompanied by synthesis of a new tubulin, perhaps specific for the mitotic apparatus. Characterization of *Artemia* tubulin thus provided an opportunity to test the multitubulin hypothesis of Fulton and Simpson.[49]

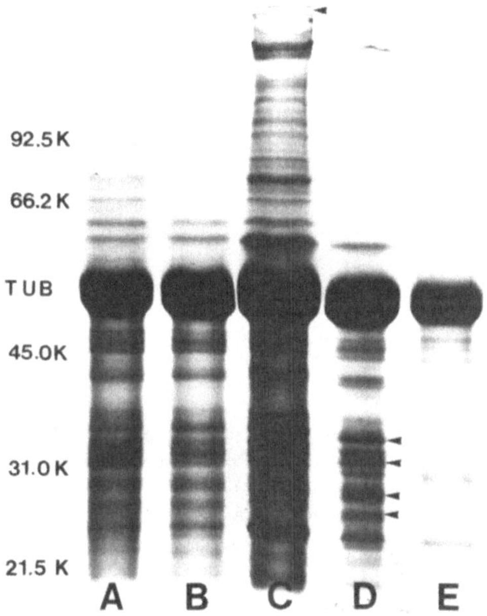

FIGURE 2. Taxol-induced assembly of *Artemia* tubulin. Microtubules formed by addition of taxol to cell-free extracts of *Artemia* were electrophoresed on SDS-polyacrylamide gels. The extracts were prepared from animals after 0 (A), 5 (B), 10 (C), 15 (D), and 24 (E) hr of postgastrula development. The arrowhead for lane C indicates high molecular weight proteins which appear on heavily loaded lanes. The arrowheads for D indicate proteins which disappear or are reduced after 24 hr of development (compare lanes D and E). Markers are expressed as mol wt $\times 10^{-3}$. TUB, tubulin. (Reprinted by permission from Rafiee, P., et al., *Biochem. Cell Biol.*, 64, 238—249, copyright© 1986.)

An extensive analysis of *Artemia* tubulin demonstrated that the major tubulin isoforms, as revealed by two-dimensional gel electrophoresis, did not change during the 36-hr period subsequent to initiation of cyst development at 28°C (Figures 4 and 7). It is possible that a change not detectable by our methods occurred and it could be revealed by use of immunological or other methods. We have shown that neural and *Artemia* tubulin is not antigenically identical but the antibodies needed for comparison of *Artemia* isotubulins are not yet available.[38]

The identical isotubulin composition regardless of the stage of development was unexpected since the members of most iostubulin families change during development. This occurs for tubulin or other small gene families, as there is a tendency for constitutive expression of some genes, and restricted expression of the remainder. Whether the *Artemia* isotubulins are the products of different genes is not known but my initial analysis using cloned *Drosophila* tubulin genes as probes suggests that *Artemia*, unlike most metazoan animals, may have only one, rather than a family of α-tubulin genes. The various α-isotubulins would thus result from post-transcriptional mechanisms. The *Artemia* β-tubulin genes, on the other hand, seem to fit a more established pattern as there are at least two, and maybe more, genes for β-tubulin. Since we have not demonstrated synthesis of the isotubulins at all stages of development, but only their presence, it is possible that *Artemia* tubulin is stable, and the tubulin synthesized at one time persists into later developmental stages. It is also uncertain whether all the available tubulins are utilized at the same time; however it does not seem likely the cells would maintain relatively large supplies of specific

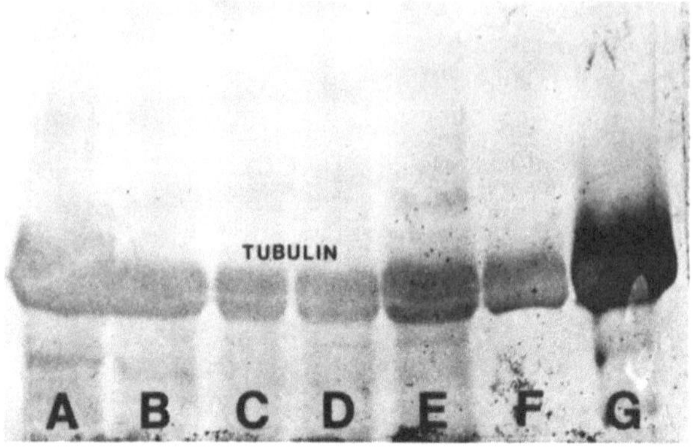

FIGURE 3. Identification of tubulin by immunoperoxidase staining. The microtubules induced to assemble with taxol at various stages of postgastrula development (A — E: 24, 15, 10, 5, and 0 hr) were blotted to nitrocellulose and stained with a polyclonal antibody to tubulin. Lane F contained biochemically purified *Artemia* tubulin while lane G contained purified bovine neural tubulin. Only that portion of the gel with proteins that reacted with the antibody is shown. The results clearly show that the major protein in taxol-induced pellets is tubulin and that *Artemia* and bovine neural tubulin are antigenically similar. (Adapted by permission from Rafiee, P., et al., *Biochem. Cell Biol.*, 64, 238—249, copyright© 1986.)

isotubulins unless there was some benefit. In total, the observations suggest the same pool of tubulins is used for all microtubule-dependent events in *Artemia*, a situation potentially contrary to the multitubulin hypothesis but with parallels in *Crithidia* and *Drosophila*.[17,50,51]

Since it is difficult to detect minor tubulin species synthesized at restricted stages of development by Coomassie blue staining of gels I have begun to approach this problem by characterization of *Artemia* mRNA. The use of cloned *Drosophila* tubulin genes has revealed α- and β-tubulin mRNA on Northern blots of *Artemia* poly(A^+) mRNA. We are purifying tubulin mRNA by hybrid selection so as to avoid problems with the translational inhibitor present in *Artemia* (see Chapter 2). Subsequent in vitro translation and fluorography may reveal products of minor mRNA which are not apparent on Coomassie blue stained gels of extracted tubulin. The potential of the work with *Artemia* mRNA is expanded when considered in concert with analysis of *Artemia* tubulin genes by hybridization of nick translated *Drosophila* tubulin coding sequences to Southern blots of restriction-digested *Artemia* DNA as just described.

C. Assembly of *Artemia* Tubulin

1. The Role of MAP in Artemia Tubulin Assembly

The next indication that the study of *Artemia* tubulin would yield important, basic information came when its assembly was examined. Contrary to neural tubulin, the assembly of *Artemia* tubulin is disrupted by glycerol and Mg^{2+}, with their presence in assembly mixtures leading to formation of morphologically abnormal microtubules (Figure 8). Further, in the absence of MAP, *Artemia* tubulin assembles more readily than does neural tubulin, a characteristic shared by other non-neural and some neural tubulins from fish. The critical concentration for assembly of *Artemia* tubulin in the absence of MAP is about 0.6 to 0.7 mg/mℓ whereas in our hands, purified bovine neural tubulin does not assemble below 2 mg/mℓ (Figures 9 and 10). The assembly of *Artemia* tubulin is stimulated by neural MAP resulting in a threefold reduction in the critical concentration. The neural MAP bind to the *Artemia* microtubules and cosediment through sucrose cushions, showing there is a relatively tight association of the tubulin and MAP (Figure 11).

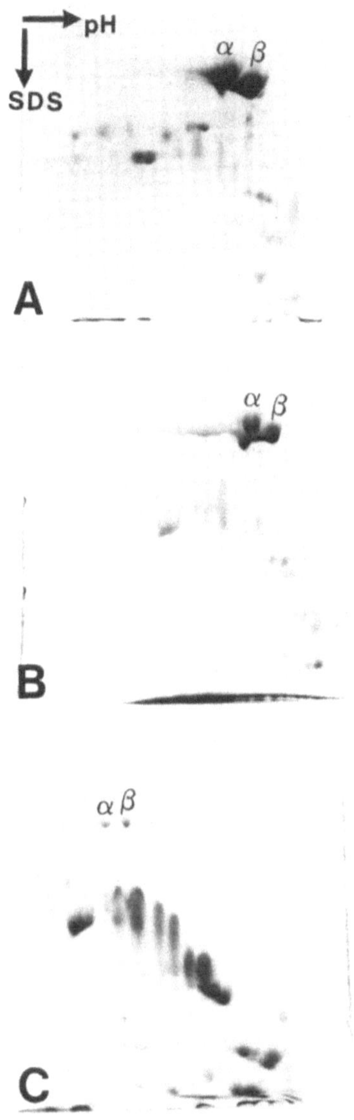

FIGURE 4. The effect of proteolytic enzymes on *Artemia* tubulin prepared from organisms after 36 hr of postgastrula development. Taxol-induced microtubules were collected and electrophoresed on isoelectric focusing/SDS polyacrylamide two-dimensional gels as indicated in panel A. (A) extracts were prepared in the presence of a protease inhibitor cocktail consisting of leupeptin as hemisulfate salt, pepstatin A, phenylmethylsulfonyl fluoride, and trypsin soybean inhibitor (type II-S) all at 10 μg/mℓ; (B) extracts were prepared in the presence of trypsin soybean inhibitor and applied immediately to the gel; or (C) were incubated at 0°C for 3 hr before application to the isoelectric focusing gel. Tubulin from organisms after 24 hr of postgastrula development was prepared at the same time as the tubulin in panel C and it showed no indication of digestion. (Reprinted by permission from Rafiee, P., et al., *Biochem. Cell Biol.*, 64, 238—249, copyright© 1986.)

Since assembly of *Artemia* tubulin is stimulated by neural MAP, I have searched for proteins from *Artemia* with similar activity. Taxol-induced microtubules, formed in cell-free *Artemia* extracts and collected by centrifugation through sucrose cushions, contain several proteins in addition to tubulin (Figure 2). The copurifying proteins possess many of

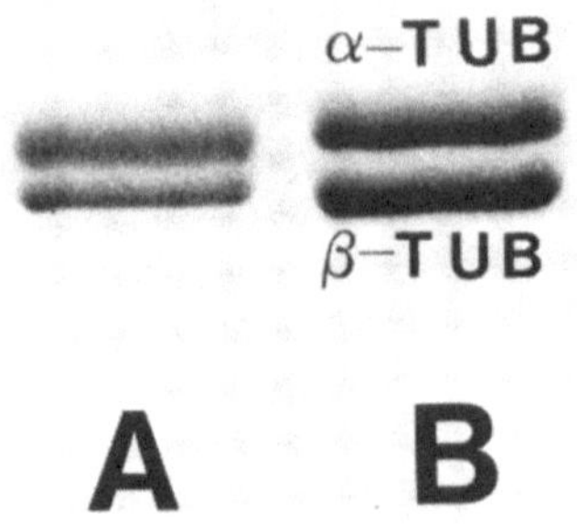

FIGURE 5. Comparison of *Artemia* and bovine neural tubulin. Biochemically purified tubulin from bovine brain (A) or *Artemia* after 15 hr of postgastrula development (B) was compared by electrophoresis on a one-dimensional SDS-polyacrylamide gel. Only that portion of the gel with the tubulin bands is shown. The separation of *Artemia* α- and β-tubulin is greater than for bovine neural tubulin. (Adapted with permission from MacRae, T. H. and Ludueña, R. F., *Biochem. J.*, 219, 137—148, copyright© 1984, The Biochemical Society, London.)

the properties of MAP from other sources, not the least of which is their ability to stimulate assembly of neural and *Artemia* tubulin. The MAP change little during development although there is some tendency for them to be reduced in complexity as the organism develops. Gel electrophoresis of proteins purified from taxol-induced *Artemia* microtubules by salt extraction and subsequent co-assembly with tubulin show that the major MAP in the organism are different in size from any other MAP reported to date. Antibodies raised against neural MAP will be used in Western blot analysis to determine if there are minor MAP in *Artemia* antigenically similar to neural MAP. It will be of interest to learn whether *Artemia* MAP, although very different in molecular weight, share similar antigenic sites with neural MAP.

2. *Effect of Tubulin-Reactive Ligands on Artemia Tubulin*

a. N,N'-ethylene-bis(iodoacetamide)

The assembly characteristics of *Artemia* tubulin, compared to other tubulins, have been examined to ascertain the effects of tubulin-reactive chemicals and drugs on *Artemia* tubulin. (Much of this work was done in collaboration with Dr. Richard Ludueña, Department of Biochemistry, University of Texas Health Science Center, San Antonio.)[44] The bifunctional agent *N,N'*-ethylene-*bis*(iodoacetamide) (EBI) forms two intrachain cross-links termed β^s and β^* with β-tubulin from brain, the products of which can be resolved on SDS-polyacrylamide gels.[63-65] The β^* cross-link in neural tubulin involves at least one sulfhydryl group critical for tubulin assembly, and its formation is inhibited by colchicine, podophyllotoxin, and nocodozole. On the other hand, the formation of β^s involves cross-linking of different sulfhydryl groups and is inhibited completely by GTP and maytansine but only partially by vinblastine. Formation of the β^* and β^s cross-links with *Artemia* tubulin is equally sensitive to drugs and the *Artemia* β^* comigrates with β^* formed with neural tubulin.[44] The electrophoretic mobility of *Artemia* β^s is, however, altered less than is the β^s for brain tubulin. Thus, it seems that the spatial interrelationships of the assembly-critical or β^* sulfhydryls and the drug binding sites are rather well conserved but the β^s sulfhydryls were less well conserved during evolution. The evolutionary implications of the results obtained by studying *Artemia* tubulin were strengthened by a concurrent examination of tubulin from the squid, *Loligo*, where the sites involved in formation of β^* cross-links were better conserved than the sites involved in formation of β^s.[44]

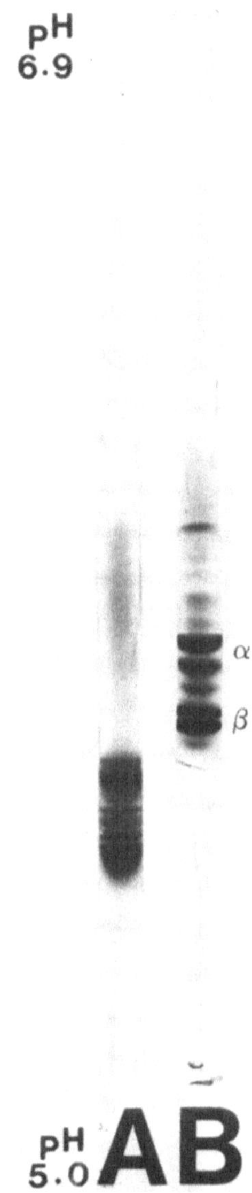

FIGURE 6. Comparison of *Artemia* and bovine neural isotubulins. Biochemically purified tubulin from bovine brain (A) or *Artemia* after 15 hr of postgastrula development (B) was electrophoresed on isoelectric focusing gels. The *Artemia* isotubulins focused in the pH range of 5.6 to 6.0 and were more basic than the bovine neural isotubulins. Upon subsequent analysis, the upper α-tubulin band has been resolved into 2 α-tubulins. (Adapted with permission from MacRae, T. H. and Ludueña, R. F., *Biochem. J.*, 219, 137—148, copyright© 1984, The Biochemical Society.)

When considering tubulin similarities across phylogenetic distances, it is noteworthy that the peptide maps of *Artemia* α- and β-tubulin, as determined by digestion with V-8 proteinase from *Staphylococcus aureus*, are very similar to those of neural tubulin (Figure 12). This was an unexpected result in light of previous peptide mapping of tubulins where substantial differences were found between the peptide maps of tubulins from the cytoplasm (i.e., nonaxoneme) of some deuterostomes and protostomes.[39,66,67] The difference in assembly characteristics and electrophoretic properties found previously also suggested that the primary

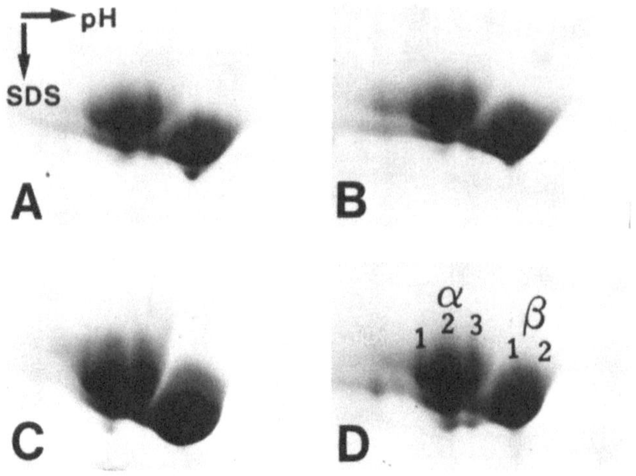

FIGURE 7. Comparison of *Artemia* isotubulins during development. Taxol-induced microtubules prepared from *Artemia* after 0 (A), 15 (B), and 24 hr (C) of postgastrula development were electrophoresed on two-dimensional gels as indicated in panel A. (D) Biochemically purified tubulin from organisms developed for 15 hr. Only the tubulin regions of the gels are shown. Results from organisms developed 5 or 10 hr were identical and are not shown. No change in the isotubulin population was detected by this method during the first 36 hr (see Figure 4) of postgastrula development. (Reprinted by permission from Rafiee, P., et al., *Biochem. Cell Biol.*, 64, 238—249, copyright© 1986.)

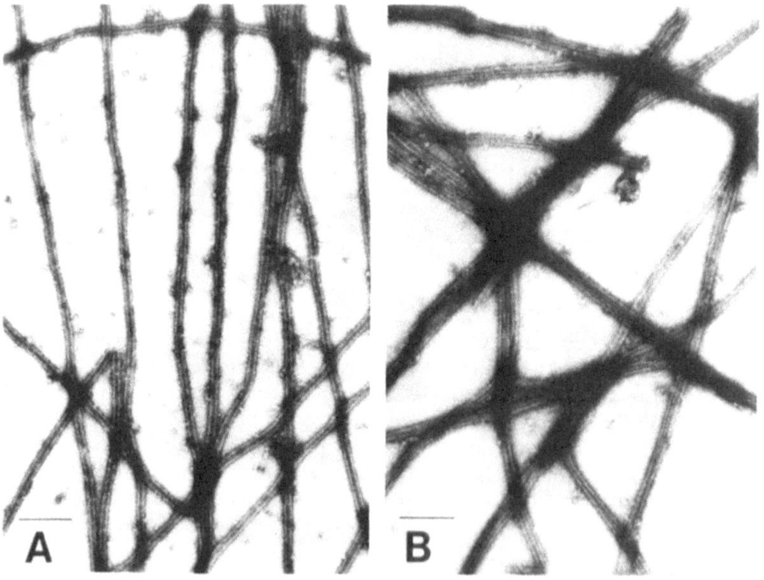

FIGURE 8. Effect of glycerol and Mg^{2+} on assembly of *Artemia* tubulin. Biochemically purified *Artemia* tubulin was assembled in the absence (A) or presence (B) of glycerol (2.5 *M*) and Mg^{2+} (11 m*M*), negatively stained with uranyl acetate, and examined in the electron microscope. The inclusion of glycerol in the assembly mixture induced formation of short, poorly formed microtubules. The bar represents 0.25 μm in both A and B. (Reprinted by permission from MacRae, T. H., *Cell Biol. Int. Rep.*, 8, 423—431, copyright© 1984.)

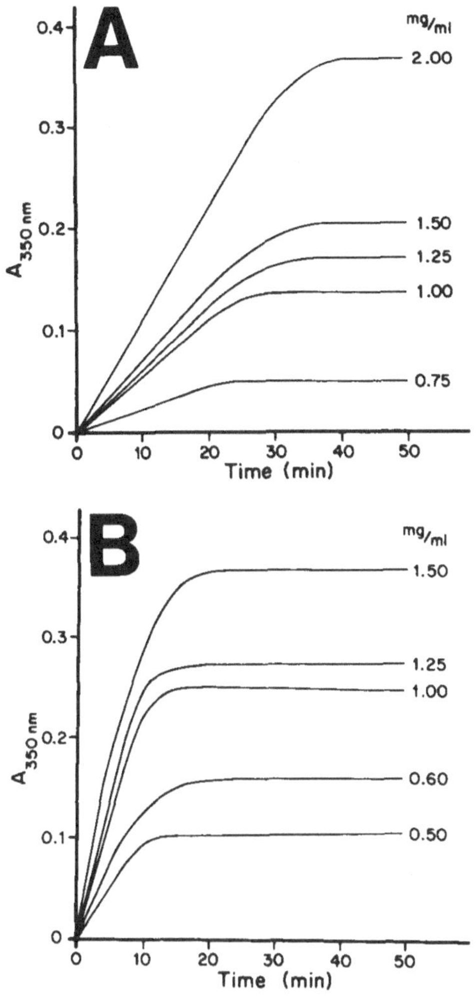

FIGURE 9. The effect of bovine neural MAP on assembly of *Artemia* tubulin. The assembly of biochemically purified *Artemia* tubulin at concentrations as indicated on the graph, in the absence (A) or presence (B) of saturating amounts of bovine neural MAP was monitored turbidimetrically. The same batch of tubulin was used for all assemblies shown here with equivalent assembly stimulation obtained from other tubulin preparations. Comparison of A and B shows that neural MAP stimulate assembly of *Artemia* tubulin. (Reprinted by permission from Rafiee, P., et al., *Biochem. Cell Biol.*, 64, 238—249, copyright© 1986.)

structure and hence the peptide maps of the tubulins from the two sources would differ considerably. From peptide map data, we are currently unable to explain why *Artemia* cytoplasmic tubulin is more similar to neural tubulin than it is to cytoplasmic tubulins from other protostomes. These findings have suggested a re-evaluation of the idea that cytoplasmic tubulins are much freer to change during evolution than are axonemal tubulins, which have constraints imposed on them by their involvement in formation of the highly organized flagellum.[39,66]

b. Vinblastine

Interaction of vinblastine with tubulin has provided interesting results related to assembly characteristics and phylogenetic interrelationships of the tubulins. When microtubules formed

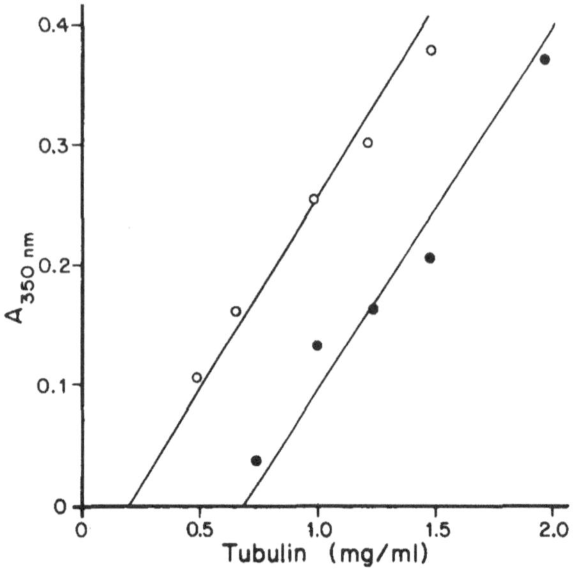

FIGURE 10. The effect of neural MAP on the critical concentration for assembly of *Artemia* tubulin. The curves were produced by plotting the maximum turbidity at each tubulin concentration tested in Figure 9 in the presence (○) or absence (●) of MAP. The critical concentration is the point where the curve intercepts the abscissa and it is reduced about 3-fold by addition of MAP to the assembly reactions. (Reprinted by permission from Rafiee, P., et al., *Biochem. Cell Biol.*, 64, 238—249, copyright© 1986.)

from bovine neural tubulin in the absence of MAP are exposed to vinblastine the microtubules disassemble completely, but when MAP are present, the microtubules are converted into coiled structures probably composed of individual protofilaments.[43] With *Artemia* tubulin, however, coils are formed even in the absence of MAP, although they are less extensive than when MAP are present (Figure 13). In all other aspects examined, the *Artemia* and neural tubulins undergo similar responses to vinblastine.[41,42] The results indicate the vinblastine and MAP-binding sites on the tubulin molecules have been conserved across large phylogenetic distances and the functional interrelationships of one site to another are partially conserved.

Vinblastine is thought to cause neural microtubule disassembly in the absence of MAP by initially disrupting the lateral protofilament-to-protofilament interactions in the microtubules. The longitudinal tubulin-tubulin interactions may be relatively unstable and the protofilaments disassemble due to the negatively charged carboxyl-terminal of the tubulins. The proposed role of the MAP in coil formation is to stabilize the longitudinal tubulin-tubulin interactions and to allow protofilament coiling to occur, a suggestion strengthened by the observation that removal of the carboxyl-terminal from tubulin stimulates vinblastine-induced coil formation in the absence of MAP.[47] Since coils form from *Artemia* tubulin in the absence of MAP, the carboxyl-terminal of *Artemia* tubulin may be less acidic than for neural tubulin and not require shielding by MAP in order to prevent disassembly. That *Artemia* tubulin is less negatively charged than bovine tubulin, as shown by isoelectric focusing, agrees with this hypothesis, but it is not known if the difference in charge is sufficient to cause the difference in tubulin response to vinblastine.

IV. CONCLUSION

In conclusion, the developmental regulation of tubulin gene expression, tubulin assembly,

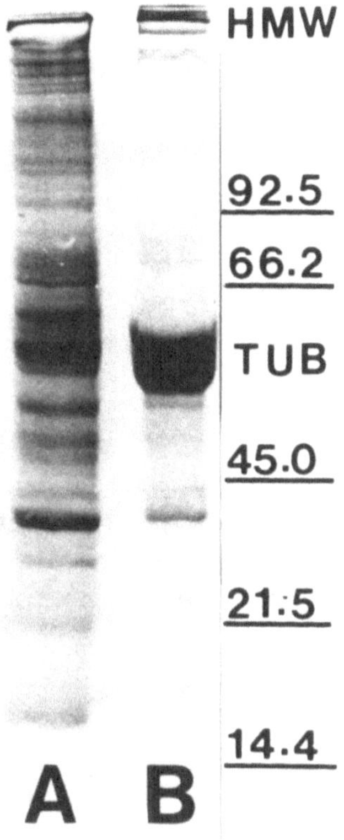

FIGURE 11. Characterization of the neural MAP that bind to *Artemia* tubulin. Biochemically purified *Artemia* tubulin and bovine neural MAP were mixed and assembly was allowed to occur. The microtubules were collected by centrifugation through sucrose cushions and analyzed on SDS-polyacrylamide gels to determine if tubulin and MAP coassembled. (A) 66 μg of bovine MAP;(B) polypeptide composition of microtubule pellets. The high molecular weight (HMW) MAP were especially prominent but the tau proteins in the molecular weight range of 60,000 to 70,000 were just barely visible. Markers are expressed as mol wt × 10^{-3}. TUB, tubulin. (Reprinted by permission from Rafiee, P., et al., *Biochem. Cell Biol.*, 64, 238—249, copyright© 1986.)

and MAP function in *Artemia* are under study in my laboratory. The most recent work involves the application of recombinant DNA methodology to these problems. Use of this approach has resulted in the cloning of two DNA sequences from *Artemia* which cross-hybridize to cloned *Drosophila* tubulin genes and has provided a preliminary estimation of the number of tubulin genes and mRNAs in *Artemia* (work in progress). Sequencing of the cloned *Artemia* genes will extend our phylogenetic comparisons and perhaps indicate modifications of tubulin amino acid sequences which lead to efficient assembly of tubulin in the absence of MAP. The potential clearly exists, through study of *Artemia* tubulin, to answer several very basic questions regarding synthesis and utilization of tubulin in eukaryotic cells, and to integrate molecular/cellular approaches in the study of *Artemia* development through analysis of the microtubule, a relatively simple cellular organelle.

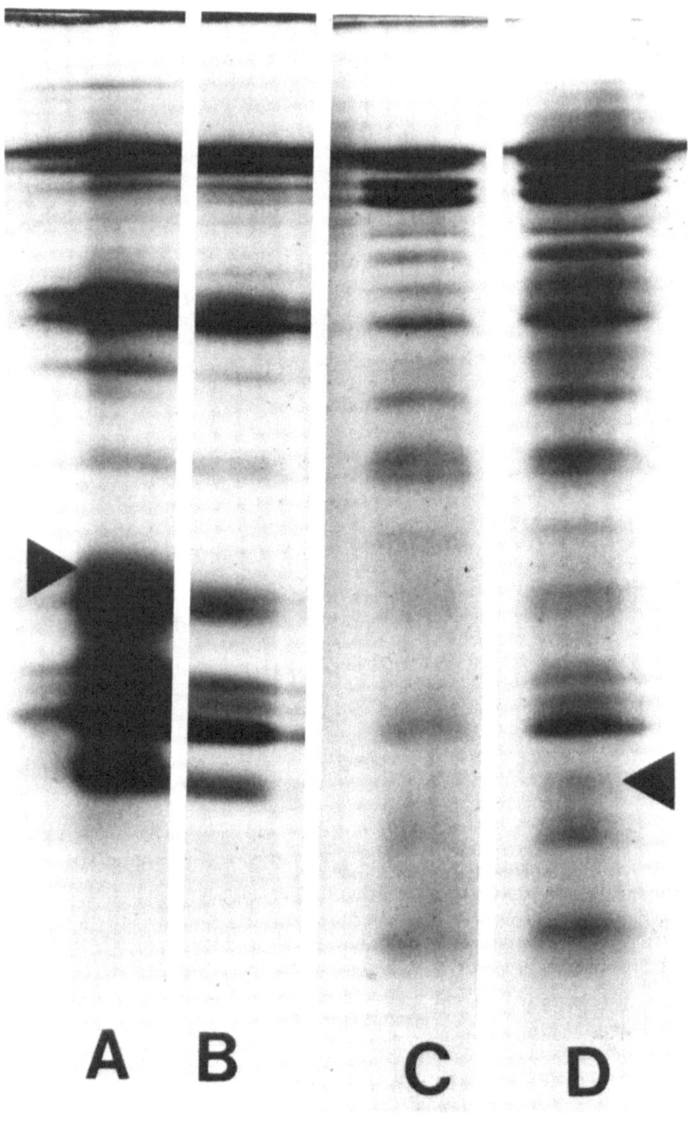

FIGURE 12. Comparison of the peptide maps of *Artemia* and bovine neural tubulin. The α- and β-tubulin subunits, separated by reduction and carboxymethylation followed by electrophoresis on 6% polyacrylamide gels, were digested with *Staphylococcus aureus* V-8 proteinase. A and B are the peptide maps of *Artemia* and bovine α-tubulin respectively whereas C and D are the peptide maps of bovine and *Artemia* β-tubulin respectively. Peptides found in the *Artemia* but not the bovine tubulin maps are indicated by arrowheads. The other differences in the β-tubulin maps are probably dependent on slightly different extents of tubulin digestion. (Adapted with permission from MacRae, T. H. and Ludueña, R. F., *Biochem. J.*, 219, 137—148, copyright© 1984, The Biochemical Society.)

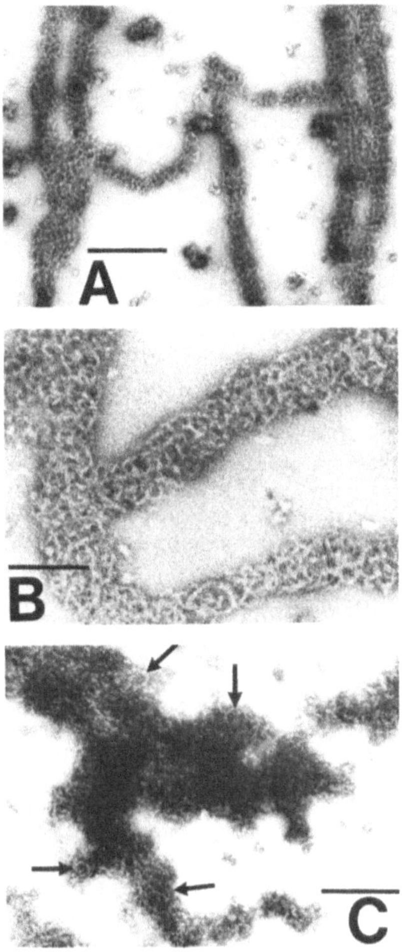

FIGURE 13. Effect of vinblastine and MAP on microtubule structure. Microtubules formed from *Artemia* tubulin in the presence (A and B) or absence (C) of bovine neural MAP were incubated with 40 μm vinblastine for 30 min, fixed with glutaraldehyde, negatively stained with uranyl acetate, and examined in the electron microscope. Extensive coils formed when MAP were present whereas aggregates with filamentous or coiled substructure (arrows in c) were present when MAP were omitted. Similar structures formed when assembly occurred in the presence of vinblastine. The bars represent 0.40 μm in A and 0.20 μm in B and C. (Adapted with permission from MacKinlay, S. A., et al., *Biochim. Biophys. Acta*, 882, 419—426, copyright© 1986.)

REFERENCES

1. **Persoone, G., Sorgeloos, P., Roels, O., and Jasper, E.,** *The Brine Shrimp Artemia,* Universa Press, Wetteren, Belgium, 1980.
2. **Hentschel, C. C. and Tata, J. R.,** The molecular embryology of the brine shrimp, *Trends Biochem. Sci.,* 1, 97, 1976.
3. **Olson, C. S. and Clegg, J. S.,** Cell division during the development of *Artemia salina, Wilhelm Roux's Arch. Dev. Biol.,* 184, 1, 1978.

4. **Miller, D. and McLennan, A. G.,** Changes in intracellular levels of Ap_3A and Ap_4A in cysts and larvae of *Artemia* do not correlate with changes in protein synthesis after heat-shock, *Nucleic Acids Res.*, 14, 6031, 1986.
5. **Vallejo, C. G., Perona, R., Garesse, R., and Marco, R.,** The stability of the yolk granules of *Artemia*. An improved method for their isolation and study, *Cell Differentiat.*, 10, 343, 1981.
6. **Morris, J. E.,** Dehydrated cysts of *Artemia salina* prepared for electron microscopy by totally anhydrous techniques, *J. Ultrastruct. Res.*, 25, 64, 1968.
7. **Thoen, C., deHerdt, E., and Slegers, H.,** Identification of the ribosomal proteins phosphorylated by the ribosome-associated casein kinase type II from cryptobiotic gastrulae of the brine shrimp *Artemia Sp.*, *Biochem. Biophys. Res. Commun.*, 135, 347, 1986.
8. **Goss, D. J. and Harrigan, T.,** Magnesium ion dependent equilibria kinetics, and thermodynamic parameters of *Artemia* ribosome dissociation and subunit association, *Biochemistry*, 25, 3690, 1986.
9. **Maassen, J. A., Schop, E. N., Brands, J. H. G. M., Van Hernent, F. J., Lenstra, J. A., and Moller, W.,** Molecular cloning and analysis of cDNA sequences for two ribosomal proteins from *Artemia*. The coordinate expression of genes for ribosomal proteins and elongation factor 1 during embryogenesis of *Artemia*, *Eur. J. Biochem.*, 149, 609, 1985.
10. **Ciesiolka, J., Nurse, K., Klien, J., and Ofengand, J.,** Conservation of RNA sequence and cross-linking of the anticodon of P site bound tRNA to the penultimate cytidine of the UACACACG sequence in *Artemia salina* 18S rRNA, *Biochemistry*, 24, 3233, 1985.
11. **Vaughn, J. C., Sperbeck, S. J., Ramsey, W. J., and Lawrence, C. B.,** A universal model for the secondary structure of 5.8S ribosomal RNA molecules, their contact sites with 28S ribosomal RNAs, and their prokaryotic equivalent, *Nucleic Acids Res.*, 12, 7479, 1984.
12. **Donceel, K., Nieuwenhuysen, P., and Clauwaert, J.,** The size and conformation of *Artemia* (brine-shrimp) ribosomal RNA free in solution, *Biochem. J.*, 205, 495, 1982.
13. **Nieuwenhuysen, P. and Clauwaert, J.,** Physicochemical characterization of ribosomal particles from the eukaryote *Artemia*, *J. Biol. Chem.*, 256, 9626, 1981.
14. **Roberts, K. and Hyams, J. S.,** *Microtubles*, Academic Press, New York, 1979.
15. **MacRae, T. H.,** Nonneural microtubule proteins: structure and function, *BioEssays*, 6, 128, 1987.
16. **Cleveland, D. W. and Sullivan, K. F.,** Molecular biology and genetics of tubulin, *Annu. Rev. Biochem.*, 54, 331, 1985.
17. **Raff, E. C.,** Genetics of microtubule systems, *J. Cell Biol.*, 99, 1, 1984.
18. **Maruta, H., Greer, K., and Rosenbaum, J. L.,** The acetylation of alpha-tubulin and its relationship to the assembly and disassembly of microtubules, *J. Cell Biol.*, 103, 571, 1986.
19. **LeDizet, M. and Piperno, G.,** Cytoplasmic microtubules containing acetylated α-tubulin in *Chlamydomonas reinhardtii*: spatial arrangement and properties, *J. Cell Biol.*, 103, 13, 1986.
20. **Piperno, G. and Fuller, M. T.,** Monoclonal antibodies specific for an acetylated form of α-tubulin recognize the antigen in cilia and flagella from a variety of organisms, *J. Cell. Biol.*, 101, 2085, 1985.
21. **Wandosell, F., Serrano, L., Hernandez, M. A., and Avila, J.,** Phosphorylation of tubulin by a calmodulin-dependent protein kinase, *J. Biol. Chem.*, 261, 10332, 1985.
22. **Gundersen, G. G. and Bulinski, J. C.,** Distribution of tyrosinated and nontyrosinated α-tubulin during mitosis, *J. Cell Biol.*, 102, 1118, 1986.
23. **Gundersen, G. G., Kalnoski, M. H., and Bulinski, J. C.,** Distinct populations of microtubules: tyrosinated and nontyrosinated alpha tubulin are distributed differently in vivo, *Cell*, 38, 779, 1984.
24. **Kobayashi, T. and Flavin, M.,** Tubulin tyrosylation in invertebrates, *Comp. Biochem. Physiol.*, 69B, 387, 1981.
25. **Vallee, R. B., Bloom, G. S., and Luca, F. C.,** Differential cellular and subcellular distribution of microtubule-associated proteins, in *Molecular Biology of the Cytoskeleton*, Borisy, D. B., Cleveland, D. W., and Murphy, D. B., Eds., Cold Spring Harbor Laboratory, Cold Spring Harbor, N. Y., 1984, 111.
26. **Riederer, B. and Matus, A.,** Differential expression of distinct microtubule-associated proteins during brain development, *Proc. Natl. Acad. Sci. U.S.A.*, 82, 6006, 1985.
27. **Vallee, R. B.,** A taxol-dependent procedure for the isolation of microtubules and microtubule-associated proteins (MAPs), *J. Cell Biol.*, 92, 435, 1982.
28. **Bloom, G. S., Luca, F. C., Collins, C. A., and Vallee, R. B.,** Use of multiple monoclonal antibodies to characterize the major microtubule-associated protein in sea urchin eggs, *Cell Motil.*, 5, 431, 1985.
29. **Kotani, S., Murofushi, H., Maekawa, S., Sato, C., and Sakai, H.,** Characterization of microtubule-associated proteins isolated from bovine adrenal gland, *Eur. J. Biochem.*, 156, 23, 1986.
30. **deGarcini, E. M., Diez, J. C., and Avila, J.,** Quantitation and characterization of tau factor in porcine tissues, *Biochim. Biophys. Acta*, 881, 456, 1986.
31. **Black, M. M., Aletta, J. M., and Greene, L. A.,** Regulation of microtubule composition and stability during nerve growth factor-promoted neurite outgrowth, *J. Cell. Biol.*, 103, 545, 1986.
32. **Lewis, S. A., Villasante, A., Sherline, P., and Cowan, N. J.,** Brain-specific expression of MAP-2 detected using a cloned cDNA probe, *J. Cell Biol.*, 102, 2098, 1986.

33. **Lewis, S. A., Sherline, P., and Cowan, N. J.**, A cloned cDNA encoding MAP1 detects a single copy gene in mouse and a brain-abundant RNA whose level decreases during development, *J. Cell Biol.*, 102, 2106, 1986.
34. **Hirokawa, N.**, 270K microtubule-associated protein cross-reacting with anti-MAP2 IgG in the crayfish peripheral nerve axon, *J. Cell Biol.*, 103, 33, 1986.
35. **Goldstein, L. S. B., Laymon, R. A., and McIntosh, J. R.**, A microtubule-associated protein in *Drosophila melanogaster:* identification, characterization, and isolation of coding sequences, *J. Cell Biol.*, 102, 2076, 1986.
36. **Aamodt, E. J. and Culotti, J. G.**, Microtubules and microtubule-associated proteins from the nematode *Caenorhabditis elegans:* periodic cross-links connect microtubules in vitro, *J. Cell Biol.*, 103, 23, 1986.
37. **Pillus, L. and Solomon, F.**, Components of microtubular structures in *Saccharomyces cerevisiae, Proc. Natl. Acad. Sci. U. S. A.*, 83, 2468, 1986.
38. **Rafiee, P., MacKinlay, S. A., and MacRae, T. H.**, Taxol-induced assembly and charcterization of microtubule proteins from developing brine shrimp *(Artemia), Biochem. Cell Biol.*, 64, 238, 1986.
39. **MacRae, T. H. and Ludueña, R. F.**, Developmental and comparative aspects of brine shrimp tubulin, *Biochem. J.*, 219, 137, 1984.
40. **Dustin, P.**, *Microtubules,* 2nd ed., Springer-Verlag, Berlin, 1984.
41. **MacKinlay, S. A., Ludueña, R. F., and MacRae, T. H.**, Vinblastine-induced aggregation of brine shrimp *(Artemia)* tubulin, *Biochim. Biophys. Acta,* 882, 419, 1986.
42. **MacKinlay, S. A., Ludueña, R. F., and MacRae, T. H.**, Temperature effects on solutions of vinblastine-induced polymers assembled from brine shrimp *(Artemia)* tubulin, *FEBS Lett.*, 203, 301, 1986.
43. **Ludueña, R. F., Fellous, A., McManus, L., Jordon, M. A., and Nunez, J.**, Contrasting roles of tau and microtubule-associated protein 2 in the vinblastine-induced aggregation of brain tubulin, *J. Biol. Chem.*, 259, 12890, 1984.
44. **Ludueña, R. F., Roach, M. C., MacRae, T. H., and Langford, G. M.**, N, N′-Ethylene-bis(iodoacetamide) as a probe for structural and functional characteristics of brine shrimp, squid, and bovine tubulins, *Can. J. Biochem. Cell Biol.*, 63, 439, 1985.
45. **Sackett, D. L. and Wolff, J.**, Proteolysis of tubulin and the substructure of the tubulin dimer, *J. Biol. Chem.*, 261, 9070, 1986.
46. **Maccioni, R. B., Serrano, L., Avila, J., and Cann, J. R.**, Characterization and structural aspects of the enhanced assembly of tubulin after removal of its carboxyl-terminal domain, *Eur. J. Biochem.*, 156, 375, 1986.
47. **Serrano, L., delaTorre, J., Ludueña, R. F., and Avila, J.**, The removal of the carboxy-terminal region of tubulin favors its vinblastine-induced aggregation into spiral-like structures, *Arch. Biochem. Biophys.*, 249, 611, 1986.
48. **Ginzburg, I., Teichman, A., Dodemont, H. J., Behar, L., and Littauer, U. Z.**, Regulation of three β-tubulin mRNAs during rat brain development, *EMBO J.*, 4, 3667, 1985.
49. **Fulton, C. and Simpson, P. A.**, Selective synthesis and utilization of flagella tubulin: the multi-tubulin hypothesis, in *Cell Motility,* Goldman, R., Pollard, T., and Rosenbaum, J., Eds., Cold Spring Harbor Laboratory, Cold Spring Harbor, N. Y., 1976, 987.
50. **Russell, D. G., Miller, D., and Gull, K.**, Tubulin heterogeneity in the trypanosome *Crithidia fasciculta, Mol. Cell. Biol.*, 4, 779, 1984.
51. **Birkett, C. R., Foster, K. E., Johnson, L., and Gull, K.**, Use of monoclonal antibodies to analyse the expression of a multi-tubulin family, *FEBS Lett.*, 187, 211, 1985.
52. **Landfear, S. M., McMahon-Pratt, D., and Wirth, D. F.**, Tandem arrangement of tubulin genes in the protozoan parasite *Leishmania enriettii, Mol. Cell. Biol.*, 3, 1070, 1983.
53. **Seebeck, T., Whitaker, P. A., Imboden, M. A., Hardman, N., and Braun, R.**, Tubulin genes of *Trypanosoma brucei:* a tightly clustered family of alternating genes, *Proc. Natl. Acad. Sci. U.S.A.*, 80, 4634, 1983.
54. **Cowan, N. J.**, Tubulin genes and the diversity of microtubule function, in *Oxford Surveys of Eukaryotic Cells,* Vol. 1, MacLean, N., Ed. Oxford University Press, London, 1984, chap. 2.
55. **Sullivan, K. F., Machlin, P. S., Ratrie, H., III, and Cleveland, D. W.**, Sequence and expression of the chicken β-3 tubulin gene. A vertebrate testis β-tubulin isotype, *J. Biol. Chem.*, 261, 13317, 1986.
56. **Sullivan, K. F. and Cleveland, D. W.**, Identification of conserved isotype-defining variable region sequences for four vertebrate β-tubulin polypeptide classes, *Proc. Natl. Acad. Sci. U.S.A.*, 83, 4327, 1986.
57. **Maccioni, R. B., Serrano, L., and Avila, J.**, Structural and functional domains of tubulin, *BioEssays,* 2, 105, 1985.
58. **Brinkley, B. R.**, Microtubule organizing centers, *Annu. Rev. Cell Biol.*, 1, 145, 1985.
59. **Brown, D. L., Stearns, M. E., and MacRae, T. H.**, Microtubule organizing centers, in *The Cytoskeleton in Plant Growth and Development,* Lloyd, C. W., Ed., Academic Press, New York, 1982, 55.
60. **Murphy, D. B., and Wallis, K. T.**, Erythrocyte microtubule assembly in vitro. Determination of the effects of erythrocyte tau, tubulin isoforms, and tubulin oligomers on erythrocyte tubulin assembly, and comparison with brain microtubule assembly, *J. Biol. Chem.*, 260, 12293, 1985.

61. **Swan, J. A. and Solomon, F.,** Reformation of the marginal band of avian erythrocytes in vitro using calf brain tubulin: peripheral determinants of microtubule form, *J. Cell Biol.*, 99, 2108, 1984.
62. **MacRae, T. H.,** Bundling of bovine and brine shrimp *(Artemia)* microtubules *in vitro, Cell Biol. Int. Rep.*, 8, 423, 1984.
63. **Ludueña, R. F. and Roach, M. C.,** Interaction of tubulin with drugs and alkylating agents. I. Alkylation of tubulin by iodo [^{14}C] acetamide and N, N'-ethylenebis (iodoacetamide), *Biochemistry*, 20, 4437, 1981.
64. **Ludueña, R. F. and Roach, M. C.,** Interaction of tubulin with drugs and alkylating agents. II. Effects of colchicine, podophyllotoxin, and vinblastine on the alkylation of tubulin, *Biochemistry*, 20, 4444, 1981.
65. **Roach, M. C., Bane, S., and Ludueña, R. F.,** The effects of colchicine analogues on the reaction of tubulin with iodo [^{14}C] acetamide and N, N'-ethylenebis(iodoacetamide), *J. Biol. Chem.*, 260, 3015, 1985.
66. **Little, M., Ludueña, R. F., Keenan, R., and Asnes, C. F.,** Tubulin evolution: two major types of α-tubulin, *J. Mol. Evol.*, 19, 80, 1982.
67. **Little, M., Ludueña, R. F., Langford, G. M., Asnes, C. F., and Farrell, K.,** Comparison of proteolytic cleavage patterns of α-tubulins and β-tubulins from taxonomically distant species, *J. Mol. Biol*, 149, 95, 1981.

Chapter 10

THE INTEGUMENT OF *ARTEMIA* DURING EARLY DEVELOPMENT

John A. Freeman

TABLE OF CONTENTS

I. INTRODUCTION

The naupliar and metanaupliar larvae of the brine shrimp, *Artemia* sp., have been employed as experimental animals in a diverse group of disciplines within the biomedical sciences. Several recent works have reviewed the findings on development in the brine shrimp.[1-3] It is clear from these studies that brine shrimp are favorable organisms for studies in developmental biology because they encyst at an early point in development, undergo synchronized development upon rehydration and hatching, and large numbers of animals can be obtained with little expense.

The majority of studies have dealt with the cellular and molecular biology of the early development where the biochemical changes during rehydration, emergence, and larval morphogenesis have been examined.[1-3] Only a few studies have considered the morphology of the developmental changes in relation to the biochemical events. Yet, considerable morphogenesis occurs in the tissues during the early larval stages.

One tissue in which morphogenesis is particularly evident is the epidermis. This tissue covers the entire exterior of the shrimp and secretes the exoskeleton. Thus, the shape of the larva and the appendages are a result of the changing form of the epidermis. Moreover, the development of the thoracic and abdominal segments, which continues during the rest of the larval period, is a result, in part, of the metameric activation of the differentiation process in the epidermal cells. In this chapter, the many functions of the epidermal cells during early development will be reviewed and examined with the aim of comparing the morphological changes with the molecular activities which have been reported for the naupliar tissues.

The region of the integument that undergoes the greatest, and, thus, the most obvious development is the thorax. The study of the thorax during the emergence and early larval periods permits the examination of the interaction of the processes of growth, cell differentiation, and pattern formation in the morphogenesis of a segment. The segments form in an anterior-posterior gradient such that the metameric body plan can be followed at several levels of maturity with the most complete segment located at the anterior end of the thorax.[4-10] In this chapter, the segment will be discussed as a developing "unit" of the body in which growth and cell differentiation in the epidermis interact to form the mature segment. Although some of the epidermal cells develop into morphologically distinct structures, all the cells will be considered as undergoing differentiation since they will attain the final, differentiated state, in which they form an integumental structure of specific shape and with specific cuticular elements.

II. MORPHOLOGY OF THE EXOSKELETON

A. Embryonic Membranes

From the early blastula stage to the point of hatching, the embryo becomes surrounded by four membranes: the cyst shell (tertiary membrane), the embryonic cuticle (membrane), the hatching membrane, and the larval (naupliar) cuticle of the first naupliar instar.[1] The encysted embryo is in the post gastrula state (GII of Benesch[8]) at the time of release from the cyst.

The outermost layer is the cyst shell. This rigid structure is only found on embryos that are ovoviviparous, and it is not secreted around embryos that develop *in utero*.[10,11] Anderson et al.[10] demonstrated that the components of the tertiary membrane are secreted by the shell glands and come to surround the blastula while it is in the uterus. The superstructure of the tertiary membrane appears to develop from the inside out after contact with the embryo. The control of cyst shell formation is unknown but it may be autocatalytic. Ultrastructural observations have shown that the tertiary membrane consists of an outer thin membrane

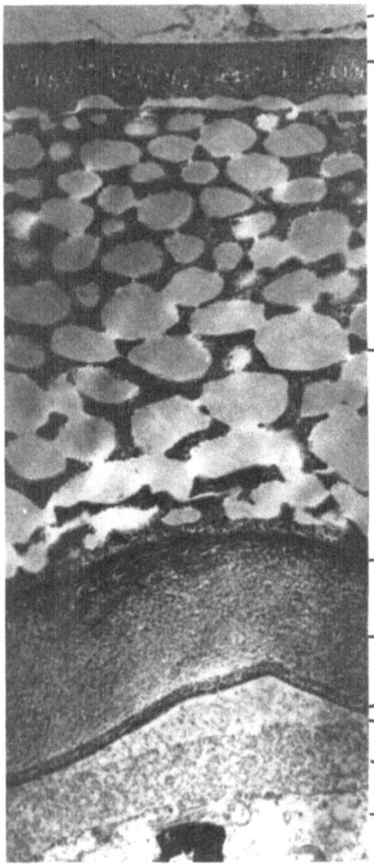

FIGURE 1. Electron micrograph of a developing brine shrimp cyst 14 hr following rehydration. Embryonic (larval) ectodermal tissues are at bottom (magnification × 6680). (From Trotman, C. N. A., et al., *Dev. Biol.*, 80, 167, 1980. With permission.)

overlaying a 1-μm thick cortical layer (Figure 1). Beneath this structure is a thick (6 μm) alveolar layer with numerous interconnected spaces. The function of the tertiary membrane is thought to be more physical than chemical since many compounds can pass through this layer.[11,12] It has also been suggested that certain compounds in this layer may act as chemical signals to direct the embryonic cells towards the developmental pathway of encystment and anhydrobiosis.[10,11]

Beneath the cyst shell is the embryonic cuticle. This layer is not as thick as the tertiary membrane but is more complex in structure. The origin of this cuticle is not known. De Chaffoy et al.[13] have suggested that the molecules for this cuticle may be provided by the female. Although this possibility has not been disproven, it is difficult to see how the secretion of the components of this layer and the control of its organization could be accomplished by the uterine epithelium alone through the tertiary membrane. Several lines of evidence support the contention that it is more likely the epidermal cells secrete the embryonic cuticle. First, it is secreted after the initial phase of gastrulation has occurred.[11,13] By this time the epidermal cells are situated on the outside of the embryo and would be capable of secreting the layers.[8] Second, the morphology of the layer resembles that of a crustacean cuticle and, in some micrographs, a lamellar orientation of the materials in the fibrous layer have been observed.[11] The uterine epithelium, however, may provide the proper physiological milieu for the construction of the layers.[13] There are three layers to the embryonic cuticle. An outer cuticular membrane 500 nm thick has a tripartite structure and covers a fibrous layer that is 1.6 μm thick and composed of chitin and protein (Figure 1). The inner cuticular membrane is 400 nm thick and shows little structural definition. The outer cuticular layer acts as a barrier to the surrounding fluids since, after its completion, the permeability of the embryo to salts is reduced.[13] In embryos that have been removed from the uterus between the early and late phases of gastrulation (GI-GII of Benesch[8]), the formation of the outer membrane is not completed and these larvae remain permeable to most precursor molecules throughout the embryonic and hatching periods.[13] Many of these larvae, however, appear to be abnormal at hatching and are unable to develop past the first nauplius stage. Trotman et al.[12] have suggested that the inner membrane acts as a semi-permeable barrier which maintains an osmotic gradient that would be necessary for uptake of water during emergence and hatching. Clearly, further research on these interesting and complex membranes must be done before the role of each in development can be understood.

If the embryonic cuticle is, in fact, secreted by the embryonic epidermal cells, then it is clear that these cells have attained the ability to secrete a highly structured exoskeleton. The release of the embryonic cuticle from the epidermal cells and the sequential production of the hatching membrane and the naupliar cuticle by these cells may be considered as an embryonic molt and, thus, the initial round of the cyclical production of the exoskeleton. Embryonic molts have been observed in both crustaceans and insects,[14,15] and in the former, may lead to the prezoea stage.[16] It would be of interest to find out if this molt is controlled by the hormones that regulate molting in the larvae and adults of crustaceans. If these hormones are found in the embryos it is possible that they are not produced by a larval molting gland (which may not yet be differentiated, see below) but, instead, may be released from stored sources, such as the yolk platelets.

The structure of the embryonic cuticle is important for controlling the milieu in which the postgastrular events occur. Although De Chaffoy et al.[13] found that only larvae maintained *ex utero* were abnormal in appearance, under certain conditions, or with cysts from some sources, abnormal nauplii which resemble those found by De Chaffoy[13] have been obtained.[89] The reason for this is not known, but it may be a result of improper membrane formation during embryogenesis.

During the emergence phase the embryonic cuticle separates from the hatching membrane such that the larva remains surrounded by the hatching membrane even as it leaves the cyst.[11-13] The mechanism that permits this separation is unknown but is of interest since, in electron micrographs, the two structures are tightly apposed. One unsolved question is the origin of the hatching membrane. This layer is produced between 14 and 19 hr after rehydration and is present in emerging embryos.[11] Morris and Afzelius[11] have considered this layer to be of embryonic origin and secreted before the larval cuticle is laid down. If so, then this layer may be the unassigned layer of Trotman et al.[12] and must be secreted in a very short period of time. The importance of the hatching membrane in the development of the nauplius has not been studied. Possible functions include: (1) to permit the thorax to reach its proper shape which cannot be done in the spherical cyst, (2) to support the larval cuticle while permitting the uptake of water, and (3) to allow the initial antennal movements to occur in a controlled environment. The completion of emergence is seen as the larva ruptures through the hatching membrane as a stage I nauplius larva.[17] A hatching enzyme has been reported and it may be the first chitino-proteolytic enzyme produced by the epidermal cells.[1]

B. Naupliar Cuticle

The developing larval cuticle can be observed during rehydration in the cyst.[11,12] It is thinner than either the cyst shell or the embryonic membrane and appears to be amorphous or slightly fibrous in structure. This form is altered during emergence since the cuticle at hatching has achieved a more defined structure. The larval cuticle is thin and divided into an outer epicuticle, which is probably tanned, and an inner procuticle that is 0.3 to 1.0 μm thick. Although these values were determined from micrographs of specialized regions of the cuticle,[18-20] observations in this laboratory have shown that the cuticle has a similar appearance in all regions of the integument (Figure 2).

The procuticle appears to be fibrous, but not lamellar, in structure. There is no division of the procuticle into an exo- and endocuticle and no membranous layer is present. Adult cuticles, however, have been shown to have a lamellar procuticle.[21] Thus, the cuticle-synthesizing mechanism appears to change and develop further during the larval period. The crustacean cuticle is normally divided into four sections, each of which is secreted during a specific phase of the molt cycle.[22,23] The lack of identifiable sections of the procuticle in *Artemia* nauplii suggests that there is less molt cycle stage dependency in the secretory process and that, perhaps, structural differentiation of discrete layers is not necessary for the skeletal function of the cuticle. These characteristics make the *Artemia* nauplii cuticle unusual, since other anostracans and small planktonic crustaceans, such as *Triops* and *Daphnia*, exhibit a layered cuticle.[24-28] Although the possibility that the conditions necessary to detect the lamellar form have been lacking in preparations observed thus far, it is more likely that this condition is somehow a modification of the cuticle secretory process. The cellular processes that take place when the new cuticle is secreted in nauplii may involve a different chemical interaction between chitin and protein. Furthermore, the strands of chitin-protein may be liberated into the apical space in a less oriented manner than those seen in adult *Artemia* or other crustaceans.[21-28]

In most crustaceans the cuticle is thicker than the epidermis and, in some cases, it is the major component of the body. Ultrastructural analyses of the cuticle in *Artemia* indicate that the cuticle is much thinner than the epidermal cells and less than 1 μm thick. It is interesting to speculate as to how much support such a thin cuticle could provide since the much thicker epidermis would probably be less deformable. The brine shrimp, then, may actually have a hydrostatic skeleton[29] where the cuticle may act to limit the expansion of the body as well as be a scaffold for further morphogenesis.

C. Naupliar Epidermal Cells

The general morphology of the epidermis of the nauplius has not been studied at the ultrastructural level. Histological preparations of paraffin sections of larvae show the epidermis to be a monolayer of cells (Figure 2).[4,7-9] The cells vary in thickness depending on their location. Cells in the head and antennal region are thin and underlie a greater area of the cuticle than in the developing thoracic and abdominal regions where the cells are cuboidal and contain nuclei that fill most of the volume of the cell.[7-9]

The difference in morphology between the differentiated cells in the cephalic region and the undifferentiated cells in the developing thoracic region is typical of developing tissues where cells that are determined for a certain fate generally tend to maintain a rather general, unspecialized morphology until the development of that structure is complete and the cells differentiate. Moreover, during embryonic growth, undifferentiated cells usually undergo more cell divisions during morphogenesis as the structure is generated. As will be discussed later, this situation is true for the thoracic and abdominal epidermal cells during the larval period. Several structures in the larva, however, have differentiated during embryogenesis. These structures will be discussed next in order to examine the specialized morphology of these cells.

1. Salt Gland

The salt gland, or neck organ, is an oval region of the integument in the dorsal head area.

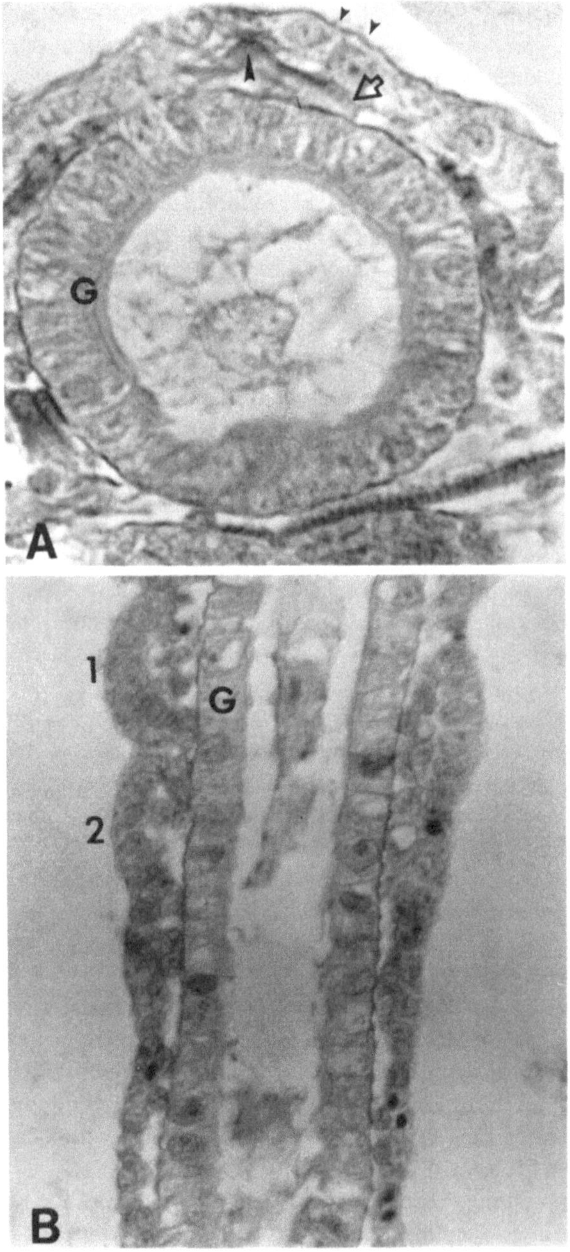

FIGURE 2. Light micrographs of paraffin sectioned larvae. (A) Cross section through the head. The salt gland is at the top. The small arrowheads point to the cuticle. The antennal muscles (open arrow) travel medially from the antennae and impinge upon the mid-dorsal region of the salt gland (black arrowhead). The gut (G) lies immediately beneath the salt gland (magnification × 800). (B) Frontal section through the thorax of a late second instar nauplius. The epidermal cells in the antero-ventral region have begun to evaginate as thoracopod buds of segments (1) and (2) (magnification × 800).

The cells in this region are much larger than the surrounding epidermal cells and, as a result, the salt gland rises above the plane of the surrounding integument to form an oval dome. Muscles from the antennae and the mandibles impinge upon the salt gland cells in three regions and, in the living larvae, appear to pull the epithelium in as they contract (Figure 2).[4]

Approximately 60 to 70 cells make up the epithelium gland. Nuclei of the cells are much larger than those in the regular epidermal cells and are polyploid (discussed below). Whole mount preparations show that the cells appear to be arranged in concentric circles.

The fine structure of these cells is similar to that observed in other transport epithelia.[18-20,30,31] The apical surface is modified into many folds (tubular tufts) which are in contact with the cuticle and are maintained even during formation of the new cuticle. In close association with these folds are numerous mitochondria. The baso-lateral borders form a tubular labyrinth of branching sinuses which fill much of the volume of the cell and approaches closely to the nucleus. A close association among the labyrinth membranes and mitochondrial complexes, glycogen granules, and yolk platelets exists. Between these two regions of the cell, septate junctions between lateral cell membranes of adjacent cells are found. There is no basement membrane. Conte[20] has considered the three zones of the cell to be compartments of a functional ion transport cell, termed a chloride cell. Mitochondrial pumps in the baso-lateral region consist of aggregates of mitochondria in close association with the membrane. Energy storage products needed for ion transport would be produced by glycogen metabolism and ATP production by the mitochondria. In addition, the yolk granules contain the pyrophosphate ester *p,p*-diguanosine-5′-*p,p*-tetraphosphate, another source of stored energy molecules. [20,32,33] These morphological features, similar to those of known transporting epithelia, indicate that the cell possesses certain cytoarchitectural modifications, such as an enlarged surface area and increased organization of the mitochondria-cell membrane complexes for production of ATP to drive the transport activities.

One indication that the cells are, in fact, active in transport is the presence of Na,K-ATPase. Several physiological studies have demonstrated that the cells are indeed active in chloride transport and have a high concentration of Na,K-ATPase activity.[34,35] Indirect methods and inhibitor studies suggest that the enzyme is present in the apical and baso-lateral membranes of the cell.[20,30,31] Peterson and colleagues[37] have successfully isolated and purified the enzyme from *Artemia* nauplii. Subsequent biochemical analyses demonstrated that both subunits contain carbohydrates and that the subunits vary in the temporal nature of their presence during development.[38-41] Moreover, the quantity of α-1 subunit increases during development as α-2 subunit decreases.[38] Recently, antibodies against both the α and β subunits have been developed and characterized.[42] With this tool the analysis of genetic regulation of the synthesis, subunit interactions during enzyme production, and the site-specific location of the molecules in the cells will be possible.

The importance of the salt gland in osmotic and ionic regulation had not been tested directly, however, until the salt glands were isolated from other epithelia.[43,44] Lowy and Conte[45] found that the glands were viable in culture. The glands contained approximately 17% of the ATPase activity in the larva, a very high amount considering the small percent of the body surface covered by the gland.

The role of the salt gland in the developmental period is to bridge the gap from the emergence period, when the osmotic conditions are regulated by internal glycerol levels, to the point when the gills, or branchiae, on the thoracopods take over the function of ionic transport.[20,35] Salt glands are resorbed before the adult state is attained. The branchiae have a similar functional morphology to the salt gland cells with the exception that the cells are smaller and more than one cell type is present.[18] Maturation of the branchial cells does not begin until the fifth or sixth larval instar, a period of approximately 5 to 6 days. During this period the ability of the larva to exist in high salt concentrations is a function of the chloride cells in the salt gland.

Development of the salt gland cells appears to occur after encystment and possibly during emergence. It is during this period that the first signs of Na,K-ATPase are detectable, and there is a gradual transition from glycerol-dependent to enzyme-dependent osmotic regulation.[35] Morphological development of the cells, however, begins immediately after completion of gastrulation (GIII of Benesch[8]). The cells can be seen as enlarged cells in the dorsal roof of the cephalon at this time. Thus, it is possible that they are morphologically distinct at the time of encystment and then further differentiate during the rehydration and emergence periods. Since these cells attain a high degree of morphological and physiological differentiation early in development, it would be interesting to examine the genetic regulation and developmental cues involved in salt gland cell morphogenesis. With recent advances in methods for the isolation and characterization of the Na,K-ATPase from the brine shrimp and the availability of antibodies against the subunits of the enzyme, the molecular probes seem to be at hand to begin the study of the salt gland cell differentiation. In this context, an examination should be undertaken of the development of the branchial epipodite cells that generate the same functional activity as the salt gland cells and take over the ion transport and osmoregulatory functions of the salt gland.

As previously noted, one function of the gland cells is to serve as muscle attachment sites for antennal and mandibular muscles. Thus, these cells carry out the function of tendinal cells found in other crustaceans.[25,46] Studies of tendinal cell morphology in other crustaceans have shown that the cytoplasm of these cells is filled with bundles of microtubules that extend from large desmosomes joining the muscle cell with the epidermal cell at the basal surface to the apical surface and even embed in the endocuticle (Figure 2).[47] Furthermore, the nuclei of tendinal cells are displaced to the periphery of the cell. Electron micrographs of the salt gland cells, however, show no indication of the typical tendinal cell morphology. The largest group of muscles converge on the center of the gland. A thin, densely staining structure oriented perpendicular to the cuticle has been shown,[19] but further examination of this structure has not been done. The structure appears to lie between two cells. Light microscopic examination shows that the basal surface of the cell[5] has densely staining regions that are not present in neighboring cells but the nucleus is centrally located and the mid-level and apical cytoplasm does not appear to be different from that of other cells[89] (Figure 2). Further examination with the electron microscope will have to be done before the functional nature of the muscle attachment regions of these cells can be determined. Tendinal cell function may still be maintained by these cells although the skeletal support may not come from the thin exocuticle but from the thick gland cells which may be sufficient to support the tension from the relatively small muscles of the head region. An alternative suggestion is that the muscles pull against each other and thus may not need an extensive anchoring structure in the salt gland cells.[4]

2. *Setal-Forming Cells*

Setae are thin extensions of the cuticular surface which increase surface area for swimming or feeding in planktonic crustaceans such as *Artemia*. The antennae and mandibles of the brine shrimp are setated at hatching. The second antenna has an enditic process with a spine-like shape that is a modified setal structure.[7] The cells that form the setae are epidermal cells that have reached the differentiated state before emergence and hatching. The fine structure of these cells has not been examined. It is expected, however, that the morphology of these specialized cells will be similar to that found for other species.[48] Light microscopic observations have shown that the nuclei of these cells are larger than the surrounding epidermal cells and they are tetraploid[89] (Figure 3). Since each setal cell forms an entire seta once during each molt cycle (every 1 or 2 days), the increased DNA content may reflect the heightened cuticlar biosynthetic activity of these cells. The cells form an array that follows a strict pattern in the antennae. Benesch[8] does not show antennal setal structures

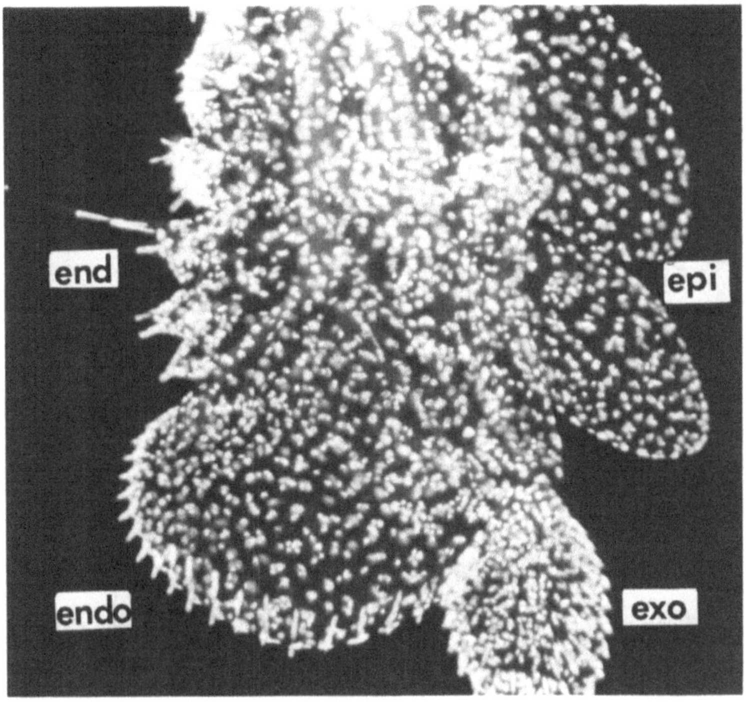

FIGURE 3. Dark field micrograph of a mature thoracopod of *Artemia*. The nuclei have been stained with the nuclear fluorochrome Hoescht 33258. Note the bright, oblong nuclei of the setal cells around the edge of the endite (end), endopodite (endo), and exopodite (exo). Setae are lacking from the epipodites (epi) (= branchiae) at the upper right (magnification × 130).

until late in the embryonic period and setal cells were not described. Thus, the development of these cells must occur during a relatively short period of time during rehydration. The distinct external and histological morphology of these cells combined with a simple but exact pattern may allow these cells to become useful models for examining pattern formation and cell differentiation during early development.

3. *Other Cephalic Epidermal Cells*

The epidermal cells in the remainder of the head region are diploid.[89] The cells are less dense than in the thoracic region and each secretes a greater area of the cuticle. Since this is typical of the adult condition and since little growth will occur in the head region,[5] we must assume that these cells have become fully differentiated by the time of hatching.

III. MOLTING AND MORPHOGENESIS

A. The Molt Cycle

For growth to occur, crustaceans must periodically build a new, larger cuticle beneath the old one and shed the older, restrictive exoskeleton. The events that take place in the integument during the period from one ecdysis to the next comprise the molt cycle and can be divided into four periods: postmolt, intermolt, premolt, and ecdysis.[22,23]

Artemia nauplii have a short, or diecdysic, molt cycle, which is typical of larvae and other, generally small, rapidly molting crustaceans.[9] The predominant phase of the molt cycle is premolt and, in comparison to many other crustaceans, the intermolt period is greatly shortened (Figure 4). Thus, the larvae appear to initiate molting activities immediately after completing the previous molt.

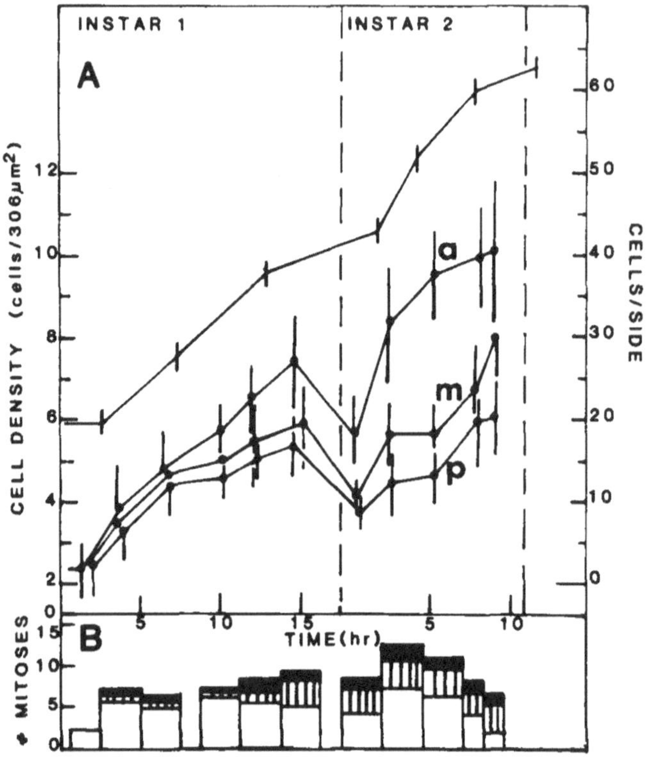

FIGURE 4. Comparison of the change in cell number in different regions of the ventral thorax during the first two instars. (A) The uppermost line (CELLS/SIDE, right ordinate) shows the increase in the number of cells in a longitudinal file of epidermal cells in the ventral thorax. The bottom three lines represent the change in cell density in the anterior (a), middle (m), and posterior (p) regions of the ventral thorax (CELL DENSITY, left ordinate). The time for both instars (abscissa) is not cumulative. (B) Number of mitoses in ventral thorax measured from the same larvae as were used for cell density determinations. Each bar is divided into number of mitoses with longitudinal (clear), transverse (vertical stripes), and oblique (black) axes. The abscissa is the same as shown for (A) (From Freeman, J. A., *J. Crust. Biol.*, 6, 37, 1986. With permission.)

Molt cycle phases for the first two instars are shown in Figure 4. The combined periods of postmolt and intermolt take up 5 to 6 hr in the first instar and 3 to 4 hr in the second instar. The remainder of each period is premolt, first determined by the onset of apolysis (D_O). New cuticle formation was first observed approximately midway through the premolt period, although future ultrastructural studies may show that secretion of the epicuticle and procuticle begins before this point and, perhaps, within 1 hr after apolysis. Comparisons of molt cycle duration of first and second instar larvae have demonstrated that the second instar is shorter than the first (Figure 4). The significance of this difference is not known, but it may be a function of the onset of feeding that begins by the third instar. The molt cycle duration increases during the third instar and then remains relatively constant. The growth rate is also constant.[5,6] Microscopical examination of the head appendages and the thorax during early development has shown that apolysis does not occur simultaneously in all regions of the integument. Setated regions on the antennae and the mandibles as well as the segmenting region of the thorax demonstrate separation of the old cuticle from the epidermis at least 1 hr before apolysis in the general integument. The functional significance of these temporal differences is thought to be related to the general complexity of the new structure to be formed (setated regions) or the morphogenetic activity that must take place before a new cuticle is synthesized. The role of early apolysis in segment morphogenesis will be discussed later (Section III.B.3.c).

Most studies of the development of *Artemia* have assumed that the larvae are hatching synchronously and, thus, that all the larvae are the same age and progressing at the same rate through the first two instars. Examination of the molt cycles of first and second instar larvae, however, has shown that the hatching process must be precisely timed to obtain synchrony in the population and that the synchrony obtained during the first instar may be lost by the onset of the third instar, 24 to 30 hr later. These considerations are important for examining molt cycle stage-dependent biochemical and morphogenetic events.

1. Hormonal Control of Molting

The process of molting in crustaceans had been shown to be under the influence of two hormones.[49-51] Ecdysone, a polyhydroxy steroid, is secreted by the Y-organ, located in the thorax. It is then metabolized to 20-hydroxyecdysone by peripheral tissues.[50] This form of the hormone is 200 times more potent than ecdysone and it is considered to be the true form of the molting hormone. During intermolt, the hormone levels are low and remain at a steady level. Following apolysis, the blood hormone levels rise to a peak late in premolt before falling to intermolt levels before ecdysis. The events of cuticle synthesis are thought to be controlled by the rate of rise and fall of the hormone levels.[23,49] Measurement of ecdysteroid levels in brine shrimp has not been done. Moreover, the presence and location of the Y-organ are also unknown. Benesch[8] did not note the location of any cells that resemble this organ, although the presence of these cells in a rudimentary form cannot be overlooked. Thus, the mechanism of molt control in *Artemia* nauplii remains elusive. We must assume that it is similar to other crustacean systems. Preliminary findings in this laboratory have shown that the onset of apolysis and a more rapid molting rate can be induced in first and second instar larvae by the presence of 20-hydroxyecdysone in the sea water. These results suggest that the tissues of the nauplii are responsive to molting hormone, and that the molt-controlling system is similar to other crustaceans.

If the Y-organ is not developed at the time of hatching, then another source of ecdysteroids must be present. A possible source may be the yolk granules. These organelles are found throughout the epidermal structures and appear to be used up during the first two instars.[1] Few yolk granules are visible in the fourth instar larvae. Ecdysteroids may be synthesized by the nurse cells in the ovary prior to embryogenesis and stored in the granules during vitellogenesis. Upon some cue, the hormone could be enzymatically cleaved and (1) released into the hemolymph, (2) metabolized by the cell into a more active form, or (3) actively stimulate morphogenesis in that cell. Such an embryonic molt-regulating system has been shown to exist in insect[15] and crustacean[14,52,53]eggs. Moreover, it is possible that this system may operate through the first instar as the Y-organ completes differentiation. Alternatively, the Y-organ may be present as a small rudiment, as is true for crab larvae, and may grow during larval development.[54] The rehydrated brine shrimp egg and nauplii would be well suited to examining the biochemistry of the onset of hormonal control mechanisms in crustaceans.

Another factor that controls the molting rate is a peptide hormone, the molt-inhibiting hormone (MIH), which is secreted by the sinus gland X-organ (SGXO) in the eyestalk.[49-51,55] Numerous studies have demonstrated that removal of the crustacean eyestalk results in an accelerated molting rate. The isolation and purification of this hormone have not been carried out, although it is thought to be a small (1000-5000 dalton) peptide.[49] During early development in *Artemia*, the eyestalks are in the early stages of morphogenesis and, in fact, the anlage of the X-organ has been described as a small group of cells in the protocerebrum in the nauplius.[8] Confirmation that these cells are the actual X-organ primordia and that the eyestalk develops a neurosecretory center for MIH production must await future studies. Since the SGXO has been described in the eyestalk of adult *Artemia*,[56] regulation of molting in the nauplius larva may be different from that in the adult. The role the molt-controlling hormones play in morphogenesis is also an area in which more research is needed. Studies done with crab larvae have shown that ecdysone and MIH appear to have little effect on morphogenesis but appear to regulate the onset of premolt and the schedule of cuticle synthetic activities.[57-58]

2. *Chitin Synthesis*

Chitin, a polymer of *n*-acetylglucosamine, is a primary constituent of the arthropod cuticle.[23] The chitin is linked to the protein part of the cuticle and the chitin-protein fibers are secreted by the epidermal cells as layers beneath the old exoskeleton during premolt and postmolt. Raineri employed a histological test to identify the chitin in the first five larval instars.[59] Although her data suggested that chitin is not secreted in all parts of the cuticle of the nauplius larva, it is likely that the test was not sensitive enough and that, in fact, chitin is present in all regions of the cuticle. The existence of a long premolt period in the molt cycle of *Artemia* suggests that cuticle secretion occurs over a period of 5 to 8 hr in the first instar and 5 hr in the second instar.[9] As noted above, the ultrastructure of the cuticle does not lend itself to analysis of the percent of the cuticle that is secreted before ecdysis. However, since the postmolt period is only 1 to 2 hr long, it is probable that most of the cuticle is synthesized during premolt.

The synthesis of chitin and other cuticular precursors has been an active area of research in crustaceans and other arthropods.[23] The first demonstration of an enzyme that is involved in chitin synthesis in *Artemia* was the work of Carey[60] who found that microsomes from day three nauplii could incorporate UDP-*n*-acetyl-^{14}C-glucosamine into chitinase-sensitive chitodextrins. A much more in-depth study of the biochemistry of chitin synthesis has been carried out by Horst.[61] He partially purified and characterized chitin synthetase from brine shrimp and found that the enzyme was membrane associated (Table 1).[61] Most of the enzyme activity appeared to be associated with plasma membranes, although Golgi membrane components were also found in the homogenate. Chitin synthetase activity increased during the larval period with a peak of activity at 72 hr of development.

Lipid-linked intermediates in chitin synthesis were shown to be active in glycosylation of proteins in cell-free extracts of early rehydrating embryos.[62] Further research has demonstrated that a dolichol-sugar intermediate is active in the synthesis of the glycopeptides.[62-64] Incubation of microsomes with dolichol resulted in the transfer of the *n*-acetylglucosamine from the UDP-n-acetylglucosamine precursor to the dolichol. In addition, up to eight sugars could be carried by the lipid intermediate.[64] Horst[65] has charactertized this enzyme, UDP-GlcNac:dolichol phosphate GlcNac-1-phosphate transferase, in brine shrimp and found that it is associated with a liposome fraction (Table 1). The oligosaccharide formed by this enzyme may act as an endogenous primer for further chitin synthesis. Alternatively, the oligosaccharide may be transferred *en bloc* to a protein acceptor. Recently, analysis of the protein acceptor has shown that certain amino acids (asparagine, arginine, threonine, lysine, glutamine, and alanine) may be associated with the chitin attachment site of the glycopeptide.[66] Thus, the enzymes and enzyme reactions involved in the synthesis of cuticular chitin have been characterized. These studies may soon provide a description of the sequence of biochemical and molecular events that occur during the formation of chitin. Future studies may probe the molt cycle stage dependency of these enzymatic events and the cellular regulation of chitin-protein fiber secretion during cuticle production.

Table 1
SUMMARY OF CHARACTERISTICS OF ENZYMES INVOLVED IN CHITIN BIOSYNTHESIS

1. Chitin Synthetase[61]
   a. Transfers n-acetylglucosamine from UDP-n-acetylglucosamine to acceptor (chitin).
   b. pH optimum of 7.0, Mg^{++}-dependent.
   c. Sensitive to diflubenzuron (Dimilin).
   d. Associated with Golgi and/or plasma membranes.
   e. Maximal activity 72 hr after incubation.
2. UDP-GlcNac: dolichol phosphate GlcNac-1-phosphate transferase[63-65]
   a. Transfers n-acetylglucosamine from UDP-n-acetylglucosamine to lipid acceptor.
   b. Sensitive to tunicamycin.
   c. Associated with microsomal fraction.
   d. Endogenous (lipid) acceptor may carry up to eight glycosyl units.

B. Segment Formation

1. General Aspects

At the time of hatching, the postcephalic region of the nauplius consists of a cone of undifferentiated epidermal cells with islands of mesodermal cells located bilaterally along the ventral surface. The cone of cells will, during the ensuing 19 instars, develop into 11 thoracic and 8 abdominal segments. The general morphology and morphometrics of these events have been extensively dealt with previously.[4-7] Of primary interest here are (1) the cellular events that occur in a region of the epidermis where a segment forms and (2) the regulation of these events.

Segment formation occurs in a sequential manner, in an anterior-posterior gradient. As Weisz[4,5] and Anderson[7] have shown, each segment region begins development before the region behind it and develops at a rate that is one instar ahead of the next region. This pattern results in the presence of segments in varying degrees of maturation adjacent to one another along the thorax. Thus, the cellular activities can be followed in several segments at once. In each segmenting region, the epidermal cells undergo two distinct but related processes, growth and differentiation. The interaction of these processes results in: (1) organization of the segment and segmental boundaries and (2) the formation of thoracopod structures (setae, gills, and tendinal cells). In addition, epidermal cells from the mid-ventral region of the thorax give rise to nerves of the segmental ganglia of the ventral nerve cord.[4,8,67]

2. Growth of the Epidermal Cells

From the point in development when encystment occurs to the point of hatching, there is little cell replication. The number of cells remains relatively constant (approximately 3900) during this period.[68-72] At hatching the replication mechanism is activated and the number of cells in the thorax increases. The epidermal cells are arranged in a hexagonal manner but the nuclei appear to be arranged in rows or files, running in an anterior-posterior direction (Figure 5).[9] Examination of cell replication during the first two instars has shown that the greatest amount of cell division is found in the antero-ventral region of the thorax (Figure 4). This is the region where the first two segments will form and where evagination of the thoracopod will occur. Increase in cell density in this region far exceeds that in the more posterior regions throughout the first two instars (Figure 4). Moreover, cell division on the dorsal surface is much less frequent than that on the ventral surface. Thus, there is a well defined region-specific difference in cell replication in the epidermis of the thorax.

In addition to the differences in the amount of growth, there is a difference in the pattern of cell division in the segmenting and nonsegmenting regions. Throughout the first instar, the cell division axes are oriented in such a manner that the daughter cells will be aligned in the same file as the mother cell (Figure 5). The directional nature of the files appears to

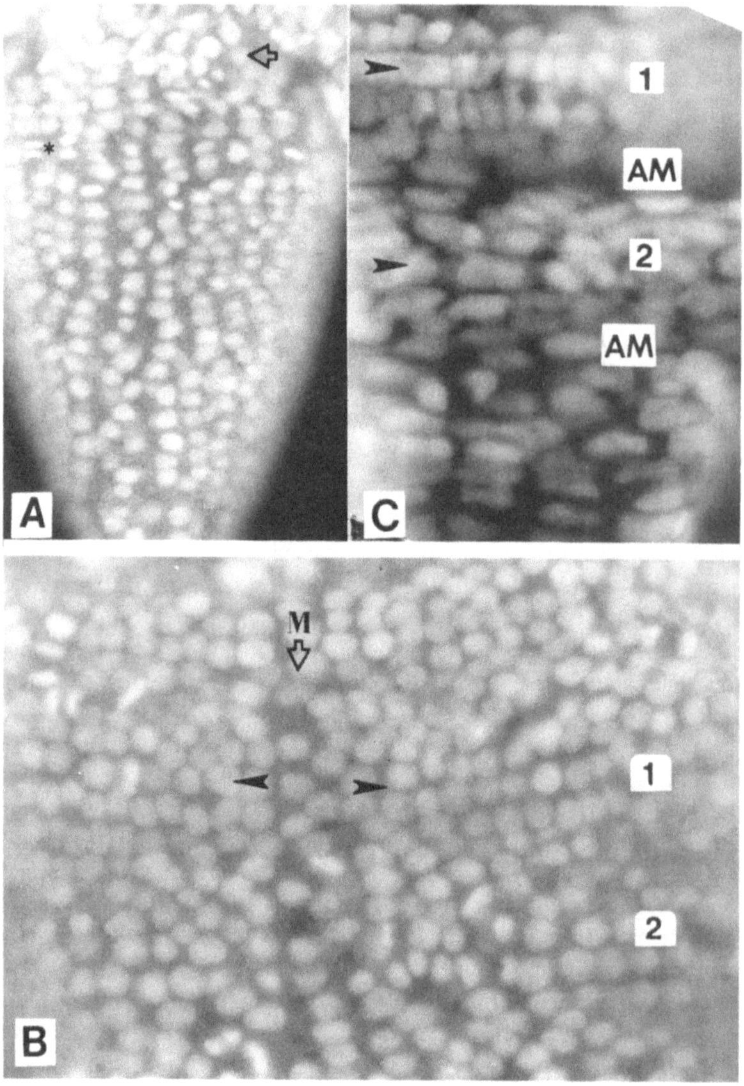

FIGURE 5. Pattern and density of the nuclei of epidermal cells in the ventral thorax of *Artemia* during the first two instars. Nuclei are stained with the nuclear fluorochrome Hoescht 33258. Anterior is at the top in all photographs. (A) Instar 1, the nuclei appear to be arranged in files oriented in an anterio-posterior direction. The star is placed between two cells that are both dividing in a longitudinal direction. Arrow denotes the junction of the cephalon and thorax (magnification × 350). (B) Anterior region of ventral thorax of instar 2 showing the arrangement of nuclei in the regions of the first and second thoracopods (1, 2). Note the transverse orientation of the files of cells of the first segment (black arrowheads). The ventral midline (M, arrowhead) separates the left and right halves of the first two segments.(C) Late instar 2 nauplii showing increased density of cells in the thoracopod bud region of segments 1 and 2 and lower cell density in the region of the arthrodial membrane (AM). The black arrow heads point to the transverse file orientation of the cells in the first two segments (B and C, magnification × 700). Note that the cells behind the second segment maintain the longitudinal file formation. All photographs are of whole mount preparations.

be maintained through such divisions in both the anterior and posterior regions of the ventral thorax. During the second instar, the pattern changes such that, in cells in the anterior region of the ventral thorax (where the first two segments form), the orientation of the division axes changes from longitudinal to transverse, while those cells in the other regions continue to divide in a longitudinal direction.[9,73] The result of this division behavior is an increase in the length of the thorax and an accumulation of cells aligned in transverse files in the antero-ventral region.

By the mid-premolt period of the second instar, the organization of the first two segments and the segment borders can be discerned. In each segment, the ventral thorax is divided into bilateral halves (Figure 5). A mid-ventral strip of cells, consisting of two rows of cells with laterally displaced nuclei demarcates the medial border of each segment. Midway through the premolt period of the second instar, the division between the first segment and the second segment can be visualized as a region of lower cell density. This region will become the arthrodial membrane, which acts as a place of articulation for the thoracopods.[9,46] Epidermal cells in the area of greatest density will evaginate ventro-laterally as the primordium of the thoracopod. These findings show that there is a spatio-temporal nature to the pattern of cell division that leads to the formation of the ventral aspect of the first two segments. Although the pattern of cell division for the other thoracic segments in later instars has not been followed, it is assumed that the same pattern repeats itself in a sequential manner as more segments are delineated towards the posterior region of the thorax.

The nature of the regulation of the pattern of epidermal cell divisions is of fundamental importance in the eventual understanding of segmentation. The pattern could be clonal in nature, whereby the cells in a region of the segment exist as a patch of contiguous cells that arose by sequential divisions of one cell. For instance, one cell could divide once in a longitudinal division and then divide twice in a transverse orientation to produce a set of cells two by four cells in dimension. If such a division pattern were to occur with cells in a band three cells wide in the anterior thorax, the results could resemble the pattern observed in the late second instar (Figure 5). An interesting question is when in development are the precursor cells for the clones defined. Benesch[8] shows a file of nine cells in the ventral thorax in nauplii during postgastrulation morphogenesis. At hatching, there are 18 cells in a file,[9] suggesting that each cell divided once before emergence, and in encysted embryos, before encystment. The clonal division pattern could have been established even at this early point in development. Alternatively, the pattern of cell divisions may not have any role in the origin of the integumental structures in each segment. In this case, the cells could be responding to information generated through a gradient of some morphogen, or to cell-cell contacts, as has been proposed for segment formation in other organisms.[74]

The importance of the division pattern in segment morphogenesis has recently been investigated.[73] Larvae were exposed to 5′-bromodeoxyuridine (BrdU), a drug known to inhibit differentiation in several developing systems.[75] When the larvae were exposed to the compound for the first 8 hr of the first instar, there was no segment differentiation by the end of the second instar (Table 2). Overall growth was not affected but the pattern of cells that was generated in the anterior-ventral thorax appeared to be totally random, with no trace of longitudinal or transverse files. These results indicate that the pattern of cell division is important for later morphogenic events and suggest that one aspect of the differentiation of the cells may be the pattern of cell division. There is clearly a need to find if there is, in fact, a set pattern of division and if attainment of the position of the cells in the pattern sets the stage for the differentiation of these cells into specific integumental structures.

Several studies on growth in crustaceans have demonstrated that cell division occurs primarily during the late intermolt-early premolt phase of the molt cycle.[76,77] In these crustaceans, this period was a relatively small part of the molt cycle. A similar stage-dependent division pattern is evident in insect larvae.[78,79] In *Artemia*, however, cell divisions in the thorax can be seen throughout the first two instars, although the number of dividing cells declines slightly just before ecdysis (Figure 4). Thus, in contrast to other crustaceans, it appears that there is no molt cycle stage-dependent cell division in *Artemia*. *Artemia* nauplii

Table 2
EFFECT OF 5′-BROMODEOXYURIDINE OR TRIFLUOROMETHYLTHYMIDINE ON SEGMENTATION (LATERO-VENTRAL EVAGINATION, ARTHRODIAL MEMBRANE FORMATION, TENDINAL CELL FORMATION, TRANSVERSE CELL PATTERN) IN *ARTEMIA* LARVAE

Treatment	Percent segmenting (#)	Treatment	Percent segmenting (#)
Bromodeoxyuridine (100 μg/mℓ)		Trifluoromethylthymidine (20 μg/mℓ)	
Control-no drug	99 (117)	Control-no drug	98 (48)
Hour 0-2	84 (68)	Hour 0-2	88 (29)
Hour 2-26	19 (103)	Hour 2-26	0 (29)
Hour 0-4	54 (52)	Hour 0-4	26 (29)
Hour 4-26	31 (51)	Hour 4-26	12 (25)
Hour 0-6	68 (93)	Hour 0-6	0 (30)
Hour 6-26	71 (81)	Hour 6-26	0 (31)
Hour 0-8	18 (81)	Hour 0-8	0 (18)
Hour 8-26	73 (39)	Hour 8-26	0 (30)
Hour 0-10	13 (63)	Hour 10-26	79 (33)
Hour 10-26	79 (34)	Hour 22-26	80 (30)
Hour 0-26	4 (153)	Hour 0-26	0 (173)

Note: The times refer to the period of exposure to the compounds. The total development period was 26 hours. Larvae in the pulse exposures were rinsed after the exposure period and then placed in drug-free medium for the remainder of the development period. Larvae were observed live and then fixed and stained with bisbenzimide to observe the cell pattern in the ventral thorax.

have a relatively long premolt stage and so the possibility that the period of cell division is extended throughout this stage cannot be overlooked. Alternatively, there may be no particular stage of the molt cycle that is specific for cell division in all crustaceans.

Weisz[5] has shown that growth in the larvae is a result of growth of the thoracic regions prior to segment delineation. To further examine the location of the cycling epidermal cells, the cell cycle characteristics were determined for cells in various parts of the naupliar integument. Preliminary findings have demonstrated that the epidermal cells in the cephalon and head appendages (antennae, mandibles) have few dividing cells. Instead, the cells appear to already be differentiated by the time of hatching. They secrete a relatively wide area of the cuticle and, in the case of the setal cells in the antennae and the salt gland cells, they become polyploid. Cells in the thorax can be divided into two groups based on cell cycle kinetics. Those in the antero-ventral region, where segmentation is taking place, are continuously cycling throughout the first two instars. Those cells in the dorsal and posterior-ventral regions, on the other hand, have a longer cell cycle period and many of the cells are in stage GII. Cell proliferation, then, is not evenly dispersed throughout the integument. The cells in the segment-forming region show the highest proliferative activity while cells in the rest of the thorax remain in a lower state of proliferation prior to entering segment-related growth. These findings have important implications for studies on DNA synthesis. Clearly, the amount of DNA synthesis will depend upon the level of segmentation-related proliferation. Future consideration of this information may explain the dramatic decline in DNA synthesis after 72 hr in culture observed by several investigators.[80,81]

Several reports have shown that cell division in encysted embryos does not resume until the later stages of emergence. Benesch[8] on the other hand, showed cell divisions and increases in cell numbers throughout the embryonic and prenauplius phases. It is not clear from these studies at what point in development mitosis stops during the encystment process. If, in fact, no divisions occur until hatching, then encystment must have occurred following the establishment of files with 18 cells in the thorax. Moreover, if DNA synthesis in encysted embryos is not re-established until the late emergence phases,[80,81] then these cells must have reached the cell cycle stage seen at hatching before encystment. These considerations lead to the possibility that the temporal nature of the cell division pattern may have been established at some point in development before encystment and that the events are suspended during encystment. Morris has suggested that the period of differentiation is the emergence period, since morphogenesis can be observed during this period.[82] The possibility that the pattern is established at this point cannot be overlooked. The answer to these questions may be found through tests using the nonencysted embryo instead of the cyst.

The molecular biology of DNA synthesis has been studied by several investigators.[1-3] In relation to the cell proliferation that takes place during the first two instars, the S phase synthesis of new DNA may begin as early as emergence.[80,81] The enzymes involved become evident during this period and remain at high levels during the first two instars. Future efforts to study the temporal nature of DNA synthesis must rely on a high degree of developmental synchrony within the larvae. If larvae are continuously hatching for a 48-hr period, the levels of DNA synthesis measured for the period, and possibly for the next 24 hr, may not accurately reflect the actual levels of synthesis during the first two instars.

Another change in the pattern of cell division is the onset of polyploidy in some epidermal cells. This process must occur through endoreduplication at some point in development. Although many of the cells become tetraploid (setal cells, gut), other cells (salt gland) reach higher ploidy levels and the level is tissue- and structure-specific.[89] When during embryogenesis the cells achieve polyploidy is not known.

In summary, the epidermal cells in the region of the first two segments begin rapid proliferation at hatching. This proliferation results in specific regions of high and low cell density that define the segment primordia by the end of the second instar. The cell division occurs in a spatio-temporal pattern that may be essential to the establishment of cell position prior to differentiation. Genetic regulation of these events may begin at hatching. *Artemia*, then, would seem to be an excellent research organism for studies aimed at determining the molecular control of segmentation and pattern formation.

3. Cell Differentiation-Thoracopod Morphogenesis

Differentiation of the epidermal cells of the first two segments begins midway through the second instar. At this time the thoracopod bud has begun to evaginate. Several cell types can be identified by their morphology at the light microscopic level. Other cell types cannot be identified until later in development. The dorsal surface of the segment consists of cells that are general epidermal cells that will eventually become anchoring points for muscles from the thoracopods. Little change occurs in these cells until the third and fourth instars. The cells on the ventro-lateral surface are the most active during early development. The differentiation of these cells is discussed below.

a. Arthrodial Membrane

The first sign of segment morphogenesis is seen in the ventral epidermis during mid-premolt of the second instar. As the density of the epidermal cells increases in the presumptive bud region, a region of lower cell density extending from the midline to the lateral edge and separating the first segment from the maxillary segment in front and the second segment behind becomes evident. The rate of cell division in these cells is less than that of the bud cells. This region, which will eventually become the arthrodial membrane, initially consists of two transverse files of cells.[9] Addition of other cells to the membrane may occur in later instars. By the end of the second instar, the nuclei of these cells appear to be displaced to the side such that the nucleus-free zones of the cells in each file are contiguous, resulting in a thin region of cytoplasm separating the first and second segments. Preliminary evidence shows that the displacement of the nuclei can be blocked by colchicine.[89] Thus, nuclear displacement may involve microtubules. Since arthrodial membranes in embryos of other crustaceans show similar morphology, there may be a necessity for the formation of a thin, nucleus-free region of the cells in order to attain flexibility in this region.[83-84]

b. Tendinal Cells

As ventro-latero evagination proceeds during the second instar, the arthrodial membrane bends severely and appears to point into the haemocoel (Figure 6). This morphology is even more distinct in the third instar. Late in the second instar, several cells in the posterior region of the segment develop cytoplasmic extensions that begin to grow towards the dorsal surface. Although it appears as though these cells stem from arthrodial membrane cells, participation of thoracopod bud cells cannot be eliminated at this time. These cells become tendinal cells, cells modified for the attachment of muscles or which act as an endoskeleton.[4,8,9,73] The identity of the fibrous proteins that provide the cell extension with a support structure is not known, but it is expected that microtubules are involved. Benesch[8] has shown these cells to be present in the developing head appendages. The function of the tendinal cells is probably to form a permanent connection of the arthrodial membrane to the dorsal surface of the segment and, thus, provide support for the membrane cells. Further growth of the thoracopod cells will force the bud outward with little deformation of the arthrodial membrane. It is also possible that this attachment establishes a scaffold for development of the many muscles that form in the thoracopod. The differentiation of the tendinal cell is blocked by prior exposure of the larvae to bromodeoxyuridine or trifluoromethylthymidine.[73] Since the drugs blocked both patterned cell division and differentiation, the molecular event that guides one process may be the same for the other one (Table 2). Alternatively, the drugs may have blocked the patterned cell divisions and the subsequent lack of a specific pattern may then have blocked the differentiation of the cells. These and other possible explanations for the effect of the drugs will form the basis for future research designed to define the interaction of cell division and cell differentiation in the process of segment formation.

c. Ventro-Lateral Evagination

The evagination process which gives rise to the bud could occur by several cellular mechanisms.[85] Some preliminary observations, however, point to the differential build-up in cell densities and the molt cycle events during the second instar as the principal forces acting to generate the evagination. Apolysis occurs in the region of the first two segments slightly before it occurs in more posterior regions of the thorax. Loss of contact with the restrictive exoskeleton may permit deformation in areas of different cell densities, which may result in the thicker region with greater cell density curving slightly out to meet the old exoskeleton. Adjacent, thinner regions of lower cell density, the arthrodial membranes, would then be forced to deform inwardly. The disparity in thickness in the two regions would help to maintain the invagination. The more posterior regions of the thorax would not be susceptible to deformation since the density of the epidermal cells remains roughly equivalent in all areas of this region (Figure 2). Cell shape change by the action of microtubules and microfilaments may also be active forces in the evagination process. Further maintenance of the evagination may be by microtubules, which would hold the shape of the cells, and/or by the tendinal cells, which would anchor the arthrodial membrane. Moreover, the shape of the newly synthesized cuticle may also act to maintain the shape of the evaginating epithelium. The series of events outlined here are hypothetical and future studies

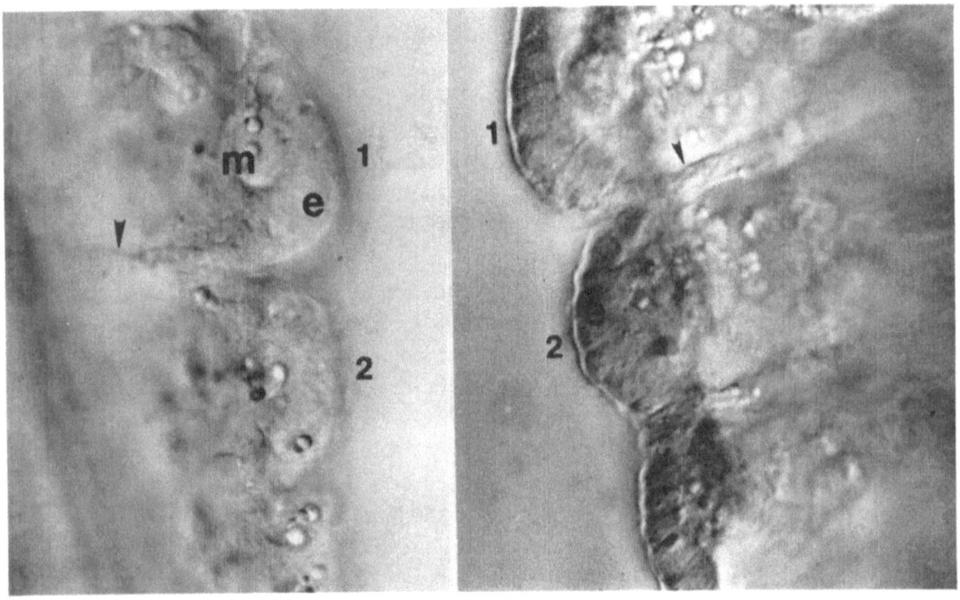

FIGURE 6. Nomarski light micrographs of living third instar larvae taken at slightly different levels through the larvae showing the evagination of the epidermis (e) on the thoracopod buds of the first and second segments and the tendinal cell extension (arrowheads). The clump of mesoderm cells is also shown (m). (Magnification × 700.)

are required to elucidate the mechanisms involved. It is clear, however, that the simplicity of the developing thoracopod and ease with which observation of the events can be done make the brine shrimp an excellent system for investigation of the processes of shape change in epithelial tissues during morphogenesis.

The molecular events which regulate the many activities of thoracopod evagination may take place during the early hours of the first instar. Evagination and increased cell density were blocked when bromodeoxyuridine or trifluoromethylthymidine was present in the sea water for the first 8 hr of the first instar (Table 2).[73] No segment organization occurred in these larvae and the results indicate that all the cellular events of segment formation were blocked. The molecular basis for genetic regulation of cell differentiation during segment morphogenesis is not understood. It is possible that all the developmental events are under the control of an initial gene product. If so, then establishment of the mitotic pattern may be the first step in a sequence of cellular activities leading to segment delineation and cell differentiation. Alternatively, the drugs may have blocked both the regulation of the mitotic pattern as well as other genetic events that would regulate differentiation of the tendinal cells, arthrodial membrane cells, and setal cells. Whatever the mechanism, it is clear that the drugs blocked the morphogenetic control of both patterned growth and differentiation.

d. Other Epidermal Cells

The ganglia and peripheral nerves of the ventral nervous system arise from epidermal cells near the ventral midline of each segment.[8] Weisz[4] and Benesch[8] have described the morphology of the ganglia. The development of the ganglia is closely tied to the process of segment formation. Differentiation of the nerves can be seen as early as two instars after the segment primordia are observed. Thus, by the third instar, the ganglia of the first segment are formed.[4,67] Completion of the ganglia is closely associated with the development of the muscles and epidermal derivatives in the thoracopod. Raineri and Falugi[67] followed the development of acetylcholinesterase (AChE) activity in embryos and larvae and found that the enzyme activity could first be observed in membranes of epidermal cells along the midline and then in the cytoplasm of neuroblasts as the structure of the ganglia evolved. AChE-positive myoblasts were observed ventrally near the ganglia and extending into the segment bud.

As the thoracopod develops, the endite (medial lobe, feeding), exite (lateral lobe, swimming), and epipodites (dorso-lateral lobes, respiration, and ion transport) are demarcated from the central body of the appendage (Figure 3). Arthrodial membranes and tendinal cells for connection of the muscles will form in each of these structures. Completion of the first thoracopod occurs several instars after hatching.[4] The epipodite or metepipodite,[8,20] develops from the dorsal-lateral region of the thoracopod and becomes a thin-walled sac. The cells have large nuclei and secrete rather large regions of the cuticle.[18,86,87] As mentioned previously, these cells have been characterized as chloride cells capable of ion transport as well as serving as gills. Although the demarcation of the epipodite has been noted as occurring as early as the fourth larval instar, the timing of the differentiation of these cells has not been followed.[8]

Other sections of the thoracopod are modified for locomotion and food capture. These structures develop setae during appendage morphogenesis. The setal cells of the thoracopod, as with those in the antennae, appear to be polyploid and modified for the complex mechanism of secretion of a new seta at every molt cycle (Figure 3).[48] The presence of the setae in these structures during later development has been shown by Benesch, but no work has been done on the development of these regions.[8] Since the setal cells are arranged in a specific pattern, studies aimed at understanding the mechanism that regulates the generation of the pattern during larval development may provide valuable clues as to how patterns are controlled.

e. Role of Mesodermal Cells

Upon evagination of the thoracopod, the mesodermal cells in the segmental clumps begin to develop into muscle cells, blood forming cells, and connective tissue cells. The morphology of the development of certain muscles has been described for early instars.[4,8] How the muscles find, and make the proper contacts with the specific tendinal cells of the developing appendage, and what role the muscle cells play, if any, in guiding the development of the epidermal structures are not known.

IV. CONCLUSIONS AND FUTURE DIRECTIONS

Artemia nauplii larvae hatch at a very early stage in development. Within two molt cycles after hatching, the primordia of the epidermal, neural, and mesodermal structures in the thorax segments are established and begin to grow and develop. Special attributes of the larvae include the rapid development time, the clarity of the cuticle, and the small number of cells that comprise each of the segment primordia. Thus, the development of one, or a group of cells can be defined early in development and followed with ease with microscopical techniques.

The larvae are particularly amenable to studies in several areas of developmental biology. These include patterned growth of an epithelium, cell differentiation, epithelial shape change, the onset and regulation of cell-specific polyploidy, ontogeny of ion transport epithelia, myogenesis, and pattern formation. Moreover, the presence of contiguous segments in a gradient of developmental stages makes the larva suitable to perturbation of the events of one segment while not affecting those segments which are either more or less, developed.

Future research on *Artemia* may take advantage of the rapid increase in the information on gene structure and gene regulation. Identification of gene products by immunofluorescence or *in situ* hybridization will permit analysis of gene regulation of segmentation. The use of growth regulators to study tissue-specific cell replication may elucidate the interaction of growth and cell differentiation. Other methods, such as tissue or organ culture of thoracopod buds or cells from a region of the bud primordia, may also be useful techniques to further understand the regulation of development in *Artemia*.

Brine shrimp are also well suited for studying the relationship of molting and morphogenesis. The amount of development of the larva is constant for some species and variable for other crustaceans.[88] Future efforts towards understanding the control mechanisms of growth, molting, and morphogenesis in crustacean larvae will have to employ a larva which has the attributes of *Artemia*.

REFERENCES

1. **Clegg, J. S. and Conte, F. P.,** A review of the cellular and developmental biology of *Artemia*, in *The Brine Shrimp Artemia*, Vol. 2, Persoone, G., Sorgeloos, P., Roels, O., and Jaspers, E., Eds., Universa Press, Wetteren, Belguim, 1980, 11.
2. **Bagshaw, J. C. and Warner, A. H.,** *Biochemistry of Artemia Development*, University Microfilms International, Ann Arbor, Mich., 1979, 239.
3. **Hentschel, C. C. and Tata, J. R.,** The molecular embryology of the brine shrimp, *Trends Biochem. Sci.*, 1, 97, 1976.
4. **Weisz, P. B.,** The histological pattern of metameric development in *Artemia salina*, *J. Morphol.*, 81, 45, 1947.
5. **Weisz, P. B.,** The space-time pattern of segment formation in *Artemia salina*, *Biol. Bull.*, 91, 119, 1946.
6. **Blake, R. W.,** The development of the brine shrimp *Artemia salina* (L): a morphometric approach, *Zool. J. Linn. Soc.*, 65, 255, 1979.
7. **Anderson, D. T.,** Larval development and segment formation in the branchiopad crustaceans *Limnadia stanlejana* (Conchostraca) and *Artemia salina*, *Aust. J. Zool.*, 15, 47, 1967.
8. **Benesch, R.,** Zur ontogenie und morphologie von *Artemia salina* L., *Jb. Anat. Bd.*, 86, 307, 1969.
9. **Freeman, J. A.,** Epidermal cell proliferation during thoracic development in larvae of *Artemia*, *J. Crust. Biol.*, 6, 37, 1986.
10. **Anderson, E., Lochhead, J. H., Lochhead, M. S., and Huebner, E.,** The origin and structure of the tertiary envelope in thick-shelled eggs of the brine shrimp, *Artemia*, *J. Ultrastruct. Res.*, 32, 497, 1970.
11. **Morris, J. E. and Afzelius, B. A.,** The structure of the shell and outer membranes in encysted *Artemia salina* embryos during cryptobiosis and development, *J. Ultrastruct. Res.*, 20, 244, 1967.
12. **Trotman, C. N. A., Mansfield, B. C., and Tate, W. P.,** Inhibition of emergence, hatching, and protein biosynthesis in embryonic *Artemia salina*, *Dev. Biol.*, 80, 167, 1980.
13. **De Chaffoy, D., De Maeyer-Criel, G., and Kondo, M.,** On the permeability and formation of the embryonic cuticle during development *in vivo* and *in vitro* of *Artemia salina* embryos, *Differentiation*, 12, 99, 1978.
14. **Lachaise, F., and Hoffmann, J. A.,** Ecdysteroids and embryonic development in the shore crab, *Carcinus maenas*, *Hoppe-Seyler's Z. Pysiol. Chem.*, 363, 1059, 1982.
15. **Hoffmann, J. A., Lageux, M., Hetru, C., Charlet, M., and Goltzene, F.,** Ecdysone in reproductively competent female adults and in embryos of insects, in *Progress in Ecdysone Research*, Hoffmann, J. A., Ed., Elsevier, Amsterdam, 1980, 431.
16. **Williamson, D. I.,** Larval morphology and diversity, in *The Biology of Crustacea*, Vol. 2, Bliss, D., Ed., Academic Press, New York, 1982, chap. 2.
17. **Myint, T.,** New details of the excystment of *Artemia salina* Leach, *Proc. La. Acad. Sci.*, 19, 24, 1956.
18. **Copeland, D. E.,** A study of salt secreting cells in the brine shrimp *(Artemia salina)*, *Protoplasma*, 63, 363, 1967.
19. **Hootman, S. R. and Conte, F. P.,** Functional morphology of the neck organ in *Artemia salina* nauplii, *J. Morphol.*, 145, 371, 1975.
20. **Conte, F. P.,** Structure and function of the crustacean larval salt gland, in *International Review of Cytology*, Danielli, J. F., Ed., Academic Press, New York, 1984, 45.
21. **Wolfe, A. F.,** A light and electron microscopic study of the frontal knob of *Artemia* (Crustacea, Branchipoda), in *The Brine Shrimp Artemia*, Vol. 1, Persoone, G., Sorgeloos, P., Roels, O., and Jaspers, E., Eds., Universa Press, Wetteren, Belgium, 1980, 117.
22. **Roer, R., and Dillaman, R.,** The structure and calcification of the crustacean cuticle, *Am. Zool.*, 25, 893, 1984.

23. **Stevenson, J. R.,** Dynamics of the integument, in *The Biology of Crustacea*, Vol. 9, Bliss, D. E. and Mantel, L. H., Eds., Academic Press, New York, 1985, chap. 1.
24. **Halcrow, K.,** Some ultrastructural features of the nuchal organ of *Daphnia magna* Straus (Crustacea: Branchiopoda), *Can. J. Zool.*, 60, 1257, 1982.
25. **Schultz, T. W. and Kennedy, J. R.,** Analyses of the integument and muscle attachment in *Daphnia pulex* (Cladocera: Crustacea), *J. Submicrosc. Cytol.*, 9, 37, 1977.
26. **Dahm, E.,** The carapace of Cladocera-a morphological comparison of Cladocera and Ostracoda, *Abh. Verh. Naturwiss. Ver. Hamburg*, 18/19, 331, 1976.
27. **Halcrow, K.,** The fine structure of the carapace integument of *Daphnia magna* Straus (Crustacea Branchiopoda), *Cell Tissue Res.*, 169, 267, 1976.
28. **Rieder, N.,** Ultrastruktur und polysaccharidanteile der cuticula von *Triops cacriformis* Bosc. (Crustacea, Notostraca) während der häutungsvorbereitung, *Z. Morphol. Tiere*, 73, 361, 1972.
29. **Barrington, E. J. W.,** *Invertebrate Structure and Function*, 2nd ed., John Wiley & Sons, New York, 1979, 174.
30. **Conte, F. P., Hootman, S. R., and Harris, P. J.,** Neck organ of *Artemia salina* nauplii, *J. Comp. Physiol.*, 80, 239, 1972.
31. **Hootman, S. R., Harris, P. J., and Conte, F. P.,** Surface specialization of the larval salt gland in *Artemia salina* nauplii, *J. Comp. Physiol.*, 79, 97, 1972.
32. **Finamore, F. J. and Clegg, J. S.,** Biochemical aspects of morphogenesis in the brine shrimp, *Artemia salina*, in *The Cell Cycle: Gene-Enzyme Interactions*, Padilla, G. M., Whitson, G. L., and Cameron, I. L., Eds., Academic Press, New York, 1969, chap. 12.
33. **Conte, F. P.,** Role of C-4 pathway in crustacean chloride cell function, *Am. J. Physiol.*, 238, R269, 1980.
34. **Ewing, R. D., Peterson, G. L., and Conte, F. P.,** Larval salt gland of *Artemia salina* nauplii. Effects of inhibitors on survival at various salinities, *J. Comp. Physiol.* 80, 247, 1972.
35. **Conte, F. P., Droukas, P. C., and Ewing, R. D.,** Development of sodium regulation and *de novo* synthesis of Na,K-activated ATPase in larval brine shrimp, *Artemia salina, J. Exp. Zool.*, 202, 339, 1977.
36. **Russler, D. and Mangos, J.,** Micropuncture studies of the osmoregulation in the nauplius of *Artemia salina, Am. J. Physiol.*, 234, R216, 1978.
37. **Peterson, G. L. and Hokin, L. E.,** Improved purification of brine-shrimp *(Artemia salina)* (Na,K)-activated adenosine triphosphatase and amino-acid and carbohydrate analyses of the isolated subunits, *Biochem. J.*, 192, 107, 1980.
38. **Churchill, L.,** Differences in phosphorylation of the two large subunits of brine shrimp Na,K-ATPase, *J. Exp. Zool.*, 231, 335, 1984.
39. **Churchill, L., Hall, C. C., Peterson, G. L., Ruoho, A. E., and Hokin, L. E.,** Photoaffinity labeling of the ouabain binding site in Na,K-ATPase in developing brine shrimp, *J. Exp. Zool.*, 231, 343, 1984.
40. **Churchill, L.,** A reversal in relative mobility of the two large subunits of brine shrimp (Na,K)-adenosinetriphosphatase, *J. Exp. Zool.*, 231, 351, 1984.
41. **Peterson, G. L., Churchill, L., Fisher, J. A., and Hokin, L. E.,** Structural and biosynthetic studies on the two molecular forms of the (Na,K)-activated adenosine triphosphatase large subunit in *Artemia salina* nauplii, *J. Exp. Zool.*, 221, 295, 1982.
42. **Fisher, J. A., Baxter-Lowe, L. A., and Hokin, L. E.,** Site of synthesis of the α and β subunits of the Na,K-ATPase in brine shrimp nauplii, *J. Biol. Chem.*, 259, 14217, 1984.
43. **Lowy, R. J.,** Isolation of salt glands from the larval brine shrimp *Artemia salina, Am. Zool.*, 24, 265, 1984.
44. **Lowy, R. J. and Conte, F. P.,** Morphology of isolated crustacean larval salt glands, *Am. J. Physiol.*, 248, R709, 1985.
45. **Lowy, R. J. and Conte, F. P.,** Isolation and functional characterization of crustacean larval salt gland, *Am. J. Physiol.*, 248, R702, 1985.
46. **Dennell, R.,** Integument and exoskeleton, in *The Physiology of Crustacea*, Vol. 1, Waterman, T. H., Ed., Academic Press, New York, 1960, 449.
47. **Koulish, S. A.,** Microtubules and muscle attachment in the integument of the Balanidae, *J. Morphol.*, 140, 1, 1973.
48. **Tchernigovtzeff, C.,** Sur l'organisation des matrices sétigères chez deux Crustacés Malacostracés, étudiée au microscope électronique *C. R. Acad. Sci. Ser. D*, 282, 727, 1976.
49. **Kleinholz, L. H. and Keller, R.,** Endocrine regulation in Crustacea, in *Hormones and Evolution*, Vol. 1, Barrington, E. J. W., Ed., Academic Press, New York, 1979, chap. 4.
50. **Chang, E. S.,** Hormonal control of molting in decapod Crustacea, *Am. Zool.*, 25, 179, 1985.
51. **Skinner, D. M.,** Interacting factors in the control of the crustacean molt cycle, *Am. Zool.*, 25, 275, 1985.
52. **McCarthy, J. F.,** Ponasterone A: a new ecdysteroid from the embryos and serum of brachyuran crustaceans, *Steroids*, 34, 799, 1979.
53. **Walgraeve, H., Van Beek, E., Criel, G., Van Brussel, K., and De Leenheer, A.,** Comparison of three separation systems for the radioimmunoassay of ecdysteroids in *Artemia* (Crustacea: Branchiopoda), *Insect Biochem.*, 16, 41, 1986.

54. **McConaugha, J.**, Identification of the Y-organ in the larval stages of the crab, *Cancer anthonyi* Rathbun, *J. Morphol.*, 164, 83, 1980.
55. **Passano, L. M.**, Molting and its control, in *The Physiology of Crustacea*, Vol. 1, Academic Press, New York, 1960, 473.
56. **Baid, I. C., and Ramaswami, L. S.**, Neurosecretory cells in *Artemia salina* L., *Experientia*, 21, 528, 1965.
57. **McConaugha, J. R.**, Regulation of crustacean morphogenesis in larvae of the mud crab, *Rhithropanopeus harrisii*, *J. Exp. Zool.*, 223, 155, 1982.
58. **Freeman, J. A. and Costlow, J. D.**, Endocrine control of spine epidermis resorption during metamorphosis in crab larvae, *Roux's Arch. Dev. Biol.*, 192, 362, 1983.
59. **Raineri, M.**, Histochemical localization of chitin in larvae of *Artemia salina* Leach (Phyllopoda), *Boll. Zool.*, 48, 139, 1981.
60. **Carey, F. G.**, Chitin synthesis *in vitro* by crustacean enzymes, *Comp. Biochem. Physiol.*, 16, 155, 1965.
61. **Horst, M. N.**, The biosynthesis of crustacean chitin by a microsomal enzyme from larval brine shrimp, *J. Biol. Chem.*, 256, 1412, 1981.
62. **Quesada-Allure, L. A.**, Early dolichol sugar synthesis during differentiation of non-growing brine-shrimp *(Artemia salina)* embryos, *Biochem. J.*, 198, 429, 1981.
63. **Horst, M. N.**, Lipid-linked oligosaccharides and their role in primer formation for crustacean chitin synthesis, *Fed. Proc.*, 42, 2164, 1983.
64. **Horst, M. N.**, The biosynthesis of crustacean chitin. Isolation and characterization of polyphenol-linked intermediates from brine shrimp microsomes, *Arch. Biochem. Biophys.*, 221, 254, 1983.
65. **Horst, M. N.**, Solubilization and partial characterization of a n-acetylglucosamine-1-phosphate transferase from larval brine shrimp, *Fed. Proc.*, 44, 1408, 1985.
66. **Horst, M. N.**, Studies on the structure and biosynthesis of the linkage regions between chitin and protein in *Artemia salina*, *Fed. Proc.*, 45, 1846, 1986.
67. **Raineri, M. and Falugi, C.**, Acetylcholinesterase activity in embryonic and larval development of *Artemia salina* Leach (Crustacea Phyllopoda), *J. Exp. Zool.*, 227, 229, 1983.
68. **Nakanishi, Y. H., Iwasaki, T., Okigaki, T., and Kato, H.**, Cytological studies of *Artemia salina*. I. Embryonic development without cell multiplication after the blastula stage in encysted dry eggs, *Annot. Zool. Jpn.*, 35, 223, 1962.
69. **Nakanishi, Y. H., Okigaki, T., Kato, H., and Iwasaki, T.**, Cytological studies of *Artemia salina*. II. Deoxyribonucleic acid (DNA) content and the chromosomes in encysted dry eggs and nauplii, *Proc. Jpn. Acad. Sci.*, 39, 306, 1963.
70. **Olson, C. and Clegg, J.**, Nuclear numbers in encysted dormant embryos of different *Artemia salina* populations, *Experientia*, 32, 864, 1976.
71. **Olson, C. S. and Clegg, J. S.**, Cell division during the development of *Artemia salina*, *Wilhelm Roux's Arch.*, 184, 1, 1978.
72. **Iwasaki, T.**, Incorporation of ^{3}H-thymidine during early developmental stages of ovoviviparous and oviparous embryos in the brine shrimp, *Artemia salina*, *Zool. Mag.*, 89, 264, 1980.
73. **Freeman, J. A.**, Effect of thymidine analogs on segmentation and cell division pattern in *Artemia* epidermis, *J. Cell Biol.*, 101, 471a, 1985.
74. **Meinhardt, H.**, Generation of structures in a developing organism, in *Developmental Order: its Origin and Regulation*, Subtelny, S. and Green, P. B., Eds., Alan R. Liss, New York, 1982, 439.
75. **Holtzer, H., Weintraub, H., Mayne, R., and Mochan, B.**, The cell cycle, cell lineages, and cell differentiation, in *Current Topics in Developmental Biology*, Vol. 7, Moscona, A. A. and Monroy, A., Eds., Academic Press, New York, 1972, 229.
76. **LeRoux, A.**, La crise mitotique dans le proventricule des zoés I et II de *Pisidia longicornis* (Linné): sa position en fonction des stades du cycle d'intermue, *Arch. Zool. Exp. Gen.*, 119, 353, 1978.
77. **Tchernigovtzeff, C.**, Multiplication cellulaire et régéneration au cours du cycle d'intermue des crustacés décapodes, *Arch. Zool. Exp. Gen.*, 106, 377, 1965.
78. **Locke, M.**, The molt/intermolt cycle in the epidermis and other tissues of an insect, *Calpodes ethlius*, *Tissue Cell*, 2, 197, 1970.
79. **Kato, Y.**, Mitotic frequency in the epidermis during early life in *Bombyx mori*, *Zool. Mag.*, 86, 250, 1977.
80. **McClean, D. K. and Warner, A. H.**, Aspects of nucleic acid metabolism during development of the brine shrimp *Artemia salina*, *Dev. Biol.*, 24, 88, 1971.
81. **Slater, J. M. and McLennan, A. G.**, DNA polymerases α and γ during pre-emergence and early larval development of *Artemia*, *Eur. J. Biochem.* 129, 415, 1982.
82. **Morris, J. E.**, Hydration, its reversibility, and the beginning of development in the brine shrimp *Artemia salina*, *Comp. Biochem., Physiol.*, 39A, 843, 1971.
83. **Dohle, W.**, Die bildung and differenzierung des postnauplialen keinstreifs von *Diastylis rathkei* (Crustacea, Cumacea). II. Die differenzierung und musterbildung des ektoderms, *Zoomorphologie*, 84, 235, 1976.

84. **Scholtz, G.,** Investigations on the formation and differentiation of the post-naupliar germ band in *Neomysis integer* Leach (Crustacea, Malacostraca, Pericarida), *Zool. Jb. Anat.*, 112, 295, 1984.
85. **Ettensohn, C. E.,** Mechanisms of epithelial evagination, *Q. Rev. Biol.*, 60, 289, 1985.
86. **Copeland, D. E.,** Salt transport organelle in *Artemia salenis* (brine shrimp), *Science*, 151, 470, 1966.
87. **Croghan, P. C.,** The mechanism of osmotic regulation in *Artemia salina* (L.): the physiology of the branchiae, *J. Exp. Biol.*, 35, 234, 1958.
88. **Costlow, J. D.,** Metamorphosis in crustaceans, in *Metamorphosis: A Problem in Developmental Biology*, Vol. 1, Etkin, W. and Gilbert, L. I., Eds., Appleton-Century-Crofts, New York, 1968, chap. 1.
89. **Freeman, J. A.,** unpublished results.

INDEX

F

G

R